Decision Support Methods for Assessing Flood Risk and Vulnerability

Ahmed Karmaoui
Southern Center for Culture and Sciences (SCCS), Morocco

A volume in the Advances in Environmental Engineering and Green Technologies (AEEGT) Book Series

IGI Global
DISSEMINATOR OF KNOWLEDGE

Published in the United States of America by
IGI Global
Engineering Science Reference (an imprint of IGI Global)
701 E. Chocolate Avenue
Hershey PA, USA 17033
Tel: 717-533-8845
Fax: 717-533-8661
E-mail: cust@igi-global.com
Web site: http://www.igi-global.com

Library of Congress Cataloging-in-Publication Data

Names: Karmaoui, Ahmed, 1982- editor.
Title: Decision support methods for assessing flood risk and vulnerability / Ahmed Karmaoui, editor.
Description: Hershey, PA : Engineering Science Reference, [2020] | Includes bibliographical references and index.
Identifiers: LCCN 2019009105| ISBN 9781522597711 (h/c) | ISBN 9781522597735 (eISBN) | ISBN 9781522597728 (s/c)
Subjects: LCSH: Flood damage--Risk assessment. | Floods--Economic aspects. | Climatic changes--Environmental aspects. | Decision support systems.
Classification: LCC HD1675 .D43 2020 | DDC 363.34/932--dc23 LC record available at https://lccn.loc.gov/2019009105

This book is published in the IGI Global book series Advances in Environmental Engineering and Green Technologies (AEEGT) (ISSN: 2326-9162; eISSN: 2326-9170)

British Cataloguing in Publication Data
A Cataloguing in Publication record for this book is available from the British Library.

Advances in Environmental Engineering and Green Technologies (AEEGT) Book Series

ISSN:2326-9162
EISSN:2326-9170

MISSION

Growing awareness and an increased focus on environmental issues such as climate change, energy use, and loss of non-renewable resources have brought about a greater need for research that provides potential solutions to these problems. Research in environmental science and engineering continues to play a vital role in uncovering new opportunities for a "green" future.

The **Advances in Environmental Engineering and Green Technologies (AEEGT)** book series is a mouthpiece for research in all aspects of environmental science, earth science, and green initiatives. This series supports the ongoing research in this field through publishing books that discuss topics within environmental engineering or that deal with the interdisciplinary field of green technologies.

COVERAGE

- Air Quality
- Industrial Waste Management and Minimization
- Green Transportation
- Cleantech
- Electric Vehicles
- Radioactive Waste Treatment
- Green Technology
- Renewable Energy
- Biofilters and Biofiltration
- Waste Management

IGI Global is currently accepting manuscripts for publication within this series. To submit a proposal for a volume in this series, please contact our Acquisition Editors at Acquisitions@igi-global.com or visit: http://www.igi-global.com/publish/.

The Advances in Environmental Engineering and Green Technologies (AEEGT) Book Series (ISSN 2326-9162) is published by IGI Global, 701 E. Chocolate Avenue, Hershey, PA 17033-1240, USA, www.igi-global.com. This series is composed of titles available for purchase individually; each title is edited to be contextually exclusive from any other title within the series. For pricing and ordering information please visit http://www.igi-global.com/book-series/advances-environmental-engineering-green-technologies/73679. Postmaster: Send all address changes to above address.

Titles in this Series

For a list of additional titles in this series, please visit:
https://www.igi-global.com/book-series/advances-environmental-engineering-green-technologies/73679

Advanced Design of Wastewater Treatment Plants Emerging Research and Opportunities
Athar Hussain (Ch. Brahm Prakash Government Engineering College, India) and Ayushman Bhattacharya (Ch. Brahm Prakash Government Engineering Colleg, India)
Engineering Science Reference • ©2019 • 350pp • H/C (ISBN: 9781522594413) • US $195.00

Advanced Multi-Criteria Decision Making for Addressing Complex Sustainability Issues
Prasenjit Chatterjee (MCKV Institute of Engineering, India) Morteza Yazdani (Universidad Loyola Andalucía, Spain) Shankar Chakraborty (Jadavpur University, India) Dilbagh Panchal (Dr. B. R. Ambedkar National Institute of Technology (NIT) Jalandhar, India) and Siddhartha Bhattacharyya (RCC Institute of Information Technology Kolkata, India)
Engineering Science Reference • ©2019 • 360pp • H/C (ISBN: 9781522585794) • US $195.00

Amelioration Technology for Soil Sustainability
Ashok K. Rathoure (Biohm Consultare Pvt Ltd, India)
Engineering Science Reference • ©2019 • 280pp • H/C (ISBN: 9781522579403) • US $185.00

Advanced Agro-Engineering Technologies for Rural Business Development
Valeriy Kharchenko (Federal Scientific Agroengineering Center VIM, Russia) and Pandian Vasant (Universiti Teknologi PETRONAS, Malaysia)
Engineering Science Reference • ©2019 • 484pp • H/C (ISBN: 9781522575733) • US $195.00

Spatial Planning in the Big Data Revolution
Angioletta Voghera (Politecnico di Torino, Italy) and Luigi La Riccia (Politecnico di Torino, Italy)
Engineering Science Reference • ©2019 • 359pp • H/C (ISBN: 9781522579274) • US $195.00

For an entire list of titles in this series, please visit:
https://www.igi-global.com/book-series/advances-environmental-engineering-green-technologies/73679

IGI Global
DISSEMINATOR OF KNOWLEDGE

701 East Chocolate Avenue, Hershey, PA 17033, USA
Tel: 717-533-8845 x100 • Fax: 717-533-8661
E-Mail: cust@igi-global.com • www.igi-global.com

Table of Contents

Detailed Table of Contents

Mohd Talha Anees, University of Malaya, Malaysia
Ahmad Farid Bin Abu Bakar, Universiti of Malaya, Malaysia
Lim Hwee San, Universiti Sains Malaysia, Malaysia
Khiruddin Abdullah, Universiti Sains Malaysia, Malaysia
Mohd Nawawi Mohd Nordin, Universiti Sains Malaysia, Malaysia
Nik Norulaini Nik Ab Rahman, Universiti Sains Malaysia, Malaysia
Muhammad Izzuddin Syakir Ishak, Universiti Sains Malaysia, Malaysia
Mohd Omar Abdul Kadir, Universiti Sains Malaysia, Malaysia

Flood can be assessed through flood vulnerability, risk, and susceptibility analysis using remote sensing, geographic information system, and hydrological modelling. In this chapter, different stages, complexities, and processes of flood vulnerability, risk, and susceptibility assessment were discussed. The study reveals that flood vulnerability should be assessed based on four aspects: physical, social, economic, and environmental. Flood risk should be assessed by three stages: risk analysis, disaster relief, and preparedness, whereas flood susceptibility assessment involves three processes. Overall, it was found that the responsible factors vary as per the local conditions, which need to be carefully analyzed and selected. Furthermore, the role of remote sensing and geographic information system in flood risk management were found important especially in flood risk mapping and in the selection of responsible flooding factors.

national development strategies. Policies will be presented to implement these recommendations for adaptation to climate variability and global CC. The authors provide general recommendations and identify challenges for the incorporation of climate change impacts and risk assessment into long-term land-use national development plans and strategies. Overall, this chapter provides an overview of the implications for CC to coastal management.

Nonlinear Muskingum method is a very efficient tool in flood routing implementation. It is possible to estimate an outflow hydrograph by a given inflow hydrograph of a flood at a specific point of the river channel. However, it turns out an optimization problem at the stage of employing this method, and it becomes important to reach the optimal model parameters so as to obtain precise outflow hydrograph estimations. Hence, it was decided to utilize five up-to-date optimization algorithms, namely, vortex search algorithm (VSA), gases brownian motion algorithm (GBMO), water cycle algorithm (WCA), flower pollination algorithm (FPA), and colliding bodies optimization (CBO). The algorithms were integrated with the nonlinear Muskingum model so as to estimate the outflow hydrograph of Wilson data, and it was deduced that WCA, FPA, and VSA perform relatively better than the models employed in the other researches before.

At-site flood frequency analysis (FFA) of extreme hydrological events under Bayesian paradigm has been carried out and compared with frequentist paradigm of maximum likelihood estimation (MLE). The main objective of this chapter is to identify the best approach between Bayesian and frequentist one for at-site FFA. As a case study, the data of only two stations were used, Kotri and Rasul, and Bayesian and MLE approaches were implemented. Most commonly used tests were applied for checking initial assumptions. Goodness of fit (GOF) tests were used to identify the best model, which indicated that the generalized extreme value (GEV) distribution appeared to be best fitted for both stations. Under Bayesian paradigm, quantile estimates are constructed using Markov Chain Monte Carlo (MCMC) simulation method for their respective returned periods and non-exceedance probabilities. For MCMC simulations, as compared to other sampler, the M-H sampling technique was used to generate a large number of parameters. The analysis indicated that the standard errors of the parameters' estimates and ultimately the quantiles' estimates using Bayesian methods remained less as compared to maximum likelihood estimation (MLE), which shows the superiority of Bayesian methods over conventional ones in this study. Further, the safety amendments under two techniques were also calculated, which also show the robustness of Bayesian method over MLE. The outcomes of these analyses can be used in the selection of better design criteria for water resources management, particularly in flood mitigation.

This chapter presents an overview of the important concepts related to flood hazard assessments and explores the use of remote sensing data from satellites to supplement traditional assessment techniques. The method presented in this chapter can be used in sectoral planning activities and integrated planning studies and for damage assessment. The chapter presents the application of flood modelling to the study area. The study area, Adyar River in Chennai, has experienced major floods in the past decade which is attributed to increased urbanization. The hydrologic model for the Adyar River Basin was set up using HEC geoHMS and was run and calibrated using observed flow in HEC-HMS. The chapter also discusses the results obtained from the IDF analysis and its application in HEC HMS to generate hypothetical storm hydrographs. Furthermore, the chapter goes on to discuss the results obtained from the hydraulic modelling such as the inundation map for the 2005 flood event and the inundation map for hypothetical storms of varying return periods.

The Ziz Watershed is located in the arid zones of South-Eastern Morocco and belongs to the large basin of Ziz-Rheris. In this basin, floods are related to natural factors and mainly to the occupation of the hydraulic public domain and the human intervention on the courses of the rivers. Increases in sediment yield are observed in many places in the Ziz, dramatically affecting water quality and reservoir management. In order to map overland sediment generation and delivery to the stream (studying the service of sediment retention), the InVEST sediment delivery ratio (SDR) model was applied. The sedimentation analysis in the Hassan Dakhil Dam, located in this watershed, shows that there is a very important erosion rate. The proof is the rapid filling of the dam. This is due to the transport of sediments in the rivers. If this situation continues at the current rate, the dam will no longer be fully operational for irrigation by 2050.

The Innaoune Watershed represents an important hydric potential of the oriental part of Morocco. However, the basin exhibits a set of hydrologic drawbacks, such as floods, erosion, and pollution. This chapter is focused on flood forecast study. In order to help managers and decision makers to adopt the appropriate land management strategies for protecting the population from flood damages, the study of the hydrological behavior and quantification of water yield are paramount. According to this perspective, the main goal of this chapter is to test the ability of the SWAT model to simulate and reproduce the hydrological behavior of the upstream of Innaouene

Watershed. The output of the model could be used to map, delineate, and forecast the floods expansion for a particular rainfall event. SWAT was performed on a daily time step from 2004 to 2012 for calibration and 2012 to 2014 for validation. The model accuracy was evaluated by measuring the Nash-Sutcliffe coefficient and R2.

Flooding has a wide range of impacts on societies and natural environments. In this sense, the city of Fez suffers from these problems reflected by the overflow of Oued El Jawaher during the rainy periods. This situation led the authors to compare between the current situation and the situation developed by the thresholds of Oued El Jawaher. HEC-RAS hydraulic model consists of 31 cross-sections, which will be used in the course of this study. The simulations will concern the current state and the developed state for flows of different frequencies. The result of the simulations confirms that the capacity of the proposed hydraulic structures is insufficient to transit and should be considered. To conclude, the development of the channel by thresholds, which serves for the creation of water plan, magnifies the risk of an overflow of the banks of the canal by the water line along with the longitudinal profile.

Climate change is a word that we have heard hundreds of times, but what is it? Is it happening or is it something made by us? There are thousands of such questions, thoughts, doubt which come to our minds as soon as we hear the words “climate change.” Even though there are hundreds of research works and many more proofs stating that the climate change is happening, there is a side which has been generally overlooked, and that is what if the climate change that we look is just something made by us. Climate change refers to long-lasting changes in temperature, clouds, humidity, and rainfall around the world. Both local and global factors cause regional climate change. This difference is significant because if a regional climate change occurs on account of local factors, then these changes can be mitigated by local actions. This chapter explores flood hazard casting prediction of climate change impressions.

Effective climate change and disaster policy communication services are vital for enhancing the adaptive resilience capacity of the vulnerable local communities in poor countries like Uganda. This chapter focuses on the effectiveness of the Ugandan national climate change and disaster policy information communication strategies in addressing national flooding disaster risks, highlights the recent trends of knowledge based responses to climate change induced floods, assesses the impact of the flood on the socio-economic well-being of local households and communities, and determines the vulnerability issues with corresponding adaptation strategies to floods in the flood prone country. Climate change flood risks have continued to exact huge socio-economic loss and damage effects due to the vulnerability and weak adaptation strategies to floods. The national meteorological services tend to forecast seasonal flood events; some flood forcing factors; and the impact of floods on social, economic, ecological, and physical infrastructure are on the rise in some parts of the country.

Ecosystem management requires biophysical and socio-economic measurement. The intervention of the government and the local community in order to combat the degradation of ecosystems must take into account the effects of the environmental hazards. This can reinforce the inhabitants' ability to adapt at local level. The impact on ecosystem and resources are numerous and complex. Consequently, a multidisciplinary evaluation is needed. In this context, a new approach was proposed, called environmental hazards assessment at local scale. It was used to evaluate the risk of several oasis resources to multiple hazards in the Middle Draa Valley. The findings show that for all resources, desertification is the biggest challenge affecting this area followed by drought, sandstorms, and then floods. This risk assessment approach can provide guidance for future assessments.

Preface

The impact of climate change is increasingly felt with an increased frequency of extreme events (droughts, fires, floods, etc.). All these phenomena are going to affect the regional economy and subsequently the national socio-economic stability. Floods are highly relevant extreme events and remain among the most dangerous and complex natural disasters in middle and low-income countries where the fragility of the environment is quite critical, despite efforts by governments.

Globally, floods cause an enormous amount of damage (more than $100 billion) in property damage and kill over 1000 people every year. The impact of floods has several negative economic and social consequences on transportation systems, water supplies, agriculture, and health, in regions at a higher risk. Human response to flood management requires measures involving integrated planning, adaptation, and recovery strategies.

The proposed book details this phenomenon, debating the associated aspects such as the impact, the risk, the vulnerability, the resilience, and the adaptation. This book, also, provides an opportunity to share experiences and methods for assessing the flood linking science and practice, through a collection of decision support methods. The explored methods have the purpose to assess flood risk and vulnerability going through the discussion of the impact, the susceptibility the exposure, the resilience, and the adaptation to flooding in different scales: urban, rural, basin, sub-catchment, and coastal areas.

Specifically, flood refers to an overflowing of water in a usually dry area. It is considered a current hazard and the most frequent worldwide. The frequency and intensity increase with climate change. A better understanding of flood hazards is needed, including its changes over time and its impact on human well-being (social and economic). In fact, flood risks are linked to social, economic, physical, and environmental components.

The book will be an essential reference for practitioners in the field and aims to:

- Establish useful and effective tools focused on the diagnosis and assessment of vulnerability to flooding risks;

- Reflect the current trends of knowledge in terms of floods;
- Help, evaluate, and measure the impact of the flood on the socio-economic well-being of vulnerable population.
- Identify the appropriate adaptation strategies to floods;
- Present theoretical frameworks and empirical research findings to be used by professionals who want to improve their understanding of the flood phenomenon.

This book simplifies and gives the tools and methodologies in several aspects linked to floods (e.g., human, economy, climate change, and cultural aspects). This will be helpful for a large audience in the above-mentioned fields. The book provides also a set of new tools and new research trends, models, review of relevant literature, analysis, etc. These can be used for education, sensitization and scientific research, and then suggest strategies for different spatial-temporal scales. The impact of floods on biophysical and socio-economic systems are numerous and complex. Consequently, multidisciplinary decision tools are needed. These tools use hydrological, topographic, climatic, and socio-economic environmental parameters which are strongly associated with flood vulnerability and risk.

In this context, several approaches and methods were proposed in this collection: The Environmental hazards assessment at a local scale (EHALS), the flood vulnerability index (FVI), Flood Frequency Analysis (FFA), likelihood estimation (MLE), and Nonlinear Muskingum method. In this latter method, five algorithms were integrated: Vortex Search Algorithm (VSA), Gases Brownian Motion Algorithm (GBMO), Water Cycle Algorithm (WCA), Flower Pollination Algorithm (FPA) and Colliding Bodies Optimization (CBO). Other tools were applied in this collection of chapters such as InVEST Sediment Delivery Ratio (SDR) model, HEC-RAS hydraulic model (with 31 cross-sections), Soil & Water Assessment Tool) (SWAT model), Nash-Sutcliffe coefficient and R^2, IDF analysis and its application in HEC HMS to generate hypothetical storm hydrographs.

The uses of these approaches have been complemented by other techniques and methods like:

- Geographic information system: mapping the service of sediment retention, delineate and forecast, the floods expansion;
- Remote sensing analysis: flood vulnerability, risk, and susceptibility analysis;
- Traditional assessment techniques in hydrological modeling;
- Flood forecast studies;
- Important concepts related to flood hazard assessments.

The results of this book can be used in the selection of better approaches in floods mitigation. The policies and recommendations have been presented to adapt to climate variability and floods. In this context, a number of challenges were identified in risk assessment into long-term land-use national and local development plans. These can help also managers and decision-makers to adopt the appropriate strategies to protect human life and property from flood damages. This should allow a better understanding of degradation, provide guidance for appropriate responses at different scales, and may contribute as an important step toward preparedness to flooding.

The book is organized in 13 chapters.

Chapter 1 introduces the concepts of Flood Vulnerability, Risk and Susceptibility Assessment using remote sensing, geographic information system and hydrological modeling. In this study, different stages, complexities and processes of flood vulnerability, risk and susceptibility assessment were discussed which are the recent concerns among researchers and decision makers. The study reveals that flood vulnerability should be assessed based on four aspects such as physical, social, economic and environmental. Flood risk should be assessed by three stages such as risk analysis, disaster relief and preparedness. Overall, it was found that the responsible factors vary as per the local conditions which need to be carefully analyzed and selected. Furthermore, the role of remote sensing and geographic information system in flood risk management were found important especially in flood risk mapping and in the selection of responsible flooding factors.

Chapter 2 presents the research question of composite indicators as Decision Support Method for Flood Analysis. In fact, floods are highly relevant extreme events, with increasing frequency at a global scale. It remains among the most dangerous and complex natural disaster mainly in middle and low-income countries. In this context, it is necessary to develop decision support tools to reduce the flood risk and increase the resilience. The chapter reviews one of the most relevant tools, the flood vulnerability index (FVI) category at a global scale. These tools use hydrological, topographic, socio-economic parameters strongly associated with flood vulnerability. The findings indicate that FVI is a flexible tool for integrated assessment of vulnerability to floods for application in different regions. Social, environmental, and physical components are the main components used in the FVI. Household and Neighborhood, Basin, Urban, Sub-catchment, and Coastal are the different levels of vulnerability analysis. This may contribute as an important step toward a preparedness to flood.

Chapter 3 described natural values, coastal landforms, and types of infrastructure that most likely to be affected by climate change (CC) and provide information for assessing inundation, erosion and recession risks for a chosen location. This chapter was focused on the land uses, the vulnerability of coastal infrastructure, and argued for effective linkages between CC issues and development planning. It also recommended the incorporation of CC impact and risk assessment into long term national development strategies. Policies will be presented to implement these recommendations for adaptation to climate variability and global CC. The authors provide general recommendations and identify challenges for the incorporation of climate change impacts and risk assessment into long-term land-use national development plans and strategies. Overall, this chapter provides an overview of the implications for CC to coastal management.

Chapter 4 explores the Nonlinear Muskingum method that is a very efficient tool in flood routing implementations. It is possible to estimate an outflow hydrograph by a given inflow hydrograph of a flood at a specific point of the river channel. However, it turns out an optimization problem at the stage of employing this method and it becomes important to reach the optimal model parameters so as to obtain precise outflow hydrograph estimations. Hence, it was decided to utilize five up-to-date optimization algorithms namely Vortex Search Algorithm (VSA), Gases Brownian Motion Algorithm (GBMO), Water Cycle Algorithm (WCA), Flower Pollination Algorithm (FPA) and Colliding Bodies Optimization (CBO). The algorithms were integrated with the nonlinear Muskingum model so as to estimate the outflow hydrograph of Wilson data and it was deduced that WCA, FPA and VSA perform relatively better than the models employed in the other researches before.

Chapter 5 presents a Flood Frequency Analysis using Bayesian Paradigm. At-site Flood Frequency Analysis (FFA) of extreme hydrological events under Bayesian paradigm has been carried out and compared with frequentist paradigm of maximum likelihood estimation (MLE). The main objective of this chapter is to identify the best approach between Bayesian and frequentist one for at-site FFA. As a case study, the authors used the data of only two stations named Kotri and Rasul and implemented Bayesian and MLE approaches. Most commonly used tests were applied for checking initial assumptions. Goodness of fit (GOF) tests were used to identify the best model, which indicated that the Generalized Extreme Value (GEV) distribution appeared to be best fitted for both stations. Under Bayesian paradigm quantile estimates are constructed using Morkov Chain Monte Carlo (MCMC) simulation method for their respective returned periods and non-exceedance probabilities. For MCMC simulations, as compared to other sampler, the authors use the M-H sampling technique to generate a large number of parameters.

Chapter 6 explores the Flood Modelling and Mapping. This chapter presents an overview of the important concepts related to flood hazard assessments and explores the use of remote sensing data from satellites to supplement traditional assessment techniques. The method presented in this chapter can be used in sectoral planning activities and integrated planning studies, and for damage assessment. The chapter presents the application of flood modeling to the study area. The study area, Adyar River in Chennai, has experienced major floods in the past decade which is attributed to increased urbanization. The hydrologic model for the Adyar river basin was set up using HEC geoHMS and was run and calibrated using observed flow in HEC-HMS. The chapter also discusses the results obtained from the IDF analysis and its application in HEC HMS to generate hypothetical storm hydrographs. Furthermore, the chapter goes on to discuss the results obtained from the hydraulic modeling such as the inundation map for the 2005 flood event and the inundation map for hypothetical storms of varying return periods.

Chapter 7 provides the quantification and evaluation of water erosion using SDR-InVEST Model in the Ziz Basin South-eastern of Morocco. The Ziz watershed is located in the arid zones of south-eastern Morocco and belongs to the large basin of Ziz-Rheris. In this basin, floods are related to natural factors and mainly to the occupation of the hydraulic public domain and the human intervention on the courses of the rivers. Increases in sediment yield are observed in many places in the Ziz, dramatically affecting water quality and reservoir management. In order to map overland sediment generation and delivery to the stream (studying the service of sediment retention), the InVEST Sediment Delivery Ratio (SDR) model was applied. The sedimentation analysis in the Hassan Dakhil dam, located in this watershed, shows that there is a very important erosion rate, the proof is the rapid filling of the dam. This is due to the transport of sediments in the rivers. If this situation continues at the current rate, the dam will no longer be fully operational for irrigation by 2050.

Chapter 8 draws conclusions based on the hydrologic modeling using SWAT: Test the Capacity of SWAT Model to Simulate the Hydrological Behavior of Innaoune Watershed in Semi-Arid Climate. This watershed presents an important hydric potential of the oriental part of Morocco. However, the basin exhibits a set of hydrologic drawbacks, such as floods, erosion, and pollution. This chapter is focused on flood forecast study. In order to help managers and decision makers to adopt the appropriate land management strategies for protecting the population from flood damages, the study of the hydrological behavior and quantification of water yield are paramount. According to this perspective, the main goal of this chapter is to test the ability of the SWAT model to simulate and reproduce the hydrological behavior of the upstream of Innaouene watershed. The output of the model could be used to map, delineate, and forecast the floods expansion for a particular rainfall event. SWAT was performed on a daily time step from 2004 to 2012 for calibration and 2012 to 2014 for validation. The model accuracy was evaluated by measuring the Nash-Sutcliffe coefficient and R^2.

Chapter 9 presents a hydraulic modeling using HEC RAS Model for Flood Protection. In fact, Flooding has a wide range of impacts on societies and natural environments. In this sense, the city of Fez (Morocco) suffers from these problems reflected by the overflow of Oued El Jawaher during the rainy periods. This situation led to compare between the current situation and the situation developed by the thresholds of Oued El Jawaher. HEC-RAS hydraulic model consists of 31 cross-sections, which will be used in the course of this study. The simulations will concern the current state and the developed state, for flows of different frequencies. The result of the simulations confirms that the capacity of the proposed hydraulic structures is insufficient to transit and should be considered. To conclude the development of the channel by thresholds which serves for the creation of water plan magnify the risk of an overflow of the banks of the canal by the water line along with the longitudinal profile.

Chapter 10 analyses the flood hazard casting and predictions of climate change impressions. Climate change is a word that we have heard hundreds of times, but what is it? Is it happening or is it something made by us? There are thousands of such questions, thoughts, doubt which come to our mind as soon as we hear the words "climate change." Even though there are hundreds of research works and many more proofs stating that the climate change is happening, still there is a side which has generally overlooked, and that is what if the climate change that we look is just something made by us. Climate change term refers to long-lasting changes in temperature, clouds, humidity, and rainfall around the world. Both local and global factors cause regional climate change. This difference is significant because if a regional climate change occurs on account of local factors, then these changes can be mitigated by local actions.

Chapter 11 focuses on Climate Change Induced Flood Disaster Policy Communication Issues for Local Community Adaptation Resilience Management in Uganda. Effective climate change and disaster policy communication services are vital for enhancing the adaptive resilience capacity of the vulnerable local communities in poor countries like Uganda. This chapter focuses on the effectiveness of the Ugandan national climate change and disaster policy information communication strategies in addressing national flooding disaster risks; highlight the recent trends of knowledge based responses to climate change induced floods; assess the impact of the flood on the socio-economic well-being of local households and communities; and determine the vulnerability issues with corresponding adaptation strategies to floods in the flood prone country. Climate change flood risks have continued to exact huge socio-economic loss and damage effects due to the vulnerability and weak adaptation strategies to floods. The national meteorological services tend to forecast seasonal flood events, some flood forcing factors, and the impact of floods on social, economic, ecologic, and physic infrastructure are on the rise in some parts of the country.

Chapter 12 investigates the environmental hazards assessment at pre-Saharan local scale. Ecosystem management requires biophysical and socio-economic measurement. The intervention of the government and the local community in order to combat the degradation of ecosystems must take into account the effects of environmental hazards. This can reinforce the inhabitants' ability to adapt at the local level. The impact on the ecosystem and resources are numerous and complex. Consequently, a multidisciplinary evaluation is needed. In this context, a new approach was proposed, called environmental hazards assessment at a local scale. It was used to evaluate the risk of several oasis resources to multiple hazards in the Middle Draa Valley. The findings show that for all resources, desertification is the biggest challenge affecting this area followed by drought, sandstorms, and then floods. This risk assessment approach can provide guidance for future assessments.

Chapter 13 assesses the Vulnerability of forest vegetation due to anthropogenic disturbances in Western Himalayan region of India. The Western Himalayan zone of India is not only threatened by rapid climatic changes, natural floods, fires but also due to anthropogenic activities. Himalayan forests are vulnerable due to climatic changes and faced severe ecological deterioration due to anthropogenic pressures. The degradation of forests due to anthropogenic disturbances is increasing because of overgrowth of population, high poverty ratio, as well as the limitations of alternative livelihood options. Further resources from forest makes it inseparable to manage forest stands without considering the importance of socio-economic status and ecological aspects of forest management to the well-being of local communities. Therefore, the Himalayan forests and the communities depending on forests should be seen as a part of an evolving. This chapter will explore the vulnerability of the knowledge towards Western Himalayan forests and community based managements of forests. Additionally, it will on an out sketch of potential sites affected through anthropogenic pressures.

Acknowledgment

Extreme events impacts on biophysics and socio-ecological systems have become common in geosciences and ecology. Through 13 chapters, this book tries to assess the impact, the vulnerability, the susceptibility, the exposure, the resilience, and the adaptation to floods, in urban, basin, sub-catchment, and coastal areas. This book provides a collection of various decision tools. These chapters were organized in three sections. Section 1 discusses the floods vulnerability, risk, exposure, susceptibility, and resilience and the section 2 investigates the Floods and the associated hydrologic events. However, the section 3 explores the climate change, natural hazards and anthropogenic impacts.

This collection of chapters is a result of the valuable participation of several researchers, students, academics, and policy makers with the important support of the IGI global team over a period of 12 months.

Special thanks go to the authors for their hard work to develop and revise the chapters. I would like to take this opportunity to thank the reviewers for their effort to improve the chapter quality.

I am grateful for the help provided by my little family, my wife Siham, my two children, Yassin and Amine. Many thanks are addressed to my dad Mimoun for his continuous encouragement.

Section 1

Flood Vulnerability, Risk, Exposure, Susceptibility, and Resilience

Chapter 1
Flood Vulnerability, Risk, and Susceptibility Assessment:
Flood Risk Management

Mohd Talha Anees
https://orcid.org/0000-0002-4751-0484
University of Malaya, Malaysia

Ahmad Farid Bin Abu Bakar
https://orcid.org/0000-0002-2731-5198
Universiti of Malaya, Malaysia

Lim Hwee San
https://orcid.org/0000-0002-4835-8015
Universiti Sains Malaysia, Malaysia

Khiruddin Abdullah
Universiti Sains Malaysia, Malaysia

Mohd Nawawi Mohd Nordin
Universiti Sains Malaysia, Malaysia

Nik Norulaini Nik Ab Rahman
Universiti Sains Malaysia, Malaysia

Muhammad Izzuddin Syakir Ishak
Universiti Sains Malaysia, Malaysia

Mohd Omar Abdul Kadir
Universiti Sains Malaysia, Malaysia

ABSTRACT

Flood can be assessed through flood vulnerability, risk, and susceptibility analysis using remote sensing, geographic information system, and hydrological modelling. In this chapter, different stages, complexities, and processes of flood vulnerability, risk, and susceptibility assessment were discussed. The study reveals that flood vulnerability should be assessed based on four aspects: physical, social, economic, and environmental. Flood risk should be assessed by three stages: risk analysis, disaster relief, and preparedness, whereas flood susceptibility assessment involves three processes. Overall, it was found that the responsible factors vary as per the local conditions, which need to be carefully analyzed and selected. Furthermore, the role of remote sensing and geographic information system in flood risk management were found important especially in flood risk mapping and in the selection of responsible flooding factors.

DOI: 10.4018/978-1-5225-9771-1.ch001

INTRODUCTION

Natural Hazard is a natural phenomenon that have the potential to affect negatively on humans and on the environment because of their location, severity and frequency. Flood is one of the common natural hazard in the world which affects loss of life, economy and agricultural activities. According to a National Geographic news, in the United States, floods causes approximately $6 billion worth of damage and killing of approximately 140 people every year. In a report issued by the Organization for Economic Cooperation and Development in 2007, it was found that coastal flooding inflict around $3 trillion damage worldwide. It is also added in the news about China's Yellow River valley where some of the world's worst floods have occurred, millions of people have died in floods during the last century (NGN, 2015). Whereas, in India and Nepal, thousands of killings were due to flooding in 2013, while the Philippines has suffered from recurring flooding that caused more than 100 fatalities every year between 2011 and 2013 (Tanoue et al., 2016). Major recent floods occur in East African Countries (Kenya, Rwanda and Somalia) in May 2018 (NASA, 2018), Japan (Southwestern Japan) flood in July 2018 (UNISDR, 2018), India (Kerala) flood in August 2018 (Kerala Government, 2018) and Jordan (Zara Maeen hot springs area) flash flood in October 2018 (UN, 2018). In future, it is expected that flooding will occur more rigorously and frequently due to rapid urbanization, climate change, land use and land cover pattern changes, poor watershed management activities and reduction of infiltration capacity due to urbanization (Nasiri et al., 2016). It is a serious issue which should be analysed to lower the impact of flooding on the humans and on the environment.

There are three most common types of flooding such as coastal, river and flash flooding while tsunami and hurricanes are also two less common types of floods witnessed and reported in the literature (Jonkman, 2005). Coastal flooding are due to the processes of waves, tides, storm surge, heavy rainfall from coastal storms or stream overflow. River flooding occurs when stream has bankfull discharge which means that the capacity of stream channel is not enough to conduct the amount of water available in the channel. Mostly, the bankfull discharge is due to high rainfall, steep slope and deposition of sediment on river bed which lowers the river's depth. Flash floods caused by high intensity of rainfall at local level due to which sudden rise in water levels. Tsunamis are long-period waves generated by disturbances in the ocean such as earthquakes, volcanic activity and undersea landslides. Hurricanes are large swirling storms which produces winds at very high speed (average speed of 120 km/h). The high speed winds damages buildings and trees which effects the humans.

Hazards associated with all the flooding can be divided into three types. First is the primary hazard that occur due to the direct contact with water. Massive amount of erosion due to flooding which weakens the foundations of structures, levees and buildings causes their collapse. Due to high flood velocity, large boulders and gravels, sediments and sometimes even the automobiles, houses and bridges are transported as the suspended load. Damages of house interior and furniture, loss of crops, deposition of sediment and mud after retreat of the flood water are also the primary effects of flooding.

Secondary effects are mainly the disruption of services and health problems due to flooding. These effects includes disruption in drinking water supply which results in disease and other health effects, gas and electricity services and transportation which effects food supply. Third, the tertiary effects are the long term changes in which river morphology shifted and modified. These damages can be minimized by proper implementation of traditional flood protection measures. Furthermore, flood prevention, flood risk analysis and management and flood risk mitigation should be implemented in parallel.

BACKGROUND

Flooding occurs from the combinations of different variety of sources whereas flood risk is the combination of probability of an event and its impact on exposed elements (Rooke, 2009). The management of flood risk is necessary to reduce the risk rather than avoiding it completely. The purpose of flood risk management is to reduce the human loss and economic cost at acceptable limit (Nasiri et al., 2016). Flood risk management include the following three stages (Schanze, 2006):

1. **Pre-flood management:** The aim of this stage is to reduce the flood risk, planning for flood mitigation and try to maintain it for long term duration.
2. **During flood event management:** During flood event, the response time is very limited and it is very difficult to take action at this stage. The effects of flood event depends on actions taken before the flood event.
3. **Post-flood management:** In this stage, recovery and rehabilitation actions on the basis of a flood event impact takes place.

In these three stages of flood risk management, pre-flood management is the most important because good and proper planning for prevention and mitigation of flood will reduce the flood risk at acceptable limit and hence reduce the flood impact. There are two approaches for pre-flood management such as structural and non-structural measures (Tingsanchali, 2012). Structural measure are the interventions in the flood risk system on the basis of hydraulic engineering structures such as dam, levees, dredging and widening of rivers and river dikes which modifies the river flow (Meyer et al., 2012; Tingsanchali, 2012). Whereas, non-structural measures are those which use knowledge and experience such as educating, warning, forecasting, reporting, assessing, land use land cover planning, emergency services and insurances to reduce risks and impacts (Abbas et al., 2015; Nasiri et al., 2016; Barcellos et al., 2017). Three stages of flood risk management and the processes associated with them are shown in Figure 1.

As per the topic of this study, the scope of this study is the assessment of vulnerability, risk and susceptibility in flood planning. Flood vulnerability, risk assessment and susceptibility are the important terms for flood analysis. Flood vulnerability is refers to the ability of community and its systems to face and manage adverse conditions, emergencies or disasters by using available skills and resources (Brito et al., 2017). Risk assessment is used to describe the overall processes and methods including hazards identification, risk analysis, risk evaluation and risk control. Flood susceptibility is refered to the estimation of the possibility of flood's impact in an area with certain intensity. In the following sections, flood vulnerability, risk and susceptibility assessment through different techniques or models (as stated in the literature) will be discussed. Furthermore, the role of remote sensing and the geographic information system (GIS) in flood vulnerability, risk and susceptibility assessment will also be discussed.

Figure 1. Flood risk management approaches and their associated processes (Ali, 2013)

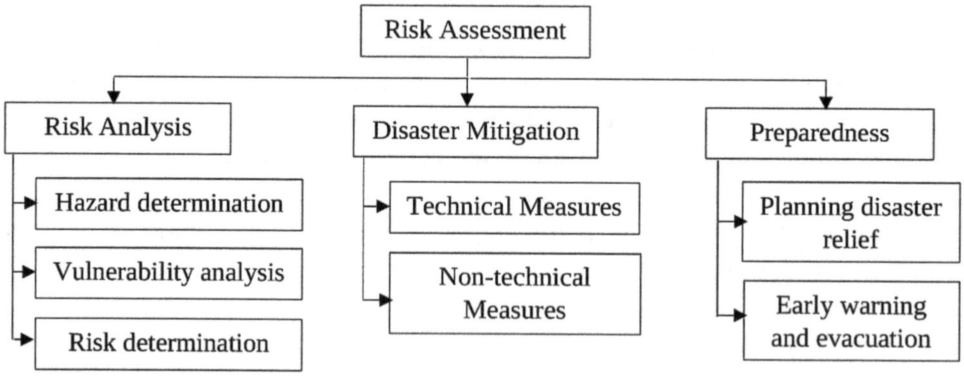

FLOOD VULNERABILITY

Flood vulnerability is one of the important component in pre-flood planning and damage assessment (Connor and Hiroki, 2005). Flood vulnerability assessment is a complex process which depends on number of regional, social, environmental, economic and political factors. Vulnerability needs to be assessed by modern techniques which should be easily understandable by officials at all levels and even community leaders (Jixi Gao, 2007). There are number of definitions of vulnerability available in the literature which are based on specific conditions and local level factors (Nasiri et al., 2016). Some definitions focused on the protection of individuals, others on the maintenance of economic activities or the protection of the environment. Some specific aspect of vulnerability depends on the type of study, on the results required (damage evaluation or urban planning project), on the kind of flood hazard (flash flood or slow-flood), on the spatial and temporal scale of study, on the specificity of the study area and on the temporality (prevention, crisis, post crisis) (Barroca et al., 2007). Different definitions given in the literature are shown in Table 1.

Flood Vulnerability Assessment

Flood vulnerability assessment is an important issue at present for the safety and protection of community and properties because of increasing rainfall intensity in warmer climates which will increase the flood frequency events (Fernandez et al., 2016). Variety of approaches are available in the literature to assess vulnerability, therefore, selection of appropriate methodology is necessary for researchers and decision makers to reduce damages and fatalities. Following are the four aspects for flood vulnerability assessment which needs to be considered (Fernandez et al., 2016):

1. The physical aspect that represents the potential of physical impact on the exposed area.
2. The economic aspect which shows the potential impact of flood on economic resources.
3. The social aspect which represents the capacity to handle, prevent and recover from flood impact by individuals or communities, and
4. The environmental aspect which shows the potential of natural environment and the ability of ecosystems to handle and recover from flood impacts.

Table 1. Vulnerability definitions stated in the literature.

Sources	Definitions
United nations (1982)	Vulnerability is a degree of damage to a certain objects at flood risk with specified amount and present in a scale from 0 to 1 (no damage to full damage)
Menoni and Pergalani (1996)	Vulnerability term is damage goods, people, buildings, infrastructures and activities in hazard condition
Mileti (1999)	Degree of the capacity to endure or recover from the impacts of a hazard during the time
WHO (2002)	Vulnerability is the degree to which a population, individual or organization is unable to anticipate, cope with, resist and recover from the impacts of disasters.
Wisner (2004)	The characteristics of an individual or group of people and their condition that affect their ability to predict, tackling, struggle, and recover from the effects of environmental threats
Adger (2006)	Susceptibility to harm from exposure to pressures related with environmental and social changes, and in lack of adaptation ability
Barroca et al. (2007)	Vulnerability is defined as the results of damages due to any possible occurrence of hazard.
Borden et al. (2007)	Distinct vulnerability means potential or sensitivity to losses or harm. Social vulnerability contains the susceptibility of society or social groups to potential losses from hazards
UNISDR (2009)	Vulnerability is the characteristics and circumstances of a community, system or asset that make it susceptible to the damaging effects of a hazard
Balica and Wright (2010)	Vulnerability is defined with interaction between Exposure, susceptibility and resilience of each community in risk condition.
Hong et al. (2015)	Vulnerability is associated with a disruptive event and quantified by the performance drop of the system under the event.
UNISDR (2017)	Vulnerability is the conditions determined by physical, social, economic and environmental factors or processes which increase the susceptibility of an individual, a community, assets or systems to the impacts of hazards.

The selection of model is very significant to understand the behaviour of natural hazard and disaster. The four aspects (mentioned above) are analysed by different models which have their own importance and significance. These models are discussed in the following section.

Physically Based Flood Vulnerability Assessment

Physically based models are widely used to understand the physical environmental processes and their impact on the built environment (Tate, 2012). Physically based models generally simulate the actual flood event with an input and output boundary conditions. The input may be either single or multiple. The simulated flood event first calibrated with the actual or in situ data to ensure that the model working properly and accurately. After calibration, future or artificial simulations are conducted to analyse the potential of physical impact on the exposed area and then validated with in situ or actual data. On the basis of estimated physical impact on the exposed area through simulation, flood vulnerability can be assessed. Based on the flood vulnerability assessment, structural measures are taken such as construction of dam, levees, river dredging and widening to minimize physical impact of flood event up to acceptable limit. However, it is difficult to find out the exact causes and effects of flood event because of the involvement of several minor or major parameters and factors. These parameters and factors varies from region to region. Although, during flood event, some important parameters and factors prevails which needs to be considered which are given in Table 2.

Physically based flood vulnerability assessment can be done on the basis of flood action in the form of hydrostatic, hydrodynamic, erosion, buoyancy and debris actions (Kelmen, 2003). In hydrostatic action of flood, lateral pressure and capillary rise are the two forms which damages the housing walls and interior of the

Table 2. Some important parameters in physically based model during flooding event

Parameters	Description
Topographic data	Topographic data are in the form of pixel based elevation values. In physical models, water moves from higher elevation value to lower elevation value. So, physical model simulate flood event on the basis of distribution of elevation values. In remote sensing, these topographic data are known in the form of Digital Elevation Model which varies from coarse resolution to fine resolution.
Slope	Slope can be derived from elevation values. The distribution of flood event is also depends on the slope distribution. It is also important in terms of soil erosion. During flood, in generally high or steep slope areas, sediment yield from steep slope to gentle slope. This results the deposition of sediment on river bed which lowers the depth of river channel and hence water overflows from channel to floodplain.
Rainfall intensity	The extent and severity of flood event depends on the intensity and duration of rainfall. Rainfall intensity may not be the only possible reason of extent and severity of flood event but it can be, in combination with elevation and slope values. The severity of flood event in low lying or gentle slope areas can also be increased due to anthropogenic activities such as land use practices and agricultural activities in high slope areas. Furthermore, rainfall intensity also varies in different climate.
River bathymetry	River bathymetry data is one of the most important parameter in flood modeling because major part of rainfall flow is through river channel. River channel overflow either due to high intensity and long duration rainfall or due to lowering down the river bathymetry depth because of sediment deposition in combination with high flow in the river channel.
Land use land cover data	Land use land cover (LULC) parameter is important in overland flow. LULC work as a resistance to overland flow. Overland flow will be high in open or uncultivated land as compared to cultivated or scrub land. Similarly, impact of flood event on settlement area make trouble for human being and community.

houses. Hydrodynamic action of flood is in the form of flood velocity, turbulence and waves. Erosion action of flood cause removal of soil from river channel sides and river bed along which the water flows. The transportation of sediment, its deposition and depositional arrangement of grain size depends on the water flow velocity, turbulence and bed roughness. Buoyancy action during flooding is a force which uplift the lighter structures such as sheds, pipes and other daily usable material. These structures and materials acts as a tool which hit to the houses and the buildings due to the buoyancy forces which causes damage, destabilisation or sometimes complete destruction (Kelmen, 2003). During flooding, sediment deposited in settlement areas and accumulate inside and outside the houses. The deposited sediment varies in size and their impact on the walls of houses depends on the flow velocity and the buoyancy forces.

The analysis of above mentioned flood actions gives the idea about how vulnerable the area is and how to suggest mitigation and protection measures. Once the plan for mitigation and protection checked out then the necessity will arise for the construction of engineering structures along or on the river as per the conditions. That structures will reduce the flood vulnerability up to acceptable limit. It is not necessary that these mitigation and protection measures will work all the time. It is also need to regularly check them by linking weather forecasting and physically based flood modeling technique to make sure the system is working properly or not.

Physically based hydrological models plays an important role in flood prediction because their accurate output based on the factors prevailing in a particular region. Basically these types of models works according to user's understanding. The user can input a particular combination of factors with their particular intensities or areas according to the condition of a particular region. The user can construct imaginary engineering structures in the models, control the intensity of rainfall, steepness of the slope and roughness of open surface which play a major role in regulating the runoff. Based on the user's input parameters and conditions, hydrological models gives runoff or flooding condition for a particular region. Therefore, hydrological models are very effective in future flood prediction.

Economically Based Flood Vulnerability Assessment

Economic vulnerability, in terms of flood, refers to risks caused by the flooding to a system of production, distribution and consumption (Dincer and Hacioglu, 2017). Economic vulnerability depends on the exposure of an area and flood characteristics. The areas in floodplain nearer to the river are more vulnerable because the areas are hit by more flood velocity as compared to the areas which are far away from the river. Additionally, flow speed during flood erode house walls which damages houses or settlement areas. Economic vulnerability, in terms of flood characteristics, depends on the flood intensity, duration and inundation area and flooding depth. The damages caused by the exposure of an area and flood characteristics are used to assess the economic loss due to flooding. On the basis of economic losses, economic vulnerability can be assessed. In several studies, indexes were developed to assess economic vulnerability which will be discuss later in the following section. To understand what parameters should be used in economic vulnerability indexes to assess economic vulnerability, types of damage effects needs to be discussed first.

The effects of damages can be categorised into direct and indirect effects (Messner and Meyer, 2006). Direct flood damages effects are the immediate effects caused by physical contact of a flood event people, their property and surrounding environment. Direct effects of flood damage includes destruction of buildings, damage of economic goods, loss of standing crops and livestock in agriculture, human fatalities, immediate health impacts and contamination of ecological systems. Indirect flood damage effects are caused by post flood event. The extent of indirect flood damage effects are more than the direct flood damage effects or actually inundated area. It includes the interruption the immediate supply of goods, loss of energy and telecommunication supplies and loss of economic production due to destruction of facilities such as roads.

For economic vulnerability assessment, damage potential of an area is very important. The damage potential of an area is refers to the maximum possible amount of damage under the extent of flood inundation of the area (Messner and Meyer, 2006). This potential damage of specific area will be helpful to understand and to estimate the expected flood damages. Those indicators having considerable impact for producing damages during flooding should be considered as a vulnerability factor. Vulnerability factors are not same all the time but these factors modify or changes as per the condition, intensity and duration of flood and inundation area of flooding. Sodhi (2016) refers to these changes as vicious cycle. He represent the economic vulnerability in a cycle which is showing in Figure 2.

Socially Based Flood Vulnerability Assessment

Social vulnerability, in terms of flooding, refers to the inability of people and communities to face the impact of flood event due to deficiencies in infrastructure (Enarson, 2007), incomplete warning information (Cutter et al., 2000), lacking of planning and preparedness or insufficient resources provided by social groups and government agencies (Yang et al., 2015). Government agencies play a very important role in reducing social vulnerability risk by planning of shifting settlement areas from floodplain to safer place, proper and regular weather forecast information and emergency preparedness planning. However, social vulnerabilities are largely ignored because of the difficulty in its assessment. Social vulnerability can be assess through

Figure 2. A cycle showing the modification of economic vulnerability assessment factors due to changes in flood event

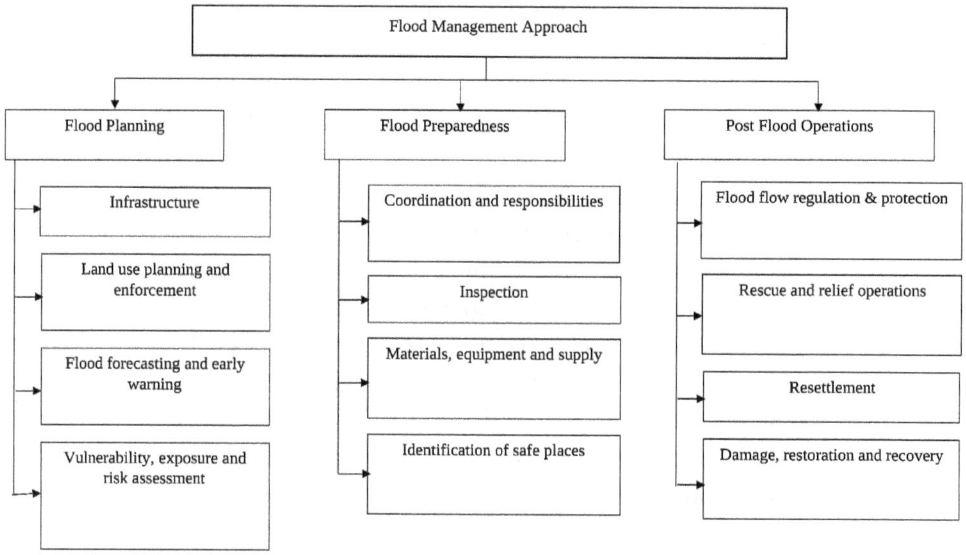

demographic data by interacting with humans and communities. It is necessary to build algorithms describing social vulnerability factors on the spatial distribution of human susceptibility to flood impacts (Tate, 2012). Validation of these algorithm can be done through independent proxy data (Frigerio et al., 2016).

Environmentally Based Flood Vulnerability Assessment

The impact of flooding, in terms of environment, occurs directly or indirectly on the ecosystem of a specific area. The restoration of environmental damages is a long term process. The Alberta Water Portal Society mentioned some direct impact of flooding which affect environment and ecosystem (Table 3).

Indirect impact of flooding on environment and ecosystem depends on the government planning and policies. There are two major indirect impacts. First, health problem in the flooding area. After flooding, the area should be clean as soon as possible otherwise various communicable diseases spread in the area. According to WHO, there are two types of communicable diseases which are as follows:

1. Water-borne diseases, such as typhoid fever, cholera, leptospirosis and hepatitis A,
2. Vector-borne diseases, such as malaria, dengue and dengue haemorrhagic fever, yellow fever, and West Nile Fever

Table 3. Direct impact of flooding on environment and their description

Direct Impact of Flooding	Description
The health and wellbeing of wildlife and livestock	Fatalities of animals in flood inundated area and livestock unable to relocate at safer place or forced to remain in polluted water until rescued. As a result, the reduction of biodiversity level, food and habitat potential in the ecosystem, creating long-term impacts for surviving wildlife.
Riverbank Erosion and Sedimentation	During high intensity and long duration of flooding, huge amount of sediment transported from one point to another point and deposited due to reduction of flood velocity. The deposition of sediment can reduce the storage capacity of reservoirs and wetlands. Huge sedimentation would lead degradation of water quality, temporarily affect municipal, industrial and recreational water supply.
Dispersal of Nutrients and Pollution	The debris in the form of trees, stones and small pieces of houses, pollutants and nutrients contained in flood water during flooding and deposited on floodplain. Pollutants in the form of oil spills from vehicles and nutrient in the form of pesticides affect water quality of river, reservoir and sea coast. It also responsible for the fatalities of coastal fishes. There is also a positive effect of nutrient deposition on landscape. Nutrient in the form of organic material and sediment carried by flood water and its deposition on landscape increase the fertility of plants and vegetation.
Replenishment of Surface and Groundwater	Replacement of surface and groundwater can benefit soil, resulting in healthy crops and pastures.
Local Landscape and Habitat	In urban areas, flooding negatively affect by damaging infrastructure, building and businesses. Whereas, in natural environment, fertility of vegetation and plantation improved by nourishing landscape.

Second is air pollution which spread due to emission of methane gas from waste deposition on landscape or flooded rice paddy fields (Minamikawa et al., 2015).

RISK ASSESSMENT

Risk is a statement of probability that an event will cause x amount of damage, or a statement of the economic impact due to an event in terms of money (Nelson, 2018).

Risk assessment is the overall process of hazard identification, risk analysis, risk evaluation, flood mitigation and preparedness. Hazard identification is the process of finding, listing and characterizing hazards while risk analysis is a process of understanding the nature of hazards and determining the level of risk. Risk evaluation is the process of comparing an estimated risk against given risk criteria to determine the significance of the risk. The process of risk assessment is shown in Figure 3.

Figure 3. Risk assessment stages and processes
(Plate, 2002).

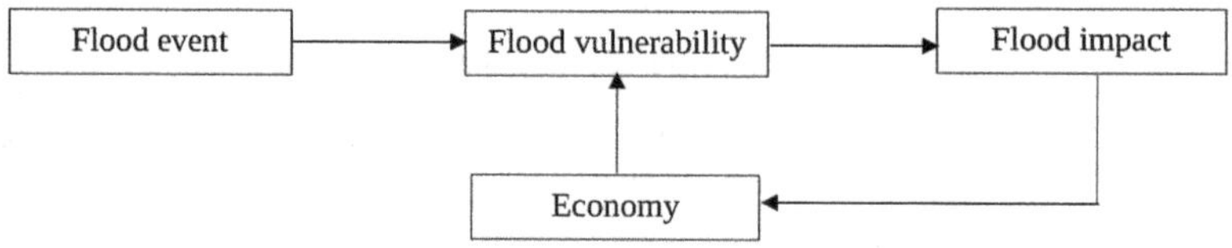

Risk Analysis

Risk analysis is the process of reviewing and identification of risks which are associated with impact and damages of a particular flood event. Flood risk is the product of two components (Balica et al., 2012):

$$Flood\, Risk = probability\, of\, flooding + consequences\, of\, flooding \quad (1)$$

Probability of flooding refers to the probability of a flood event in terms of intensity and severity occurring or being simulated in a given period of time. Whereas, consequences of flooding depends on the water depth, flow speed, duration, effects due to wave action and water quality. Consequences also depends on the vulnerability of people, property and the environment affected by a flood.

Flood hazard is the potential loss of life, property and economy due to future flood events (Guideline). Flood hazard occurrence depends on the severity of flooding which is effected by flooding behaviour, topography and emergency management. Flood hazard identification involve four processes:

1. First, the collection of past and present flood occurrences which will be helpful in simulating future flood occurrences and its spatial distribution (Rahmati and Pourghasemi, 2017).
2. Second, selection of flood conditioning factors such as morphometric parameters (elevation, slope aspect, slope angle, drainage density) and topographic parameters (wetness index, plan curvature, geology, land use, distance from rivers and soil texture) to make flood prone area map (Rahmati and Pourghasemi, 2017).
3. Third, validation of simulated flood event with past and present flood occurrences which is helpful to assess flood risk to people.
4. Fourth, quantification of flood hazard by the product of flood depth (d) and velocity (v). Quantification of flood hazard reflect the behaviour of flooding which is useful in the estimation of impact caused by flooding.

Risk to People (RP) can be calculated by (Surendran et al., 2008):

$$RP = N_z \times FHR \times AV \times PV \tag{2}$$

Where, N_z is nuber of people within the hazard zone, FHR is flood hazard risk, AV is area vulnerability which is a function effective flood warning, starting speed of flooding and nature of the area, PV is people vulnerability which is a function of presence of those people who are unable to rescue themselves at the time of flooding such as very old, infirm, disabled and long-term sick. Flood hazard risk can be calculated by (Surendran et al., 2008):

$$FHR = d \times \left(v + n\right) + DF \tag{3}$$

Where, n is a constant (0.5) and DF is debris factor (0, 0.5, 1 depending on probability that debris will lead to a hazard).

Vulnerability are of different types, as discussed above, but in general, vulnerability can be expressed as (Balica et al., 2009):

$$Vulnerability = Exposure + Susceptibility - Resilience \tag{4}$$

Exposure refers to the elements such as structures, population, agriculture, businesses and assets which is directly in contact during flooding. Susceptibility refers to those elements in exposure which have the probability to being influence by a flood event. Resilience is the ability of a system which include social, economic, physical and environmental components to recover from a flood event. Several studies assess vulnerability by using flood vulnerability index (Cornor, 2005; Pappenberger, 2007; Balica et al., 2009; Balica et al., 2013; Fernandiz et al., 2016; Ibrahim et al., 2017). Balica et al. (2009) proposed a general a flood vulnerability index (FVI) equation on the basis of exposure (E), susceptibility (S) and resilience (R) which is given as:

$$FVI = \frac{E \times S}{R} \tag{5}$$

Exposure and susceptibility are placed in numerator because these are the indicators of increasing flood vulnerability while resilience is in denominator because it represent decrease in flood vulnerability. Balica et al. (2013) modify the Eq. (5) on the basis of social, economic, physical and environmental vulnerability which is given as:

$$Total\,FVI = \frac{\left(\frac{E \times S}{R}\right)_{Social} + \left(\frac{E \times S}{R}\right)_{Economic} + \left(\frac{E \times S}{R}\right)_{Environmental} + \left(\frac{E \times S}{R}\right)_{Physical}}{4} \quad (6)$$

The indicators used in social, economic, environmental and physical vulnerability factors by previous studies are shown in Table 4.

Direct and indirect damages discussed above which include loss of life, damage estimation and number of houses in inundated area. Jonkman et al., (2005) estimate mortality by the following formula:

$$Mortality = \frac{Number\ of\ fatalities}{Total\ number\ of\ affected} \quad (7)$$

Balica et al., (2013) mentioned the formulae of damage estimation (ED) and number of houses in inundated area (NH) which are given as:

$$ED = MV \times Y \times A \times DI \quad (8)$$

$$NH = \frac{IA \times PD}{FM} \quad (9)$$

Where MV is market value, Y is yield per unit area, A is area of cultivation, DI is damage impact factor, IA is inundated area, PD is population density and FM is average number of family per household.

Several other studies assess flood vulnerability by different methods. Nasiri et al. (2016) reviewed four methods to assess the vulnerability which are curve method, disaster loos data method, computer modelling methods and indicator based method. They found that the indicator-based approach gives more precise vision of overall flood vulnerability in each area in comparison to other approaches. Lee and Choi (2018) used composite indicator to assess the vulnerability. They used three components and each component had three proxy variables. First is the exposure; which includes maximum daily rainfall, rainfall intensity and heavy rainy days. Second is the sensitivity; which includes population density, urban area ratio and basin slope. Third is the coping; which include river improvement ratio, pumping station capacity and civil servant ratio. Similarly methods for the vulnerability assessments vary from region to region must be carefully identified.

Table 4. Vulnerability factors and their indicators

Vulnerability Factors	Indicator
Social	• Population density • Population age • Demographic pressure • Unsafe settlement • Access to basic service • Disability • Poverty level • Literacy rate • Attitude • Decentralisation • Community participation • Child mortality rate • Elderly and child population rate • TV penetration rate • Employment • Family Income • Ethnicity and gender • Immigrant status
Economic	• Local resource base • Diversification • Small Businesses • Accessibility • Local gross domestic product • Growth per capita income • Agriculture • Industry • Cost of flood damages • Cost of structure damages
Environmental	• Area under forest • Degraded land • Overused land • Unpopulated land area • Types of vegetation • Deforestation
Physical	• Slope • Precipitation • Flood duration and intensity • Flood return period • Proximity to river • Soil moisture and evaporation rate • Flow velocity • Flood water depth • Sedimentation load • Flood extent • Polluted load from flood

(Connor et al., 2005; Bollin and Hidajat, 2006; Gao et al., 2007; Balica et al., 2009; Tate et al., 2012; Fernandez et al., 2016; Armenakiz et al., 2017; Brito et al., 2017; Sodhi, 2016)

Disaster Mitigation

After the assessment of flood risk, mitigation measures are proposed with the existing best methods. In the beginning of disaster mitigation process, exact causes of flooding need to be understand clearly and vulnerability of elements which are at risk. Then based on causes of flooding and risk, technical and non-technical measures are suggested. Technical measures depends on the system which is using in specific area for a particular event. For instance, if a study is conducted to know the effects of upstream anthropogenic activities on downstream in terms of flooding and drainage flow. The reason of this effect can be possible only through the physically based modelling which has the ability to give exact causes and effect of flooding. Good technical measures are those which integrate protection of rural and urban areas, through coordination with different organizations in the system and by including bridges and culverts designed to pass more than the design flood (Plate et al., 2002). Whereas, non-technical measure are those which are virtual and can be mitigate without any technical solution. For example, different government and non-government agencies help people after flooding by providing them necessary items to survive during and post flooding. Based on technical and non-technical solutions, structural and non-structural measures are taken which are discussed in the above sections.

Preparedness

Preparedness include two stages. First is the planning disaster relief and second is early warning and evacuation. Disaster relief operations are complex which need careful planning to improve disaster preparedness. Improved preparedness system can help in saving lifes, reduce the suffering of survivors and enable communities to restart normal life more quickly (Wisetjindawat et al., 2014). According to (Guidline), following are the principle actions that need in planning system for flood risk management:

1. In the earliest stage, identification of flood hazard and potential flood risk from national, regional, local and site specific.
2. The infrastructure to act for disaster relief operation should be developed in those areas where little or no flood hazard to minimize the risk.
3. In case of no alternative, the infrastructure should be developed in lower risk areas in flood hazard region.
4. If necessary, a precautionary approach should be applied to reflect uncertainty in flooding data, risk assessment techniques and the ability to predict the future climate and performance of existing flood defences.

5. Development of planning with the help of flood risk mapping.
6. Identification of land required for flood protection scheme such as camping for survivals and proper goods and medicine supplies. Some land also required to store flood water.

After principle actions planning, the responsible authorities should apply sequential approach to ensure that the development is leading towards lowering the flood risk. The sequential approach should be applied according to the flowing stages (ADRG, 2014):

1. Identification of low vulnerable flood zone to avoid flood risk.
2. Ensure the type of development proposed is at least vulnerable stage (means less disruption during flood) and is being consider at strategic reasons.
3. Ensure the reduction of flood risk at acceptable limit, and
4. Ensure the emergency planning measures are in correct place.
5. Ensure effective flow of information and coordination to provide basic needs and essential facilities in case flood damage (Ali, 2013).

In case of early warning system and evacuation, following points should be followed (Ali, 2013):

1. The flood forecasting and early warning system should be functional and regularly being checked.
2. Community based early warning system should be in correct place and functional.
3. Proper training, rehearsals, drills and strict vigilance should be exercised and sufficient resources are deployed to strengthen structure at river bank.
4. Safe places should be identified for evacuation.
5. Basic necessities and temporary shelter are arranged.
6. Transport for evacuation should be available.

FLOOD SUSCEPTIBILITY

Flood susceptibility is also an important term to assess the damages (as discussed above) which are associated with flooding. Flood susceptibility is defined as the probability of a flood event occurring in an area on the basis of local terrain conditions (Santangelo et al., 2011). Three processes are associated with flood susceptibility analysis. First is the identification of flood risk factors as per the local terrain conditions. A number of flood risk factors discussed above main those which are associated with physically based flood vulnerability assessment. Second is correlation of flood risk factors with flood inundation of a particular flood event. Third is the assessment of flood susceptibility on the basis of resulting relationships produced by the correlation.

Flooding cannot be analysed from one point or multiple point. It must be analysed through spatial and temporal analysis. For that purpose, Remote Sensing and Geographic Information System (GIS) is the most useful techniques for spatial and temporal analysis. Flood susceptibility, in general, analysed through flood susceptibility map which is also developed through remote sensing and GIS.

ROLE OF REMOTE SENSING AND GIS IN FLOOD RISK MANAGEMENT

Remote sensing and GIS are quick and more efficient techniques for capturing, storing and combining, manipulating, retrieving, analysing and displaying the information for the determination of potential hazard areas (Samanta et al., 2018). Remote sensing and GIS techniques have the ability to manipulate and analyse all relevant flooding information in order to easily separate flood hazard zones from low risk to high risk and flood related damages (Patel and Srivastava, 2013; Pourghasemi et al. 2014; Samanta et al., 2018). Several satellite data are available in which the most useful one is Landsat Thematic Mapper imageries of 30m resolution is the main source of data for monitoring flood and delineating the boundary of inundation (Joy, 2004). Joy (2004) also discussed which band of Landsat data is useful in flooding studies and about the other satellite data sets. He also discussed about the image processing required in satellite data sets to reduce uncertainties in the data such as cloud covering.

Satellite data are in the form of pixels which have a specific dimensions. These dimensions varies from one satellite data to the other. Digital Elevation Model (DEM) is one of the most important remote sensing data which has elevation values in each pixels. DEM of different resolutions, but 30 m resolution is available in public domain for developing countries, generally used in flood simulation through hydrological modelling. From DEM, slope, elevation, contour and drainage map can be generated in GIS environment. While Landsat data is useful in delineating land use land cover maps, distance from river, vegetation density and impervious surfaces. All these elements are the flooding factors and also discussed in the above sections.

Remote sensing and GIS can be used to delineate flood inundation maps. Based on flood inundation maps, zones of low, moderate and high risks and vulnerabilities are identified. Remote sensing and GIS can be used:

1. To identify suitable places to setup infrastructure for flood preparedness planning (as discussed above),
2. To identify the areas where structural measures need to applied,
3. To select safe routes for better transportation during flooding,
4. To demarcate the flooding extent based on future flood simulation techniques through hydrological modeling and
5. To identify or select those areas where vulnerability factor's indicator data need to be collected (indicators as stated in Table 4).

Remote sensing and GIS are also very important for flood susceptibility assessment because flooding covers large areas which can be analysed more easily by aerial view with the help of multiple spatial layers. In recent studies, Dottori et al. (2016) assessed flood vulnerability and susceptibility based on a mathematical index formula including simple to complex hazard scenarios. Whereas, Samanta et al. (2018) assessed flood vulnerability and susceptibility using ten independent variables, namely land use/land cover, elevation, slope, topographic wetness index, surface runoff, landform, lithology, distance from the main river, soil texture and soil drainage through remote sensing and GIS.

Several studies have been done for flood risk assessment by including land cover mapping (Sande et al., 2003), maximum flood discharge, flood intensity, flow velocity and water depth (Vojtek and Vojteková, 2016), mapping flood affected areas, estimate the damages and design the hydraulic works (Kheder, 2014). Armenakis et al. (2017) used spatial data layer and social indicators for flood risk assessment in urban areas. Spatial layers include physical components of the urban system, such as infrastructure elements while social indicator include factors such as age, single-parent families, marital status, education, income, and minority populations. All these studies used remote sensing and geographic information system (GIS) to make the maps, to identify the nature of flooding and to estimate and analyse various indicators. Vojtek and Vojteková (2016) used hydrological modelling in addition to other techniques to estimate flow velocity and water depth. However, parameters used for flood risk assessment vary from one region to other region which needs to be carefully selected for a particular region.

CONCLUSION

Flood is one of the common natural hazard in the world. The recent concern of researchers that how to reduce impact and damages due to flooding. The factors responsible for flooding vary from one region to another. Several studies were conducted for the assessment of flood risk, vulnerability and susceptibility. Number of policies and guidelines were also developed as per the particular region. Because of complexities in the assessment of flood risk, vulnerability and susceptibility assessment, present study was conducted to analyse the responsible factors and their indicators.

In terms of flood vulnerability, it was found that there are four aspects to assess flood vulnerability. First is the physically based assessment in which past, present and future flood situation can be simulated. The parameters which are needed in physically based assessment were identified and discussed. Physically based assessment can give information about the flood inundation area, flood depth, flood velocity and the impact due to land use land cover changes in a particular region. Second is the socially based assessment in which damages related to people and communities are included. Socially based assessment gives an estimation of loss of lives, migration of people from one place to other place and types of people affected by flooding. Third is the economic based assessment which gives an estimation about the economic lost either community level or state level or national level. This estimation are needed in making disaster mitigation and preparedness planning. Final and fourth is the environmentally based assessment which gives an idea that how much ecological system affected by flooding. It includes the assessment of water quality reduction, coastal water quality reduction and emission of methane gas either from waste material or rice paddy fields.

In terms of flood risk, three stages are involved. First is the risk analysis which include hazard identification, vulnerability analysis and risk determination. In hazard identification, researchers and policy makers identify the source and behaviour of hazard and the factors responsible for a specific flood event. Vulnerability analysis are conducted on the basis of exposure of elements, their susceptibilities and resilience of the elements. Whereas, risk determination involve the probability and consequence of flooding. Several formulae related to flood risk analysis are also mentioned. Second is the disaster mitigation which include technical and non-technical measures. Technical measures are related to technical failure of a specific system while non-technical measures are referred to as virtual and can be mitigate without any technical solution. Third is the preparedness which include planning disaster relief and early warning and evacuation. This is the necessary stage in which proper planning and policy making processes are involved to reduce impact and flood risk at acceptable limit.

The processes responsible for flood susceptibility assessment are identification of flood risk factors, correlation of flood risk factors with flood inundation of a particular flood event and the assessment of flood susceptibility on the basis of resulting relationships produced by the correlation. Additionally, the role and importance of remote sensing and GIS in flood risk management processes and policies were also discussed.

ACKNOWLEDGMENT

This research is gratefully acknowledge the School of Physics and School of Industrial Technology of Universiti Sains Malaysia and Department of Geology of Universiti of Malaya for providing the required research facilities for this work. Special thanks to the University of Malaya for providing the research facilities under the grant GPF017B-2018 to carry out this work. This research received no specific grant for this publication from any funding agency in the public, commercial, or not-for-profit sectors.

REFERENCES

Abbas, A., Amjath-Babu, T. S., Kächele, H., & Müller, K. (2015). Non-structural flood risk mitigation under developing country conditions: An analysis on the determinants of willingness to pay for flood insurance in rural Pakistan. *Natural Hazards*, *75*(3), 2119–2135. doi:10.100711069-014-1415-x

Adger, W. N. (2006). Vulnerability. *Global Environmental Change*, *16*(3), 268–281. doi:10.1016/j.gloenvcha.2006.02.006

Ali, A. (2013). *Indus basin floods: Mechanisms, impacts, and management.* Mandaluyong City, Philippines: Asian Development Bank.

Armenakis, C., Du, E. X., Natesan, S., Persad, R. A., & Zhang, Y. (2017). Flood Risk Assessment in Urban Areas Based on Spatial Analytics and Social Factors. *Geosciences, 7*(4), 123.

Australian Disaster Resilience Guideline (ADRG). (2014). *Guideline 7-3: Technical flood risk management guideline: Flood hazard, Australian Institute for Disaster Resilience CC BY-NC*. Australian Institute for Disaster Resilience.

Balica, S., & Wright, N. G. (2010). Reducing the complexity of the flood vulnerability index. *Environmental Hazards*, *9*(4), 321–339. doi:10.3763/ehaz.2010.0043

Balica, S. F., Douben, N., & Wright, N. G. (2009). Flood vulnerability indices at varying spatial scales. *Water Science and Technology*, *60*(10), 2571–2580. doi:10.2166/wst.2009.183 PMID:19923763

Balica, S. F., Popescu, I., Beevers, L., & Wright, N. G. (2013). Parametric and physically based modelling techniques for flood risk and vulnerability assessment: A comparison. *Environmental Modelling & Software*, *41*, 84–92. doi:10.1016/j.envsoft.2012.11.002

Barcellos, P. D. C. L., da Costa, M. S., Cataldi, M., & Soares, C. A. P. (2017). Management of non-structural measures in the prevention of flash floods: A case study in the city of Duque de Caxias, state of Rio de Janeiro, Brazil. *Natural Hazards*, *89*(1), 313–330. doi:10.100711069-017-2966-4

Barroca, B., Mouchel, J. M., Bonierbale, T., & Hubert, G. (2008). *Flood Vulnerability Assessment Tool (FVAT). In Day Water: An Adaptive Decision Support System for Urban Stormwater Management* (pp. 121–127). London, UK: Published by IWA Publishing.

Bollin, C., Hidajat, R., & Birkmann, J. (2006). Community-based risk index: Pilot implementation in Indonesia. In Measuring vulnerability to natural hazards: Towards disaster resilient societies (pp. 271-289). United Nation University Press.

Borden, K. A., Schmidtlein, M. C., Emrich, C. T., Piegorsch, W. W., & Cutter, S. L. (2007). Vulnerability of US cities to environmental hazards. *Journal of Homeland Security and Emergency Management*, *4*(2), 1547–7355. doi:10.2202/1547-7355.1279

Connor, R. F., & Hiroki, K. (2005). Development of a method for assessing flood vulnerability. *Water Science and Technology*, *51*(5), 61–67. doi:10.2166/wst.2005.0109 PMID:15918359

Cutter, S. L., Mitchell, J. T., & Scott, M. S. (2000). Revealing the vulnerability of people and places: A case study of Georgetown County, South Carolina. *Annals of the Association of American Geographers*, *90*(4), 713–737. doi:10.1111/0004-5608.00219

de Brito, M. M., Evers, M., & Höllermann, B. (2017). Prioritization of flood vulnerability, coping capacity and exposure indicators through the Delphi technique: A case study in Taquari-Antas basin, Brazil. *International Journal of Disaster Risk Reduction*, *24*, 119–128. doi:10.1016/j.ijdrr.2017.05.027

Dincer, H., & Hacioglu, U. (2017). Investigating the Economic Vulnerability Factors of Emerging Markets after the Global Financial Crisis of 2008 with a Hybrid Multi-Criteria Decision-Making Approach. In Handbook of Research on Economic, Financial, and Industrial Impacts on Infrastructure Development (pp. 313-333). IGI Global. doi:10.4018/978-1-5225-2361-1.ch016

Dottori, F., Martina, M. L. V., & Figueiredo, R. (2016). A methodology for flood susceptibility and vulnerability analysis in complex flood scenarios. *Journal of Flood Risk Management, 11*, S632–S645. doi:10.1111/jfr3.12234

Enarson, E. (2007). Identifying and addressing social vulnerabilities. *Emergency Management: Principles and Practice for Local Government, 2*, 257-278.

Fernandez, P., Mourato, S., Moreira, M., & Pereira, L. (2016). A new approach for computing a flood vulnerability index using cluster analysis. *Physics and Chemistry of the Earth Parts A/B/C, 94*, 47–55. doi:10.1016/j.pce.2016.04.003

Frigerio, I., Ventura, S., Strigaro, D., Mattavelli, M., De Amicis, M., Mugnano, S., & Boffi, M. (2016). A GIS-based approach to identify the spatial variability of social vulnerability to seismic hazard in Italy. *Applied Geography (Sevenoaks, England), 74*, 12–22. doi:10.1016/j.apgeog.2016.06.014

Gao, J., Nickum, J. E., & Pan, Y. (2007). An assessment of flood hazard vulnerability in the Dongting Lake Region of China. *Lakes and Reservoirs: Research and Management, 12*(1), 27–34. doi:10.1111/j.1440-1770.2007.00318.x

Hong, L., Ouyang, M., Peeta, S., He, X., & Yan, Y. (2015). Vulnerability assessment and mitigation for the Chinese railway system under floods. *Reliability Engineering & System Safety, 137*, 58–68. doi:10.1016/j.ress.2014.12.013

Ibrahim, N. F., Zardari, N. H., Shirazi, S. M., Mohd Haniffah, M. R., Mat Talib, S., Yusop, Z., & Mohd Yusoff, S. M. A. (2017). Identification of Vulnerable Areas to Floods in Kelantan River Sub-basins by using Flood Vulnerability Index. *International Journal of GEOMATE, 12*(29), 107–114.

Jonkman, S. N. (2005). Global perspectives on loss of human life caused by floods. *Natural Hazards, 34*(2), 151–175. doi:10.100711069-004-8891-3

Joy, S. (2005). *Application of GIS and remote sensing in flood management: A case study of West Bengal, India* (Doctoral dissertation). National University of Singapore. Retrieved from ScholarBank@NUS Repository.

Kelman, I. (2003). *Physical flood vulnerability of residential properties in coastal, eastern England* (Doctoral dissertation). University of Cambridge.

Kerala Government. (2018). *Stand With Kerala Kerala Flood Relief*. Retrieved from https://donation.cmdrf.kerala.gov.in

Kheder, K. (2014). Application of Remote Sensing and GIS for Flood Risk Assessment: A case study of Al Kharj Valley Saudi Arabia. *International Journal of Scientific & Engineering Research*, *5*(5), 452–459. doi:10.14299/ijser.2014.05.001

Lee, J. S., & Choi, H. I. (2018). Comparison of Flood Vulnerability Assessments to Climate Change by Construction Frameworks for a Composite Indicator. *Sustainability*, *10*(3), 768.

Menoni, S., & Pergalani, F. (1996). An attempt to link risk assessment with land use planning: A recent experience in Italy. *Disaster Prevention and Management: An International Journal*, *5*(1), 6–21. doi:10.1108/09653569610109523

Messner, F., & Meyer, V. (2006). Flood damage, vulnerability and risk perception – Challenges for flood damage research. Part 4: Vulnerability and flood damages. In *Flood risk management: hazards, vulnerability and mitigation measures* (pp. 149–168). Amsterdam: Springer with NATO Public Diplomacy Division. doi:10.1007/978-1-4020-4598-1_13

Meyer, V., Priest, S., & Kuhlicke, C. (2012). Economic evaluation of structural and non-structural flood risk management measures: Examples from the Mulde River. *Natural Hazards*, *62*(2), 301–324. doi:10.100711069-011-9997-z

Mileti, D. (1999). *Disasters by design: A reassessment of natural hazards in the United States*. Joseph Henry Press.

Minamikawa, K., Takahashi, M., Makino, T., Tago, K., & Hayatsu, M. (2015). Irrigation with oxygen-nanobubble water can reduce methane emission and arsenic dissolution in a flooded rice paddy. *Environmental Research Letters*, *10*(8), 1-11.

Nasiri, H., Yusof, M. J. M., & Ali, T. A. M. (2016). An overview to flood vulnerability assessment methods. *Sustainable Water Resources Management*, *2*(3), 331–336. doi:10.100740899-016-0051-x

National Aeronautics and Space Administration (NASA). (2018). *Dramatic Flooding in Eastern Africa*. Retrieved from https://earthobservatory.nasa.gov/images/92130/dramatic-flooding-in-eastern-africa

National Geographic News (NGN). (2015). *Impact of flooding*. Retrieved from https://www.nationalgeographic.com/environment/natural-disasters/floods

Nelson. (2018). *Natural Disasters & Assessing Hazards and Risk*. Retrieved from https://www.tulane.edu/~sanelson/Natural_Disasters/introduction.htm

Pappenberger, F., Beven, K., Frodsham, K., Romanowicz, R., & Matgen, P. (2007). Grasping the unavoidable subjectivity in calibration of flood inundation models: A vulnerability weighted approach. *Journal of Hydrology (Amsterdam)*, *333*(2-4), 275–287. doi:10.1016/j.jhydrol.2006.08.017

Patel, D. P., & Srivastava, P. K. (2013). Flood hazards mitigation analysis using remote sensing and GIS: Correspondence with town planning scheme. *Water Resources Management*, *27*(7), 2353–2368. doi:10.100711269-013-0291-6

Plate, E. J. (2002). Flood risk and flood management. *Journal of Hydrology (Amsterdam)*, *267*(1-2), 2–11. doi:10.1016/S0022-1694(02)00135-X

Pourghasemi, H. R., Moradi, H. R., Aghda, S. F., Gokceoglu, C., & Pradhan, B. (2014). GIS-based landslide susceptibility mapping with probabilistic likelihood ratio and spatial multi-criteria evaluation models (North of Tehran, Iran). *Arabian Journal of Geosciences*, *7*(5), 1857–1878. doi:10.100712517-012-0825-x

Rahmati, O., & Pourghasemi, H. R. (2017). Identification of critical flood prone areas in data-scarce and ungauged regions: A comparison of three data mining models. *Water Resources Management*, *31*(5), 1473–1487. doi:10.100711269-017-1589-6

Rooke, D. (2009). What is flood risk management? Why is it needed and how is it alleviated and managed? *Environmental Law & Management*, *21*(3), 140–141.

Samanta, S., Pal, D. K., & Palsamanta, B. (2018). Flood susceptibility analysis through remote sensing, GIS and frequency ratio model. *Applied Water Science, 8*(2), 66.

Santangelo, N., Santo, A., Di Crescenzo, G., Foscari, G., Liuzza, V., Sciarrotta, S., & Scorpio, V. (2011). Flood susceptibility assessment in a highly urbanized alluvial fan: The case study of Sala Consilina (southern Italy). *Natural Hazards and Earth System Sciences*, *11*(10), 2765–2780. doi:10.5194/nhess-11-2765-2011

Schanze, J. (2006). Flood risk management–a basic framework. In *Flood risk management: hazards, vulnerability and mitigation measures* (pp. 1–20). Amsterdam: Springer with NATO Public Diplomacy Division.

Sodhi, M. S. (2016). Natural disasters, the economy and population vulnerability as a vicious cycle with exogenous hazards. *Journal of Operations Management*, *45*(1), 101–113. doi:10.1016/j.jom.2016.05.010

Surendran, S., Gibbs, G., Wade, S., & Udale-Clarke, H. (2008). *Supplementary note on flood hazard ratings and thresholds for development and planning control purpose–Clarification of Table 13.1 of FD2320/TR2 and Figure 3.2 of FD2321.* Environment Agency and HR Wallingford.

Tanoue, M., Hirabayashi, Y., & Ikeuchi, H. (2016). Global-scale river flood vulnerability in the last 50 years. *Scientific Reports*, *6*(36021), 1–9. PMID:27782160

Tate, E. (2012). Social vulnerability indices: A comparative assessment using uncertainty and sensitivity analysis. *Natural Hazards*, *63*(2), 325–347. doi:10.100711069-012-0152-2

Tingsanchali, T. (2012). Urban flood disaster management. *Procedia Engineering*, *32*, 25–37. doi:10.1016/j.proeng.2012.01.1233

United Nations Office for Disaster Risk Reduction (UNISDR). (2009). *2009 UNISDR terminology on disaster risk reduction.* Retrieved from https://www.unisdr.org/we/inform/publications/7817

United Nations Office for Disaster Risk Reduction (UNISDR). (2017). *Vulnerability.* Retrieved from https://www.unisdr.org/we/inform/terminology

United Nations Office for Disaster Risk Reduction (UNISDR). (2018). *Magnitude of Japan rains and floods sign of growing challenge to disaster risk management.* Retrieved from https://www.unisdr.org/archive/59297

United Nations (UN). (1982). Assessment of Floodplain Vulnerability during Extreme Mississippi River Flood 2011. *Proceedings of the seminars on flood vulnerability analysis and on the principles of floodplain management for flood loss prevention,* 48(5), 2619-2625.

United Nations (UN). (2018). *Deeply Saddened by Flash Flooding Deaths in Jordan, Secretary-General Pledges United Nations Support for Rescue, Relief Efforts.* Retrieved from https://www.un.org/press/en/2018/sgsm19318.doc.htm

Van der Sande, C. J., De Jong, S. M., & De Roo, A. P. J. (2003). A segmentation and classification approach of IKONOS-2 imagery for land cover mapping to assist flood risk and flood damage assessment. *International Journal of Applied Earth Observation and Geoinformation*, *4*(3), 217–229. doi:10.1016/S0303-2434(03)00003-5

Vojtek, M., & Vojteková, J. (2016). Flood hazard and flood risk assessment at the local spatial scale: A case study. *Geomatics, Natural Hazards & Risk*, *7*(6), 1973–1992. doi:10.1080/19475705.2016.1166874

Wisetjindawat, W., Ito, H., Fujita, M., & Eizo, H. (2014). Planning disaster relief operations. *Procedia: Social and Behavioral Sciences*, *125*, 412–421. doi:10.1016/j.sbspro.2014.01.1484

Wisner, B., Blaikie, P., Cannon, T., & Davis, I. (2004). The challenge of disasters and our approach. In At Risk: Natural Hazards, people s vulnerability and disasters (pp. 3-48). Routledge.

World Health Organization (WHO). (2002). *Vulnerable groups*. Retrieved from https://www.who.int/environmental_health_emergencies/vulnerable_groups/en

Yang, S., He, S., Du, J., & Sun, X. (2015). Screening of social vulnerability to natural hazards in China. *Natural Hazards*, *76*(1), 1–18. doi:10.100711069-014-1225-1

Chapter 2
Composite Indicators as Decision Support Method for Flood Analysis:
Flood Vulnerability Index Category

Ahmed Karmaoui
https://orcid.org/0000-0003-3881-4029
Southern Center for Culture and Sciences, Morocco

Abdelkrim Ben Salem
https://orcid.org/0000-0002-2283-5928
Cadi Ayyad University, Morocco

Guido Minucci
Politecnico di Milano, Italy

ABSTRACT

Floods are highly relevant extreme events with increasing frequency at a global scale. They remain among the most dangerous and complex natural disasters in middle and low-income countries. In this context, it is necessary to develop decision-support tools to reduce the flood risk and increase the resilience. The chapter reviews one of the most relevant tools, the flood vulnerability index (FVI) category at a global scale. These tools use hydrological, topographic, socio-economic parameters strongly associated with flood vulnerability. The findings indicate that FVI is a flexible tool for integrated assessment of vulnerability to floods for application in different regions. Social, environmental, and physical components are the main components used in the FVI. Household and neighborhood, basin, urban, sub-catchment, and coastal are the different levels of vulnerability analysis.

DOI: 10.4018/978-1-5225-9771-1.ch002

INTRODUCTION

Globally, among all extreme events, a flood is one of the most devastating natural hazard (Anees et al., 2018) the most frequent, and the most deadly (Pulvirenti et al., 2011). Furthermore, Climate change may increase the occurrence of flood and make it more common (IPCC, 2012) whereas unplanned rapid urbanization and changes in land use may worsen the effects of a flood event in many ways. In this context, there is an international demand as the one posed by the Sendai Framework for Disaster Risk Reduction to better understanding disaster risk in all its dimensions of vulnerability, capacity, exposure of persons and assets, hazard characteristics and the environment.

There is a rising need to better understand and assess what are the vulnerability factors, which increase the susceptibility of a community to the impact of hazards (UNISDR 2009), in order to support decision-makers in mitigating and adapting to the impact of flooding hazards (Douben, 2006). In the last twenty years, the management plan of floods integrated more and more of variables and dimensions. The composite indicators constitute multidimensional tools to evaluate the risk and vulnerability to floods. The vulnerability assessment is the primordial step towards risk reduction and resilience to disasters (Birkmann, 2006). There is widespread recognition that flood vulnerability index family provides useful methods to evaluate flood in urban areas (Villordon, & Gourbesvill 2016; Karmaoui et al., 2016), in coastal zones (Balica et al., 2012), in sub-catchment and basin scale (Balica et al., 2009), and at Neighborhood level (Fernandez et al., 2016). This kind of method was developed for arid zones (Karmaoui & Balica 2019), and Mountains level (Meraj et al., 2015).

Whether using and combining statistical analysis, remote sensing, or geographic information system, the composite indicators are important decision-making tools since they facilitate the analysis of the relative state of the overall system by reflecting (or trying to reflect) the socio-economic, environmental and physical condition of a geographic region. This study was carried out to find the widely used indices for different scales using different dimension (types of parameters). A comparison of these numerical models was also carried out.

The objective of this review is to provide current methods linked to flooding assessments applied worldwide. The research was oriented toward flood vulnerability. The focus was on the flood vulnerability index categories.

THE USED METHOD

This chapter reviews composite indicators and standardized approaches available in literature. A literature review was carried out adopting keyword-based search approach through online publishers such as ScienceDirect, web of science, and through Google Scholar database in order to identify the most relevant index developed in the last twenty years regarding flood vulnerability (Table 1).The focus was on the composite indicator, the flood vulnerability index categories. The keywords used were: flood vulnerability a flood vulnerability index.

FLOOD VULNERABILITY

The concept of vulnerability was defined as "the conditions determined by physical, social, economic and environmental factors or processes, which increase the susceptibility of a community to the impact of hazards" (UNISDR 2009). In other words, vulnerability is influenced by several factors including human settlements conditions, infrastructure, authorities policy and abilities, social imbalances, economic patterns, etc. Analyzing vulnerability should be so targeted to recognize

Table 1. Flood vulnerability approaches used in this review

Index/Title	Reference
Household flood vulnerability index	Hashim et al., (2018)
FVI usingPrincipal Component Analysis & Cluster Analysis	Fernandez et al., (2016)
Flood vulnerability assessment index	Baeck et al., (2014)
Flood Vulnerability Index	Connor & Hiroki, (2005)
Runoff potential of the watersheds to assess the flood vulnerability downstream	Meraj et al., (2015)
Integrated multi-criteria flood vulnerability approach	Lee et al., (2013)
Urban settlements' vulnerability to flood risks	Salami et al., (2017)
Community Based Flood Vulnerability Index	Villordon&Gourbesvill (2016)
Flood Vulnerability Index	Rashetnia, (2016)
Flood Vulnerability Index in pre-Sahara	Karmaoui et al., (2016)
Flood Vulnerability Index	Balica et al., (2009)
Coastal Cities Flood Vulnerability Index	Balica et al., (2012)
Social-ecological index for measuring flood resilience: A composite index approach	Kotzee&Reyers, (2016)

(source: the authors)

which actions can be taken to reduce vulnerability before the possible harm is realized. Regarding this, researchers have developed many methods to assess flood vulnerability. However, despite increased knowledge about vulnerability, flood risk is still very widespread. This leads to questions being raised about (i) the effective use of knowledge, (ii) the fact that knowledge is used effectively in some respects but is overwhelmed by increases in vulnerability and in population, wealth, and poverty (White et al. 2001) and (iii) the efficiency of vulnerability assessments and their effect on flood mitigation and adaptation (Khan, 2012) .

FLOOD VULNERABILITY INDEX

The Flood Vulnerability Index (FVI) is an index for assessing vulnerability to flooding disasters that can be applied at the river basin level. The Flood Vulnerability Index can be an important policy-making tool for both assisting governments in priority setting and improving decision-making practices to reduce vulnerability in different spatial levels (Balica et al., 2009). In addition, it helps to raise public awareness and guiding international organizations in directions of involvement.

Flood Vulnerability Index Structure

Table 2 depicts the most relevant indices at different levels: Household and Neighborhood, urban, sub-catchment, basins, and coastal areas within different dimensions. According this study, several types were considered: Attitudinal, Institutional, Hydro-climatic, politico-administrative, Socio-behavioral, Meteorological, Hydrologic-topographical, Countermeasure, Morphometry, Land cover, Hydro-geological, Infrastructural, Physiographic, Land use, Anthropogenic, Access to services.

These components can be summarized in four main sub-indexes, the physical, social, economic and environmental. The major differences lie in the number and type of variables in each category, the unit used and the type of scale, and the indicator aggregation method.

The FVI can be determined using multidimensional characteristics called in most of the cases sub-indexes or components. The most used types are social, economic, physical, and environmental. The application or the applicability requires several sites or areas to normalize and compare data.

In order to develop an FVI, there are mainly four steps: Data acquisition and management, the selection of indicators, the normalization, data transformation, the weighting, and the aggregation of indicators, and the output presentation.

Table 2. Flood vulnerability approaches

Scale/Level	Index/Title	Components	Reference
Household	Household flood vulnerability index	• Social, • Economic • Physical • Environmental	Hashim et al., (2018)
Neighborhood	FVI using Principal Component Analysis & Cluster Analysis	• Social, • Economic • Physical • Environmental	Fernandez et al., (2016)
Urban	Oasean flood Vulnerability Index	• Climatic • Physiographic • Land use • Anthropogenic • Economic • Access to services	Karmaoui & Balica, (2019)
	Flood Vulnerability Index in pre-Sahara	• Social, • Economic • Physical • Environmental	Karmaoui et al., (2016)
	Urban settlements' vulnerability to flood risks	• Physical or environment • Economic • Social • Attitudinal • Institutional	Salami et al., (2017)
	Community Based Flood Vulnerability Index	• Hydro-climatic • Social • Economic • politico-administrative • Socio-behavioral	Villordon, &Gourbesvill (2016)
	Flood Vulnerability Index	• Hydrological • Economic • Social	Rashetnia, (2016)
	Analysis of applicability of Flood Vulnerability Index in pre-Sahara	• Social, • Economic • Physical • Environmental	Karmaoui et al., (2016)
Sub-catchment	Oasean flood Vulnerability Index	• Climatic • Physiographic • Land use • Anthropogenic • Economic • Access to services	Karmaoui & Balica, (2019)
	Flood Vulnerability Index	• Social, • Economic • Physical • Environmental	Balica et al., (2009)
	Analysis of applicability of Flood Vulnerability Index in pre-Sahara	• Social, • Economic • Physical • Environmental	Karmaoui et al., (2016)

continued on the following page

Data Acquisition and Management

Input data can be collected from a variety of sources, including household surveys, documents, government and ministries, scientific acquired articles, books and technical reports. There are three main types of data to use on a composite indicator:

Table 2. Continued

Scale/Level	Index/Title	Components	Reference
Basin	Flood vulnerability assessment index	• Meteorological • Hydrologic-topographical • Socioeconomic • Flood-defence vulnerability	Baeck et al., (2014)
	Flood Vulnerability Index	• Social, • Economic • Physical • Environmental	Balica et al., (2009)
	Flood Vulnerability Index	• Climate • Hydro-geological • Socio-Economic • Countermeasure	Connor & Hiroki, (2005)
	Runoff potential of the watersheds to assess the flood vulnerability downstream	• Morphometry • Land cover • Slope	Meraj et al., (2015)
	Integrated multi-criteria flood vulnerability approach	• Social • Economic • Hydrologic	Lee et al., (2013)
Coastal	Coastal Cities Flood Vulnerability Index	• Hydro-geological • Social • Economic • Politico-administrative	Balica et al., (2012)
	Social-ecological index for measuring flood resilience: A composite index approach	• Social • Economic, • Infrastructural • Ecological	Kotzee&Reyers, (2016)

Source: (The authors)

- Data which are available and published by reputable organizations;
- Data collected by the respective governments or could easily be collected;
- Data which are difficult to obtain, but could be produced or reasonably approximated;

Selection of Relevant Indicators

The best indicator is the aspect which allows identifying the level of vulnerability of an area. The selection of the relevant indicators is the first step to do. There are several techniques to select suitable indicators:

- Surveys or expert's opinions
- Statistical methods (simple correlation and Principal Component Analysis)
- Literature review

Generally, floods are due to meteorological conditions and the effects are exacerbated many causative factors. According to (Anees et al., 2018), a flood is due to topographic and climatic changes. However, Kaspersen et al., (2015) specified that precipitation with land cover changes is the main causes. Of all components (types of indicators), physical, social, economic and environmental are the most used to reflect the vulnerability levels to floods.

The indicators are numerous used in different contexts at different scales. The most used are precipitation and flood occurrence (Lee et al., 2013; Balica et al., 2009; Connor & Hiroki, 2005) and the population density (Antwi et al., 2015 ; Lee et al., 2013 ; Balica et al., 2009 ; Mavhura et al., 2017 ; Sebald, 2010 ; Antwi et al., 2015). In addition to past experience and warning system indicators (Lee et al., 2013; Balica et al., 2009), slope (Meraj et al., 2015 ; Lee et al., 2013 ; Connor & Hiroki, 2005) and flow (Meraj et al., 2015 ; Lee et al., 2013 ; Balica et al., 2009) are key variables linked to flood vulnerability. For land use category, agriculture, pasture, bare land, and forest (Villordon & Gourbesvill, 2016; Meraj et al., 2015), and reserves are considered as indicators of vulnerability (Lee et al., 2013; Balica et al., 2009 ;Villordon, & Gourbesvill, 2016 ; Sebald, 2010). Consequently, areas with a high risk of floods and high density of population and of economic activities are more vulnerable than others.

Normalization of Indicators

In order to compare indicators with deferent units, scales, values intervals, the normalization of used data indicators is needed. Generally two formulas can be used in this step:

- Maximum score

$$FVINormalized = \frac{\left[FVI\ Site\right]}{\left[FVI \max\right]} \tag{1}$$

- Extreme method

$$FVINormalized = \frac{\left[Xi - Min\ Xi\right]}{\left[Max\ Xi - Min\ Xi\right]} \tag{2}$$

Where: the ith indicator in a site j, Xij = numerical value of the ithindicator in site j, Min and Max Xi = minimum and maximum value of the ith indicator across all sites.

Transformation of Negative Values

To avoid the negative association of some indicators to flood vulnerability, a transformation using the equation 3can be done.

$$FVINormalized = 1 - \frac{\left[FVI\ specific\ site\right]}{\left[FVI \max\right]} \tag{3}$$

Weighting the Selected Indicators

This step allows assigning degree of importance of the selected indicators to reflect their real Weight. Delphi and AHP methods are the main methods to assign weights.

Aggregation of Indicators

Aggregation refers to the method of addition or multiplication of the indicators scores into FVI components and into the total FVI.

Output Presentation

After calculation, the FVI scores of the used sites are obtained. The results contain the individual scores of all the indicators, components, and the final score of the whole FVI. It should be pointed out that it is very rare to be able to collect all the data of the selected indicators for a defined area. This gap does not alter the calculation, because FVI only takes into account the data entered. The results can be presented in the form of graphs (in two and three dimensions), tables (individual and collective scores), and in the form of maps (geo-referenced). The results can be compared according to the sites, components and the chosen scenarios.

DISCUSSION AND CONCLUSION

Floods are due to the unsteady condition of climate and remain among the most dangerous natural disasters. Floods cause severe social-economic and physical effects (Albano et al. 2014). Globally, a number of floods negatively affected millions of people and caused the death of several hundreds of thousands of lives. The damage increases with the increased numbers of people, material and places exposed to floods. The complexity of this extreme event necessitates developing decision tools to analyze, manage, and compare its vulnerability. Among the most relevant approaches, FVI constitutes a holistic method for flood management and policy. This study was built on earlier work of flood vulnerability indices mentioned in table 2. FVI is a powerful tool for policy and decision-makers to prioritize investments and, by favoring an increase of public awareness, makes the decision making process more transparent. This tool allows identifying the areas with high flood vulnerability. It may guide the decision-making process towards a better way of dealing with floods by societies that take into account a long-term flood management policy approach. However, the development of this composite indicator requires a lot of information, data, and time. It is worth noting that the capacity to build the FVI index can be lessen by availability, completeness and accuracy of data at spatial and temporal scales as well as by issues of reliability of the data and sources of data and costs of data collection. This kind of indices is affected by the selection of the appropriate indicators, the complexity of the phenomena (the multi-dimensionality) and quality of input data. Balica et al., (2009) reported a list of a large number of possible indicators linked to flooding vulnerability (See the appendix, table 3). A dozen to thirty indicators and even more must be validated and classified into sub-indexes or categories. In this chapter, 15 FVI were selected and classified following the scale. Economic, social, environmental, and physical components are the main components used in the FVI of this study. Household and Neighborhood, Basin, Urban, Sub-catchment, and Coastal are the different levels of vulnerability analysis.

The development of this composite indicator is associated with several constraints as follows:

- Complexity and election of indicators: simplify using minimum relevant indicators benefiting from statistical analysis.
- Data quality and missing data can be resolved involving decision makers to provide data
- Qualitative aspect of the used method: This can be corrected by combining by remote sensing methods
- Time and interpretation: can be resolved through effective collaboration.

The FVI is a flexible framework for the integrated assessment of vulnerability to floods for application in different regions. This may contribute as an important step toward preparedness and response to the flood. It can be used also as an educational tool to aware of flood vulnerability.

REFERENCES

Anees, M. T., Abdullah, K., Nawawi, M. N. M., Ab Rahman, N. N. N., Piah, A. R. M., Zakaria, N. A., ... Omar, A. M. (2016). Numerical modeling techniques for flood analysis. *Journal of African Earth Sciences, 124*, 478–486. doi:10.1016/j.jafrearsci.2016.10.001

Baeck, S. H., Choi, S. J., Choi, G. W., & Lee, D. R. (2014). A study of evaluating and forecasting watersheds using the flood vulnerability assessment index in Korea. *Geomatics, Natural Hazards & Risk*, *5*(3), 208–231. doi:10.1080/19475705.2013.803268

Balica, S. F., Douben, N., & Wright, N. G. (2009). Flood vulnerability indices at varying spatial scales. *Water Science and Technology*, *60*(10), 2571–2580. doi:10.2166/wst.2009.183 PMID:19923763

Balica, S. F., Wright, N. G., & van der Meulen, F. (2012). A flood vulnerability index for coastal cities and its use in assessing climate change impacts. *Natural Hazards*, *64*(1), 73–105. doi:10.100711069-012-0234-1

Birkmann, J. (2006). Measuring vulnerability to promote disaster-resilient societies: Conceptual frameworks and definitions. *Measuring vulnerability to natural hazards: Towards disaster resilient societies, 1*, 9-54.

Connor, R. F., & Hiroki, K. (2005). Development of a method for assessing flood vulnerability. *Water Science and Technology*, *51*(5), 61–67. doi:10.2166/wst.2005.0109 PMID:15918359

Douben, J. K. (2006). Characteristics of River floods and Flooding: A Global Overview, 1985-2003. *Journal*, *59*, 59–521.

EMDAT. (2017). *OFDA/CRED International Disaster Database, Universitécatholique de Louvain – Brussels – Belgium*. Retrieved from https://ourworldindata.org/natural-disasters

Fernandez, P., Mourato, S., Moreira, M., & Pereira, L. (2016). A new approach for computing a flood vulnerability index using cluster analysis. *Physics and Chemistry of the Earth Parts A/B/C*, *94*, 47–55. doi:10.1016/j.pce.2016.04.003

Hashim, M. S., Hassan, S., & Bakar, A. A. (2018) Developing a Household Flood Vulnerability Index: A Case Study of Kelantan. *Sch. J. Econ. Bus. Manag.* Doi:10.21276jebm.2018.5.7.4

IPCC. (2012). *Managing the Risks of Extreme Events and Disasters to Advance Climate Change Adaptation. Cambridge University Press.*

Jorgensen, S. E. (1992). *Integration of ecosystem theories: a pattern.* Kluwer Academic Publishers Dordrecht. doi:10.1007/978-94-011-2682-3

Karmaoui, A. & Balica, S. (2019). A New Flood Vulnerability Index Adapted for the Pre-Saharan Region. *International Journal of River Basin Management.* Doi:1 0.1080/15715124.2019.1583668

Karmaoui, A., Balica, S. B., & Messouli, M. (2016). Analysis of applicability of flood vulnerability index in Pre-Saharan region, Morocco. *Natural Hazards and Earth System Sciences, 2016.* doi:10.5194/nhess-2016-96

Kaspersen, P.S., HøeghRavn, N., Arnbjerg-Nielsen, K., Madsen, H., & Drews, M. (2015). Influence of urban land cover changes and climate change for the exposure ofEuropean cities to flooding during high-intensity precipitation. *Proc. Int. Assoc. Hydrol. Sci. (IAHS), 370*, 21-27.

Kellermann, P., Schobel, A., Kundela, G., & Thieken, A.H. (2015). Estimating flood damage to railway infrastructure - the case study of the March River flood in 2006 at the Austrian Northern Railway. *Nat. Hazards Earth Syst. Sci., 15*(11), 2485-2496. . doi:10.5194/nhess-15-2485-2015

Khan, S. (2012). Vulnerability assessments and their planning implications: A case study of the Hutt Valley, New Zealand. *Natural Hazards, 64*(2), 1587–1607. doi:10.100711069-012-0327-x

Kotzee, I., & Reyers, B. (2016). Piloting a social-ecological index for measuring flood resilience: A composite index approach. *Ecological Indicators, 60*, 45–53. doi:10.1016/j.ecolind.2015.06.018

Kundzewicz, Z.W., Pinskwar, I., & Brakenridge, G.R. (2003). Large floods in Europe, 1985-2009. *Hydrol. Sci. J., 58*, 736.

Leiter, A.M., Oberhofer, H., & Raschky, P.A. (2009). Creative disasters? Flooding effects on capital, labour and productivity within European firms. Environ. *Resour. Econ., 43*(3), 333-350. . doi:10.100710640-009-9273-9

Meraj, G., Romshoo, S. A., Yousuf, A. R., Altaf, S., & Altaf, F. (2015). Assessing the influence of watershed characteristics on the flood vulnerability of Jhelum basin in Kashmir Himalaya. *Natural Hazards, 77*(1), 153–175. doi:10.100711069-015-1605-1

Pulvirenti, L., Chini, M., Pierdicca, N., Guerriero, L., & Ferrazzoli, P. (2011). Flood monitoring using multi-temporal COSMO-SkyMed data: image segmentation and signature interpretation. *Rem. Sens. Environ., 115*, 990-1002.

Rashetnia, S. (2016). *Flood vulnerability assessment by applying a fuzzy logic method: a case study from Melbourne* (Doctoral dissertation). Victoria University.

Salami, R. O., von Meding, J. K., & Giggins, H. (2017). Urban settlements' vulnerability to flood risks in African cities: A conceptual framework. *Jàmbá: Journal of Disaster Risk Studies*, *9*(1), 1–9. doi:10.4102/jamba.v9i1.370 PMID:29955335

UNISDR. (2009). *Terminology on Disaster Risk Reduction*. Available at http://www.unisdr.org/we/inform/publications/7817

Villordon, M. B. B. L., & Gourbesville, P. (2016). Community-based flood vulnerability index for urban flooding: Understanding social vulnerabilities and risks. In *Advances in Hydroinformatics* (pp. 75–96). Singapore: Springer. doi:10.1007/978-981-287-615-7_6

White, G. F., Kates, R. W., & Burton, I. (2001). Knowing better and losing even more: The use of knowledge in hazards management. *Global Environmental Change Part B: Environmental Hazards*, *3*(3), 81–92. doi:10.1016/S1464-2867(01)00021-3

KEY TERMS AND DEFINITIONS

Exposure: Refers to the position of human and properties to flood risk.

Indicator: The aspect which allows identifying the level of vulnerability of an area.

Resilience: Is the system capacity to rebalance after a perturbation.

Susceptibility: Defined as system characteristics such as awareness and preparedness.

Vulnerability: Related to exposure, sensitivity, and resilience.

APPENDIX

Table 3. Relationship between components and indicators

Flood Vulnerability	Exposure	Geographic Scale	Susceptibility	Geographic Scale	Resilience	Geographic Scale
Social Components	Population density	R.S.U	Past experience	R.S.U	Warning system	R.S.U
	Population in flood area	R.S.U	Education (literacy rate)	R.S.U	Evacuation routes	R.S.U
	Closeness to inundation area	R.S.U	Preparedness	R.S.U	Institutional capacity	R.S.U
	Population close to coast line	R.S.U	Awareness	R.S.U	Emergency service	R.S.U
	Population under poverty	R.S.U	Trust in institutions	R.S.U	shelters	R.S.U
	% of urban area	R.S.U	Communication penetration rate	R.S.U		
	Rural population	R.S	hospitals	R.S.U		
	Cadastre survey	S.U	Population with access to sanitation	R.S.U		
	Cultural heritage	S.U	Rural population who access to WS	R.S		
	% of young & older	S.U	Quality of water supply	S.U		
	Slums	U	Quality of energy supply	S.U		
			Population growth	S.U		
			Human health	S.U		
			Urban planning	U		
Economic components	Land use	R.S.U	Unemployment	R.S.U	Investment in counter measures	R.S.U
	Proximity to river	R.S.U	Income	R.S.U	Infrastructure management	R.S.U
	Closeness to inundation areas	R.S.U	Inequality	R.S.U	Dams & storage capacity	R.S.U
	% of urbanized area	R.S	Quality of infrastructure	R.S.U	Flood insurance	R.S.U
	Cadastre survey	S.U	Years of sustaining health life	R.S.U	Recovery time	R.S.U
			Urban growth	S.U	Past experience	S.U
			Child mortality	S.U	Dikes/levees	S.U
			Regional GDP/Capita	S		
			Urban planning	S		

continued on the following page

Table 3. Continued

Flood Vulnerability	Exposure	Geographic Scale	Susceptibility	Geographic Scale	Resilience	Geographic Scale
Environmental components	Ground WL	R.S.U	Natural reservations	R.S.U	Recovery time to floods	R.S.U
	Land use	R.S.U	Years of sustaining health life	R.S.U	Environmental concern	R.S.U
	Over used area	R.S.U	Quality of infrastructure	R.S.U		
	Degraded area	R.S.U	Human health	S.U		
	Unpopulated land area	R.S	Urban growth	S.U		
	Types of vegetation	R.S	Child mortality	S.U		
	% of urbanize area	R.S				
	Forest change rate	R				
Physical components	Topography (slope)	R.S.U	Buildings codes	U	Dams & storage capacity	R.S.U
	Geography	R.S.U			Roads	R.S.U
	Geology	R.S.U			Dikes /levees	S.U
	Heavy rainfall	R.S.U				
	Flood duration	R.S.U				
	Return periods	R.S.U				
	Proximity to river	R.S.U				
	Soil moisture	R.S.U				
	Evaporation rate	R.S.U				
	Temperature (yearly average)	R.S.U				
	River discharge	R.S.U				
	Frequency of occurrence	R.S.U				
	Flow velocity	S.U				
	Storm surge	S.U				
	Tidal	S.U				
	Flood water depth	S.U				
	Sedimentation load	S.U				
	Coast line	S.U				
	Coastal bathymetry	S.U				

Where: **R**, represents River Basin Scale; **S**, represents Sub-catchment Scale; and **U**, represents Urban Scale (Balica*et al.* 2009)

Chapter 3
Impacts of Climate Change on Coastal Communities

Isahaque Ali
https://orcid.org/0000-0002-7113-2882
Universiti Sains Malaysia, Malaysia

Rameeja Shaik
https://orcid.org/0000-0002-3525-6853
GITAM University, India

Maruthi A. Y.
Krishna University, India

Azlinda Azman
Universiti Sains Malaysia, Malaysia

Paramjit Singh
Universiti Sains Malaysia, Malaysia

Jeremiah David Bala
Universiti Sains Malaysia, Malaysia

Adeleke A. O.
Universiti Sains Malaysia, Malaysia

Mohd Rafatullah
Universiti Sains Malaysia, Malaysia

Norli Ismail
Universiti Sains Malaysia, Malaysia

Akil Ahmad
Universiti Sains Malaysia, Malaysia

Kaizar Hossain
https://orcid.org/0000-0002-3903-6161
Universiti Sains Malaysia, Malaysia

ABSTRACT

Earth and coastal ecosystems are not static, and they usually respond to environmental changes, mostly anthropogenic and climatic. Here, the authors described natural values, coastal landforms, and types of infrastructure that are most likely to be affected by climate change (CC) and provide information for assessing inundation, erosion, and recession risks for a chosen location. In this chapter, the authors focused on the land uses, the vulnerability of coastal infrastructure, and argued for effective linkages between CC issues and development planning. They also recommended the incorporation of CC impact and risk assessment into long-term national development strategies. Policies will be presented to implement these recommendations for adaptation to climate variability and global CC. The authors

DOI: 10.4018/978-1-5225-9771-1.ch003

provide general recommendations and identify challenges for the incorporation of climate change impacts and risk assessment into long-term land-use national development plans and strategies. Overall, this chapter provides an overview of the implications for CC to coastal management.

INTRODUCTION

Nearly 1.2 billion of people in the world (23% world`s people) live with 100 km of the coastal areas and by 2030 (Small & Nicholls, 2003) it will be at 50%. These people are unprotected to definite hazard such as hurricanes, tsunamis, coastal flooding and transmission of marine related infection diseases (Adger et al., 2005). Currently, an estimated report says, about ten million of world people facing coastal flood every year because of storm landfall typhoons and surge and near future 50 million could be at risk by 2080 due to climate change and growing population masses (Nicholls, 2004).The climate change has occurred to changes in flooding, temperature and precipitation that make more vulnerable to the people of coastal areas. Additionally, the rise in sea level and wave heights will also affect the lives of coastal people. Both straight impact (frequent storm surges and faster coastal erosion) and secondary effects (loss of coastal resources such as aquaculture and loss of critical physical infrastructure, along with decays in associated ecological, economic, subsistence and cultural values) will have socio-economic and physical impacts on coastal societies. At present the coastal societies already face a numerous difficult problems that make challenging for the policy of climate change. In demographic viewpoint, the current people of coastal societies are becoming gradually elder that results of high numbers of internal migration of elderly people with youth out migration (Ali *et al.*, 2016). In financial standpoint, the coastal people is constantly considered by high average unemployment rates, vulnerable financial conditions, including low incomes and stress on services throughout the months of summer due to tourism. Many regions of the coastal zones like England, are facing severe level of multiple deprivation like high levels of deficit related to remunerations, education, employment, skills and training. This might be due to lot of issues, comprising the reliance of naturally poor-skilled, less wages in industrial sectors for example which are related to tourism, that often also need part time and seasonal employees. Seasonal of works also creates it challenging for personnel to development in terms of educations or profession development, as each term of employment may be with a dissimilar company (Hossain *et al.*, 2016). The physical segregation of numerous seaside peoples can also frequently act as a barrier to financial progress. High levels of scarcity and the relative segregation of some coastal societies are the physical hazard of climate effects. The Impact of climate change on a particular agricultural crop, such as bananas, may lead to lost profits, unemployment of farm workers, foreclosures on mortgages, and loss of human, financial and social capital within communities. The disadvantages and hazardous of these regions in the coastal societies are the focus of this research. Here we also demonstrates how the impacts of climate change manifest across the triple-bottom line i.e. economic, environment and social. The principal objective of this chapter is to recognize current climate change vulnerabilities especially to backward coastal communities and make policy recommendations to mitigate these encounters. It also aims to make endorsements to enhance coastal people`s ability to adapt to the climate change and aimed to draw in a range of perceptions on the matter from societies and local and national stakeholders.

Climate Change Impact on Coast

The coastal communities are to be more at risk to climate change than inland areas due to changes in flooding, precipitation and temperature, rise in sea level, wave heights, and coastal erosion. The disadvantages communities identify, as those are vulnerable of the physical impacts of climate change and undergoes from severe level of deprivation of environmental isolation (Zsamboky *et al.*, 2011) and direct and indirect impacts are shown in **Table 1**. The coastal areas are highly vigorous

Table 1. The impacts of direct and indirect climate change on coastal communities

Direct Impact	Indirect Impact	Event
Inundation of low-lying residential areas	Economic- Loss of property assets Environmental- Salinity impacts on native vegetation and coastal water aquifers Social- Loss of social capital if communities are dispersed through relocation	Sea level
Damage to critical infrastructure such as a sewer pumping station	Economic- Repair costs Environmental- Increased nutrients entering waterways Social- Increased risk of disease and infection	Flooding
Extreme waves overtopping esplanades and reserves	Economic- Loss of tourism revenue Environmental: Foreshore vegetation damage Social- Loss of recreational amenity	Storm surge
Uprooting of trees	Economic- Damage to homes and cars Environmental- Loss of mature vegetation and habitat Social- Reduced shade protection in summer	Wind
Bushfire in a coastal national park	Economic- Reduced purchases by visitors for local businesses Environmental- Habitat fragmentation Social- Loss of amenity and recreational values	Fire

and geo-morphologically multifaceted systems that respond in numerous behaviors to dangerous weather events. Among the most dangerous and harmful of natural disasters coastal floods one of them (Douben, 2006). Climate change is the reason of faster sea-level rise with accelerated erosion, rising water tables, increasing occurrence of cyclones, storm surges, increased frequency of flood and increased saltwater intrusion, salt-water intrusion into aquifers as the sea rises, flooding of coastal wetlands and marshes, changes to water availability and quality, ocean acidification (due to higher concentrations of carbon dioxide in the atmosphere),lower oxygen levels in wetlands, changes in habitat and species distributions, etc. (Prakasa & Murty, 2005).

Coastal Erosion

Coastal erosion has direct impact on regular income source and income generating activities and schooling for children, health facilities. In addition, coastal people suffered from the lack of drinking water, sanitation facilities and emergency health care services, and in addition to food crisis and lack of employment (Islam *et al.*, 2012).In recent years, one of the major contribution of the climate change is rapid siltation of the river, which is intensifying bank erosion during the monsoon. Due to the climate change, riverbank erosion occurs increasingly and has longstanding effects that are difficult to manage for the coastal people (Alam, 2016). Scholars established that riverbank erosion has severe effects on financial, physical, political and social circumstances, causing in terrible hazards in lives and livelihood in coastal regions (Ahmed 2015; Alam, 2016). The erosion of riverbank is the catastrophe accounting for the severe losses in the coastal communities (Penning-Rowsell *et al.*, 2013). The coastal households are more unprotected to regular flooding and water logging due to nearness of the coastal with river bank erosion that create more vulnerable to them (*Alam et al.*, 2017). The coastal households are more vulnerable to riverbank erosion and other climate-induced hazards and forced into a low livelihood (Alam 2016; Ahmed 2015). Climate change is a key factor in coastal erosion and erosion is a complex procedure that has variety of impacts on coastal communities (Masselink & Russell, 2013). The 3,700 km coastline of England and Wales 28% is experiencing erosion greater than 10 cm per year and a large proportion of the coastline of the UK and Ireland is currently suffering from erosion (17% in the UK; 20% in Ireland) (Masselink & Russell, 2013). The riverbank erosion causes long-term slum dwellers in urban areas. The tremendous of them live in temporary housing systems that are in a very poor condition. Riverbank erosion generally creates much more suffering than floods.

Sea Level Rise

The worldwide percentage of sea level rise estimated information over than 15 (1993-2008) year is 3.5 mm per year (Nicholls & Cazenave, 2010) and few of the study suggest that the rate of sea level rise is rushing (Church & White, 2006; Rahmstorf, Perrette, & Vermeer, 2012). The rise in world sea level by 2100 will be in the range of 18–38 to 26–59 cm (Meehl, 2007) and other studies mentioned 0.75–1.90 m (Vermeer & Rahmstorf, 2009), 0.72–1.60 m (Grinsted Moore, & Jevrejeva, 2009), and 0.5 and 2 m (Nicholls et al., 2011). The latent impacts of sea-level rise are substantial for the broader coastal ecosystems (Kumar, 2006). The temperatures of sea water have changed at extraordinary rates in last 10-15 years (Philippart *et al.*, 2011) and it is expected to increase sea surface temperatures by 1° to 2 °C during the 21st century (Furevik *et al.*, 2002). Due to climate change, the temperature is raising that reasons sea level also rises and impact of coastal communities and deltas in around the world. Rising sea level will cause salinity intrusion, river bank erosion, land erosion, flood and crop failure, fisheries destruction, loss of diversity and damage of infrastructure in the coastal areas. Additionally, raising sea level it will destroy costal resources, water resources, agriculture and eco systems (Sarwar, 2005). Raising sea level will change the position of the river that affecting a change in fish environment and breeding ground. Moreover, the coastal fisheries also affecting by flooding, salinity and increasing cyclones due to climate change (Sarwar, 2005).

Storminess

The intra-annual precipitation regimes have been already become more extreme and thrilling at, local, regional and global scales (Easterling et al. 2000; Groisman et al. 2005; Knapp et al. 2008). Willmott and Legates (1991), reported that the uncertainty of climate information poses challenges for the analysis of observed rain data because the heaviest areas of precipitation may fall between recording stations. The extreme precipitation changes maybe more reliable for regions with dense networks because of the small radius of correlation for many intense rainfall events (Groisman et al. 2005). Very less literature sources are available worldwide concerning the extreme precipitation, especially about rainstorm effects on water resources and terrestrial ecosystems (Clarke and Rendell 2007; Curtis et al. 2007; Zolina et al. 2009). A large number of coastal communities are hazard from flooding- especially, when tides combine with storm surges of higher river flows (Mcgranahan, Balk, & Anderson,

2007). Climate change will enhance the hazard of equally. Due to climate change rising sea level will escalation of floods, and stronger tropical storm and marginalized income group on flood plains will more vulnerable (Mcgranahan, Balk, & Anderson, 2007). The coastal areas are more affected by storm surges and that make them more vulnerable in the communities (Philippart et al., 2011).

Climate Change Impact on Coastal Communities

The impacts of climate change hazards are visible in many coastal areas around the world (Spalding *et al.*, 2014) and it is significant sea level rise, while its degree and frequency of change is likely vary over broad and to be hard to identify locally (Cazenave & L lovel, 2010; Han *et al.*, 2010). The populations are growing faster in the coastal areas and around 10% of world population living in the coastal areas (Mcgranahan, Balk & Anderson, 2007). In the period of 1994 and 2004, nearly one-third of the 1562 floods, half of the 120,000 lives killed and 98% of the 2 million people affected by flood in Asia due to climate change (Mcgranahan, Balk, & Anderson, 2007). The experience from tsunami of Indian Ocean of 2004, lost of live s of 200,000 and millions more houses destroyed (Mcgranahan, Balk, & Anderson, 2007).

Health and Well Being

Climate change will hazards and challenges for the coastal population, with their psychological and physical health, and for coastal financial and native industry like agriculture, fisheries and tourism. Additionally, it will also affects people to entrance and quality of basic daily goods and services like food, water, health and emergency care, for them. The budgets of emergency action, prevention and recovery might be a severe burden to coastal communities and local authorities in regions with already inadequate resources (Zsamboky *et al.*, 2011). Rising sea levels enhance the risk of health vulnerability like cholera and diarrhea and it is infectious diseases of human beings and very communal in coastal area around the world (Sarwar, 2005). The growing density of population and spreading salinity, diarrhea and cholera microbes are spreading in the coastal communities (Sarwar, 2005). The coastal areas are always densely populated (Ericson et al. 2006). Increase movement of coastal people has extended infection diseases such as Acquired Immune Deficiency Virus (AIDS) and Human Immunodeficiency Virus (HIV) high in some coastal fishing communities and climate change is certain to worsen vulnerability to vector borne disease (cholera and malaria)(Allison & Seeley, 2004). On the other hand, increased in the saline water in the coastal areas will decrease fresh water and scarcity of fresh water will force to drink polluted water leading diarrhea, cholera and other vector borne diseases (Sarwar, 2005).

Livelihood and Economic Costs

The increasing salinity in the coastal communities will decline agriculture and food production that causing severe malnutrition for the coastal people (Sarwar, 2005). The higher tropical levels in worldwide have reduced big fish (Myers & Worm, 2003). A numbers of people has been displaced from their houses becoming environmental refugees due to climate change (Sarwar, 2005). The coastal states in the United States of America (USA) contributed more than $6.6 trillion of Gross Domestic Products (GDP) has of the nation`s total GDP, from less than 10% of the land area (NOAA, 2013). Due to climate change, this achievement soon will be translated heavy financial and social vulnerability and make unsustainable coastal development (Spalding *et al.*, 2014). The concern is high in the poor communities, who are direct depended on coastal ecosystems for basic needs such as livelihoods, where capacity to cope with climate change is far less than developed country (ISDR, 2009). The attention of the populations and financial matters on and near coast has severe environmental influences. It is projected that nearly one-third of coastal mangrove forest and one fifth of coral reefs have already lost. Additionally, many areas of the world wide fish has reduced significantly (Mcgranahan Balk & Anderson, 2007). The study also suggests that climate change will also effect of coastal communities livelihood those are dependent of fishing and tourism. Additionally, lower income people or unemployment were seen greater of risk to their lives and well-being of climate change (Zsamboky *et al.*, 2011).

Access, Quality and Choice of Goods and Services

Large number of coastal populations facing of important financial challenges such as older people, youth out migration and inward migration of elderly, low quality housing, transitory populations, seasonal employment and poor income (CCA, 2010). The social impacts of climate change on the coastal people will effects health due to increase in severe event such as flooding and heatwave and specially, very older to be worst affected and those who have existing health hazards. Among flood victims, mental health impacts are particular prevalence and severe events with climate change (flooding and storm) are likely to effects on public infrastructure such as health and emergency basic needs and communication systems of coastal people (Zsamboky *et al.*, 2011). Due to the climate change of coastal people, their agriculture production is less and low earning decrease their buying capacity to obtain basic needs and services. Consequently, lack of food forced them into becoming food insecurity with limited health care services that lead to them numerous diseases and poor health. This vulnerability pushes them into vicious circle of poverty. Majority of the people in coastal Bangladesh do not have access to proper nutrition, housing settlement and healthcare facilities. Poverty situation is further deteriorated because of natural hazards and calamities every year. Nearly twelve million people live in poverty in the coastal regions of Bangladesh (Dasgupta, 2016).

IMPLICATIONS FOR THE ADAPTATION OF COASTAL COMMUNITIES

Ability to Adapt

Adaptations pursues to decline the severe impact of climate change on living organisms, socio-economic including human and environment. The capacity to adjust and cope is a purpose of income/wealth, scientific and technological knowledge, skills, infrastructure, information, policy and management (Chatterjee & Huq, 2002). To mitigate to the potential disaster during the Asian tsunami 2004, individuals and communities take adaptive policies that include the mobilizations of assets, social capital and social networks (Adger et al., 2005). Social resilience, robust governance systems including institutional for mutual actions are essential wealth for buffering the effects of dangerous natural hazards and promoting social reformation such as those associate with climate change (Adger et al., 2005). The main aim of the adaptation is that to reduce vulnerability and form resilience among the coastal communities (UNFCCC, 2011) and this has led to increasing calls of cooperation into coastal adaptation policy (World Bank, 2009). There is emergency need for action of adaptation at local, national and international arena, particularly noted that sea level rise and coastal erosion are serious threats for many coastal regions and islands (Spalding *et al.*, 2014).

Understanding of Climate Change Risks and Its Awareness

To understand the climate change process, driving forces and meaningful managing and adaptive polices remained inadequate. Community practitioners and communities faces main difficulties to concreate planning due to lack of awareness, knowledge and access to forecasts during climate jargon (Leary *et al.*, 2008). In some coastal communities, people have low levels of knowledge on climate change and capacity to empower them to survive. Assimilating climate change information in to action learning process calls for a skillful balance of a possibly top-down agenda with community awareness and agency building (Tschakert, & Dietrich, 2010).

The measurement and nature of the hazard to be identify in order to motivate the costal communities and timely actions (Mcgranahan, Balk, & Anderson, 2007). A significant additional further aspect of adaptation is to improve thought and consciousness of costal erosion hazard through education, training, monitoring, policy and planned initiatives (Masselink, & Russell, 2013). The absence of active communication of climate change results to poor level of public awareness and understanding of the hazards of coastal communities and also low level preparation for the effects (Zsamboky *et al.*, 2011,). In addition to climate change impacts, an increasing trend in demand for both agricultural and domestic usages will be expected with regard to population growth (Okkan and Kirdemir, (2018).

Recommendations for Adaptation to Climate Change

Adaptation of coastal societies should be a prime policy priority and it is likely to integrated national and international policy including coastal erosion and flood management and disaster planning (Zsamboky *et al.*, 2011). In the aspect of climate change, adaptation is usually to adapt in natural or human systems in response to real or predictable climate stimuli or their impacts that harmful or exploits for the beneficial opportunities (McCarthy et al. 2001). One of the significant ideas have developed from few decades for sustainable coastal management is that adaptation to reduce the vulnerability of coastal erosion. It is of particular important to formulate enduring adaptation policy for the full range of future climate change outcomes (Nicholls et al. (2011). A concrete and holistic framework of climate adaptation is very urgent to protect of nature of coastal communities (Spalding *et al.*, 2014). To minimize the impact of climate change it requires multiple governance systems that can enrich the capacity to handle with insecurity and surprise by organizing different sources of resilience (Adger et al., 2005). Support should be provided to the coastal people long-term employment to reduce poverty and to enhance local and regional governance to manage the risk and continue to guide policy for future climate change (Tompkins, 2005). In some cases, enhance strong leadership and changes of social customs within administrations are obligatory to implement adaptive good governance of coastal social-ecological systems (Adger et al., 2005). Additionally, opportunity should be provided to vulnerable coastal communities to engage to apply indigenous knowledge and contribute directly in developing and applying adaptive policy to cope climate change (Spalding *et al.*, 2014). Further, salinity accepting production should be introduced like agriculture, fisheries and coastal forestry (Sarwar, 2005). Here we presented in details on Climate change vulnerability and for adaptation to improve costal environment in Table 2.

CONCLUSION

The coastal communities already face severe deprivation and connected with socio-economic challenges that make them more helpless to the impacts of climate change. The measurement and nature of the hazard has been identified in order to motivate the coastal communities and timely actions. Coastal erosion, sea level rise and extreme weather will gradually effects transport, housing, financial sectors including fisheries, agricultural farm, tourism potentially worsening deprivation in severe disadvantages coastal areas and future limiting coastal people`s capacity to adapt climate change. We recommended an array of short- and long-term measures, numbers of which can be used in individual to reverse the human impact of projected trends towards planning and implementing. And alos recognized current climate change vulnerabilities especially to backward coastal communities and make policy recommendations to mitigate these encounters. This will enhance the coastal people`s ability to adapt to the climate change and aimed to draw in a range of perceptions on the matter from societies and local and national stakeholders.

Table 2. Climate change vulnerability and for adaptation to improve costal environment

Vulnerability	Affect / Risk	Impact Detail	Possible Adaptation Measure
Beaches and sea-level rise	Loss of sand and vegetation; beach erosion	While beach morphology is complex and erosion is not only a result of sea-level rise, rising seas are likely to exacerbate erosion and also reduce the size of the beach.	Encourage businesses/ developers to minimise built structures close to the beach and plan for some form of managed retreat and adaptive access points to beaches.
Beaches and extreme events.	Loss of sand; beach erosion	The loss of sand is a major problem for the tourism industry (e.g. on the Gold Coast) and extreme weather events (e.g. tropical cyclones and their wider effects) exacerbate natural erosion rates, leading to amenity value decline and – in the worst case – loss of access to beach or dangerous beach profiles. Sea level rise and extreme weather events combine to increase erosion risk.	Soft (e.g. sand bags) and hard (sea wall, groynes etc.) structures for beach protection. Ecosystem-based protection of coastline. Sand pumping to replenish eroded sand. Avoid structures on the beach that exacerbate erosion.
	Debris on beaches	Debris on the beach affects aesthetic values and can be a hazard.	Clearing of beach.
Beaches and higher temperatures.	Tourist comfort Decreased and heat hazard	Increasing temperatures pose a challenge for tourism in destinations that are already classified as hot or potentially 'uncomfortable'.	Provide more shaded areas; social marketing on health risks of heat exposure. Consider planned changes in seasonality (i.e. shift away from current peak seasons to shoulder seasons) – local government to work with regional and state tourism organisations.
Adaptation of wetlands.	Saltwater intrusion	Saltwater intrusion of Fresh water wetlands may lead to substantial ecosystem changes, but also impact on visitor expectations of the natural area.	Obtain expert engineering and environmental advice on measures needed to protect significant freshwater habitats from saltwater intrusion.
	Loss of habitat	Encroaching seas reduce the size of wetlands and other coastal ecosystems that are the basis of nature based tourism.	Increase size of habitats, for example through restoration of wetlands, system repair, land swaps, compensation projects, development moratorium, etc.
Nature-based tourism and ecosystem changes.	Change in species composition and Ecosystem resilience	Fast changes in multiple physical parameters (e.g. temperature, precipitation, ocean currents, wind, salinity) will affect some species that cannot cope with new conditions.	Identify baseline parameters to effectively monitor the environmental changes. Continue to improve data management and the technical capacity of park staff. Undertake economic impact assessment of climate change impacts on key nature-based tourism activities (e.g. Penguin colony in Phillip Island).
Coastline settlements and infrastructure	Loss of assets	As a result of several factors including sea level rise, higher storm surges and larger spring tides will result in increased risk of coastal erosion and flooding/ inundation.	Encourage developers to: • build critical infrastructure (e.g. power houses) further from the beach; • Keep distance of buildings from the beach; • Raise structures to a minimum height

cotinued on the following page

Table 2. Continued

Vulnerability	Affect / Risk	Impact Detail	Possible Adaptation Measure
Coastline settlements and infrastructure and extreme events	Lack of warning resulting in unnecessarily high impacts	Many tourism operators have a very good understanding of the local weather; however, research also shows that many are insufficiently aware of early warning systems and have no systems in place themselves to respond to warnings.	Connect tourism operators into an early warning system (e.g. contact tree, text message), and encourage them to stablish weather information routines: • Check weather forecasts and warnings daily • Develop policies for dealing with warnings • Consider seasonal forecasts • Develop tourist-targeted warning systems (e.g. mobile app)
	Safety of visitors and staff	The health and safety of tourists and staff depend partly on the ability of destinations, and individual operators, to adapt planning and management practices to address the current and anticipated impacts of climate change, including the prevention of, and recovery from, weather- and climate related disasters.	Tourism operators need evacuation plans, including: • Clear plan and signage for guests • Staff training and regular drills • Multi-hazard planning (e.g. fire, cyclone, strong wind etc.) • First aid kits, medically trained staff at hand • Sufficient emergency water and food
	Damage to Assets	More severe storms will put waterfront and coastal infrastructure at risk, including marinas, jetties, boat ramps, roads, restaurants, accommodation, and other buildings.	Conduct an infrastructure risk assessment to identify assets at risk from both chronic stresses (e.g. saltwater intrusion) and additional climate change impacts and extreme weather events. Keep distance of buildings from the beach by implementing a minimum distance away from the high tide mark.
	Rising insurance costs	These increased risks to infrastructure will cause increases in the cost of insurances.	Encourage businesses to have sufficient insurance cover, including potential innovative forms of insurance such as index insurance. Businesses need a continuity plan and should invest into product diversification (e.g. non-coastal products in their portfolio). Public sector can offer emergency assistance packages if needed.
	Insufficient recognition of good practice	Many tourism businesses have some form of quality label/certification. These systems could also include a business 'resilience health check' to help a business make changes to become more resilient to climate and disaster risks.	Public-private sector partnerships to develop a risk- certification scheme (e.g. Ecotourism Australia, Earth Check).
	Reputational damage for businesses	Immediate and appropriate communication is essential and strategies (including templates) need to be prepared before the onset of a disaster	Disaster communication plan for before and after an event, tailored to different audiences, including overseas wholesalers, travel agents, airlines etc.
	Impact on destination image	Extreme events can affect a destination for a long time, especially when impacts are handled poorly. Concerted efforts for recovery are essential to 'be back in business' as soon as it is possible and appropriate.	Crisis management and communication plan (e.g. by the destination marketing organisation), media training for key staff. Development of an integrated (with other sectors) reconstruction and recovery plan

cotinued on the following page

Table 2. Continued

Vulnerability	Affect / Risk	Impact Detail	Possible Adaptation Measure
Coastline tourism infrastructure and higher temperatures.	Increased demand for air conditioning (costs and increased CO_2 emissions)	Tourism and community Infrastructure will be affected by hotter temperatures and changes in demand for energy, especially during peak times.	Ensure that tourism operators offer cool Spaces/buildings: Designs for hotels and visitor centres that create shade and cool buildings will be increasingly important in the hotter times ahead (e.g. natural air flow). Energy efficient systems e.g. ocean thermal energy conversion. Thermal energy from air conditioning exhaust vents to be used in resorts for heating water systems. Use of solar energy to benefit from renewable energy source to run air conditioning at zero emissions.
	Reduced tourist comfort	Tourists may alter their destination choices to avoid uncomfortably hot climates.	Minimise use of air conditioning, but provide shaded areas with natural air flow. Offer activities during less hot times during the day, e.g. morning and late afternoon. Networking with other businesses offers advantages in reducing impact of adverse weather conditions (e.g. heatwave).
Coastline settlements and infrastructure and changes in rainfall.	River Flooding	Inundation of coastal tourism infrastructure and settlements, and more frequent or severe flooding. Impacts can be exacerbated by storm effects and storm surges (e.g. in port areas).	Increases in the standard for drainage capacity for new transportation infrastructure and major rehabilitation projects, improvements in monitoring of conditions.
	Water shortage (drought)	Tourist businesses are often high water consumers. There is increasing need to contribute to water savings initiatives, both to reduce operational costs and as part of broader destination stewardship.	Maximising use of water efficient equipment (e.g. low flow shower heads). Staff training (e.g. housekeeping) and guest education (e.g. towel programs). Minimising water losses (e.g. leakage, evaporation from pools etc.) Building rainwater capture and storage tanks. Recirculation of recycled water to irrigate garden areas.

ACKNOWLEDGMENT

The authors are grateful to Universiti Sians Malaysia- Penang, Malaysia for providing facilities for the research.

REFERENCES

Adger, W. N., Hughes, T. P., Folke, C., Carpenter, S. R., & Rockstrom, J. (2005). Social- ecological resilience to coastal disasters. *Science*, *309*(5737), 1036–1039. doi:10.1126cience.1112122 PMID:16099974

Ahmed, I. (2015). *People of many rivers: Tales from the riverbanks*. Dhaka, Bangladesh: The University Press Limited.

Alam, G. M. M. (2016). *An assessment of the livelihood vulnerability of the riverbank erosion hazard and its impact on food security for rural households in Bangladesh* (PhD thesis). School of Commerce, University of Southern Queensland, Australia.

Alam, G. M. M., Alam, K., & Shahbaz, M. (2016). Influence of institutionalaccess and social capital on adaptation choices: Empirical evidence from vulnerable rural households in Bangladesh. *Ecological Economics*, *130*(C), 243–251. doi:10.1016/j.ecolecon.2016.07.012

Alam, G. M. M., Alam, K., Shahbaz, M., & Clarke, M. L. (2017). Drivers of vulnerability to climatic change in riparian char and river-bank households in Bangladesh: Implications for policy, livelihoods and social development. *Ecological Indicators*, *72*, 23–32. doi:10.1016/j.ecolind.2016.06.045

Ali, I., Hatta, Z., & Azman, A. (2014). Transforming the local capacity on natural disaster risk reduction in Bangladeshi communities: A social work perspective. *Asian Social Work and Policy Review*, *8*(1), 34–42. doi:10.1111/aswp.12023

Ali, I., Hatta, Z., Azman, A., Islam, S., Mazumder, A. H., & Hossain, K. (2016). Community participation in disaster management: A case study of Bangladesh. *Indian Journal of Ecology*, *43*(2), 463–472.

Allison, E. H., & Seeley, J. A. (2004). HIV and AIDS among fisherfolk: A threat to responsible fisheries? *Fish and Fisheries*, *5*(3), 215–234. doi:10.1111/j.1467-2679.2004.00153.x

Cazenave, A., & Llovel, W. (2010). Contemporary sea level rise. *Annual Marine Science Review*, *2*(1), 145–173. doi:10.1146/annurev-marine-120308-081105 PMID:21141661

CCA. (2010). *Coastal regeneration handbook*. CCA.

Chatterjee, R., & Huq, S. (2002). A report on the inter-regional conference on adaptation to climate change. *Mitigation and Adaptation Strategies for Global Change*, *7*(4), 403–406.

Church, J. A., & White, N. J. (2006). A 20th century acceleration in global sea-level rise. *Geophysical Research Letters*, *33*(1), 1–4. doi:10.1029/2005GL024826

Clarke, M., & Rendell, H. (2007). Climate, extreme events and land degradation. In Climate and Land Degradation. Springer. doi:10.1007/978-3-540-72438-4_7

Curtis, S., Salahuddin, A., Adler, R. F., Huffman, G. J., Gu, G., & Hong, Y. (2007). Precipitationextremes estimated by GPCP and TRMM: ENSO relationships. *Journal of Hydrometeorology*, *8*(4), 678–689. doi:10.1175/JHM601.1

Dasgupta, S. (2016). *Climate change poses urgent threat to poor of coastal Bangladesh*. Academic Press.

Douben, N. (2006). Characteristics of river floods and flooding: A global overview, 1985– 2003. *Irrigation and Drainage*, *55*(1), S9–S21. doi:10.1002/ird.239

Easterling, D. R., Evans, J. L., Groisman, P. Y., Karl, T. R., Kunkel, K. E., & Ambenje, P. (2000). Observed variability and trends in extreme climate events: A brief review. *Bulletin of the American Meteorological Society*, *81*(3), 417–425. doi:10.1175/1520-0477(2000)081<0417:OVATIE>2.3.CO;2

Furevik, T., Drange, H., & Sorteberg, A. (2002). *Anticipated changes in the Nordic Seas marine climate: Scenarios for 2020, 2050 and 2080*. Academic Press.

Grinsted, A., Moore, J. C., & Jevrejeva, S. (2009). Reconstructing sea level from paleo and projected temperatures 200 to 2100 AD. *Climate Dynamics*, *34*(2), 461–472.

Groisman, Knight, Easterling, Karl, Hegerl, & Razuvaev. (2005). Trends in intense precipitation in the climate record. *J. Climate, 18*, 1326–1350.

Han, W., Meehl, G. A., Rajagopalan, B., Fasullo, J. T., Hu, A., Lin, J., ... Yeager, S. (2010). Patterns of Indian Ocean sea-level change in a warming climate. *Nature Geoscience*, *3*(8), 546–550. doi:10.1038/ngeo901

Hossain, K., Quaik, S., Ismail, N., Rafatullah, M., Ali, I., Hatta, Z., ... Rameeja, S. (2016). Climate change-perceived impacts on agriculture, vulnerability and response strategies for improving adaptation practice in developing countries (South Asian Region). *International Journal of Agricultural Research*, *11*(1), 1–12. doi:10.3923/ijar.2016.1.12

ISDR. (2009). *Global assessment report on disaster risk reduction*. Geneva, Switzerland: United Nations.

Islam, M. S., Hasan, T. M., Chowdhury, S. I. R., Rahaman, M. H., & Tusher, T. R. (2012). Coping techniques of local people to flood and river erosion in char areas of Bangladesh. *Journal of Environmental Science & Natural Resources*, *5*(2), 251–261. doi:10.3329/jesnr.v5i2.14827

Knapp, A. K., Beier, C., Briske, D. D., Classen, A. T., Luo, Y., Reichstein, M., ... Weng, E. (2008). Consequences of more extreme precipitation regimes forterrestrial ecosystems. *Bioscience*, *58*(9), 811–821. doi:10.1641/B580908

Kumar, P. K. D. (2006). Potential vulnerability implications of sea level rise for the coastal zones of Cochin, Southwest Coast of India. *Environmental Monitoring and Assessment, 123*(103), 333–344. doi:10.100710661-006-9200-2 PMID:16738753

Leary, N., Adequwon, J., Barros, V., Burton, I., Kukarni, J., & Lasco, R. (Eds.). (2008). *Climate change and adaptation*. London, UK: Earthscan.

Masselink, G., & Russell, P. (2013). Impacts of climate change on coastal erosion. *MCCIP Science Review, 2013*, 71–86.

McCarthy, J. J., Canziani, O., Leary, N. A., Dokken, D. J., & White, K. S. (Eds.). (2001). *Climate change 2001: Impacts, adaptation and vulnerability. IPCC working group II*. Cambridge, UK: Cambridge University Press.

McGranahan, G., Balk, D., & Anderson, B. (2007). The rising tide: Assessing the risks of climate change and human settlements in low elevation coastal zones. *Environment and Urbanization, 19*(1), 17–37. doi:10.1177/0956247807076960

Meehl, G. A. (2007). Global climate projections. In Climate change 2007: The Physical Science Basis. Contribution of Working Group I to the Fourth Assessment Report of the Intergovernmental Panel on Climate Change (pp. 433-497). Cambridge University Press.

Myers, R. A., & Worm, B. (2003). Rapid worldwide depletion of predatory fish communities. *Nature, 423*(6937), 280–283. doi:10.1038/nature01610 PMID:12748640

Nicholls, R. J. (2004). Coastal flooding and wetland loss in the 21st century: Changes under the SRES climate and socio-economic scenarios. *Global Environmental Change, 14*(1), 69–86. doi:10.1016/j.gloenvcha.2003.10.007

Nicholls, R.J. (2011). Sea-level rise and its possible impacts given a 'beyond 4o C world' in the twenty-first century. *Philosophical Transaction of Royal Society A, 369*(1934), 161-181.

Nicholls, R.J., & Cazanave, A. (2010). *Sea level rise and its impact on coastal zones*. Academic Press.

NOAA. (2013). *State of the coast*. National Oceanic and Atmospheric Administration. Retrieved 25 October 2018, from http://stateofthecoast.noaa.gov/

Okkan & Kirdemir. (2018). Investigation of the Behavior of an Agricultural-Operated Dam Reservoir Under RCP Scenarios of AR5-IPCC. *Water Resour Manage, 32*, 2847–2866.

Penning-Rowsell, E. C., Sultana, P., & Thompson, P. M. (2013). The 'last resort'? Population movement in response to climate-related hazards in Bangladesh. *Environmental Science & Policy*, *27*(S1), 44–59. doi:10.1016/j.envsci.2012.03.009

Philippart, C. J. M., Anadón, R., Danovaro, R., Dippner, J. W., Drinkwater, K. F., Hawkins, S. J., ... Reid, P. C. (2011). Impacts of climate change on European marine ecosystems: Observations, expectations and indicators. *Journal of Experimental Marine Biology and Ecology*, *400*(1-2), 52–69. doi:10.1016/j.jembe.2011.02.023

Prakasa, R., & Murty, B.S. (2005) Estimation of flood vulnerability index for delta. *Geoscience and Remote Sensing Symposium*, *5*(25–29), 3611–3614.

Rahmstorf, S., Perrette, M., & Vermeer, M. (2012). Testing the robustness of semi-empirical sea level projections. *Climate Dynamics*, *39*(3-4), 861–875. doi:10.100700382-011-1226-7

Sarwar, M. G. M. (2005). Impacts of sea level rise on the coastal zone of Bangladesh. *Science*, *328*(5985), 1517-1520.

Small, C., & Nicholls, R. J. (2003). A global analysis of human settlement in coastal zones. *Journal of Coastal Research*, *19*(3), 584–599.

Spalding, M. D., Ruffo, S., Lacambra, C., Meliane, I., Hale, L. Z., Shepard, C. C., & Beck, M. W. (2014). The role of ecosystems in coastal protection: Adapting to climate change and coastal hazards. *Ocean and Coastal Management*, *90*(March), 50–57. doi:10.1016/j.ocecoaman.2013.09.007

Tompkins, E. L. (2005). Planning for climate change in small islands: Insights from national hurricane preparedness in the Cayman Islands. *Global Environmental Change*, *15*(2), 139–149. doi:10.1016/j.gloenvcha.2004.11.002

Tschakert, P., & Dietrich, K. A. (2010). Anticipatory learning for climate change adaptation and resilience. *Ecology and Society*, *15*(2), 1–23. doi:10.5751/ES-03335-150211

UNFCCC. (2011). *Report of the Conference of the Parties on Its Sixteenth Session, Held in Cancun from 29 November to 10 December 2010. Part Two: Action taken by the Conference of the Parties at its sixteenth session* (FCCC/CP/2010/7/Add.1). UNFCCC.

Vermeer, M., & Rahmstorf, S. (2009). Global sea level linked to global temperature. *Proceedings of the National Academy of Sciences of the United States of America*, *106*(51), 21527–21532. doi:10.1073/pnas.0907765106 PMID:19995972

Willmott, C. J., & Legates, D. R. (1991). Rising estimates of terrestrial and global precipitation. *Climate Research, 1*, 179–186. doi:10.3354/cr001179

World Bank. (2009). *Convenient solutions to an inconvenient truth: ecosystem based approaches to climate change*. Washington, DC: The World Bank.

Zolina, O., Simmer, C., Belyaev, K., Kapala, A., & Gulev, S. (2009). Improving estimates of heavyand extreme precipitation using daily records from European rain gauges. *Journal ofHydrometeorology, 10*(3), 701–716. doi:10.1175/2008JHM1055.1

Zsamboky, M., Fernández-Bilbao, A., Smith, D., Knight, J., & Allan, J. (2011). *Impacts of climate change on disadvantaged UK coastal communities*. Academic Press.

Section 2
Floods and Associated Hydrologic Events

Chapter 4
General Review of Calibration Process of Nonlinear Muskingum Model and Its Optimization by Up-to-Date Methods

Umut Kırdemir
Dokuz Eylul University, Turkey

Umut Okkan
Balikesir University, Turkey

ABSTRACT

Nonlinear Muskingum method is a very efficient tool in flood routing implementation. It is possible to estimate an outflow hydrograph by a given inflow hydrograph of a flood at a specific point of the river channel. However, it turns out an optimization problem at the stage of employing this method, and it becomes important to reach the optimal model parameters so as to obtain precise outflow hydrograph estimations. Hence, it was decided to utilize five up-to-date optimization algorithms, namely, vortex search algorithm (VSA), gases brownian motion algorithm (GBMO), water cycle algorithm (WCA), flower pollination algorithm (FPA), and colliding bodies optimization (CBO). The algorithms were integrated with the nonlinear Muskingum model so as to estimate the outflow hydrograph of Wilson data, and it was deduced that WCA, FPA, and VSA perform relatively better than the models employed in the other researches before.

DOI: 10.4018/978-1-5225-9771-1.ch004

INTRODUCTION

Flood routing techniques are user-friendly tools in estimation of downstream hydrograph by given upstream hydrograph and practically utilized in engineering issues such as designing flood control facilities and used for warning the people living in the flood regions in order to prevent them from loss of life and property as well. They are separated into two categories as hydraulic and hydrological models. Hydraulic models are built upon Saint Venant equations in which principle of mass and momentum conservation are taken into consideration. The use of momentum conservation principle introduces the flood routing processes into tedious struggle. Moreover, hydraulic models require thorough data measurement of multiple variables in the related river reach (Papaioannou et al., 2017). In view of these conditions, the hydrological methods such as Muskingum model are sufficient such that linear reservoir concept based on only principle of mass conservation is taken into consideration (Barati, 2011).

Floods are natural hazards which come to existence frequently around the world such that they have disastrous effects on human life, animals, agricultural fields, etc. Hence, it is crucial to implement planning studies about flood risk managements and vulnerability. The studies about flood risk managements deal with flood hazard prediction, its effects on human society and how to minimize flood risk (Schanze, 2006; Nikoo et al., 2016). In this respect, the methods such as flood routing specifies characteristics of a river channel by modeling the temporally or spatially varying flood wave. The flow rate is measured at a cross-section of the channel and it is predicted at an outlet point. Thus, this method enables decision makers to take measure right after a storm which may give rise to flood events. In the literature, Muskingum model is one of the well-known flood routing methods frequently used in the literature.

The linear Muskingum model in which only streamflow measurement data are utilized is the most frequently used method in natural channels and rivers among all hydrolgical models with regard to its ease of use. The model is presumed to capture the river characteristics in the routing process by means of its inherent parameters. A linear storage-discharge relationship is set up in the linear Muskingum model, however, it is highly possible to experience the linear Muskingum model to be insufficient in many flood events. That is, the nonlinear structure of streamflow with respect to time gives rise to inadequacy in estimation of outflow by linear model. Hence, a nonlinear Muskingum model in which a nonlinear storage-discharge relationship is involved in the process is more appropriate to use at this stage. Following the this idea, a three-parameter model which is frequnetly utilized in the literature is derived. The parameters to be estimated are storage time constant, absolute channel storage and an exponential constant which captures the nonlinearity in the model.

With the increase of parameter number and degrees of freedom, the calibration process of the model turns into more complex form when compared to the linear model. In this respect, it is considered that parameter estimation of the nonlinear model becomes more striving while the precision of downstream hydrograph estimations increases. Hence, it is required to use an efficient optimization procedure to estimate model parameters.

When the literature is reviewed, it can be seen that various methods have been employed by different researchers and they have tried to derive the best solution until now by means of different optimization techniques. In their works, they usually utilized Wilson's data which was released by Wilson (1974) and they made comparisons of their results with the those of previous studies. In the beginning of the studies, derivative-based methods were chosen for parameter estimation, however, the metaheuristic optimization methods in which initial parameter conditions do not effect the prediction performance were employed recently. Considering the gradual improvements in estimation of Muskingum model parameters so far, in this chapter, a general review of this calibration process will be presented in detail. Five more up-to-date optimization algorithms, whose common trait is that they were developed after the year 2010, will be introduced to the prospective readers and the calibration performances will be compared thoroughly. The optimization methods which will be used in the study are namely, Vortex Search Algorithm (Dogan and Olmez, 2015), Gases Brownian Motion Optimization (Abdechiri et al., 2013), Water Cycle Algorithm (Eskandar et al., 2012), Flower Pollination Algorithm (Yang, 2012), and Colliding Bodies Optimization (Kaveh and Mahdavi, 2014). These methods will be employed for Wilson's data which were extensively utilized in the literature to compare the performances of various parameter calibration techniques. The model performances were evaluated through nonparametric statistical tests such as Kruskal-Wallis and Dunn tests and it will be obtained an opportunity to compare up-to-date methods in terms of parameter estimation performances. To the best of our knowledge, the algorithms employed in this study have not been used in order to estimate nonlinear Muskingum model parameters before. The study was constituted in five main parts. The methodology about Muskingum flood routing and literature review of parameter estimation of nonlinear Muskingum model were presented in the next section. The up-to-date optimization techniques employed for model calibration were intorduced to the readers in the third section. In the fourth and fifth sections, the results and conclusions of the study were addressed, respectively.

MUSKINGUM FLOOD ROUTING AND PAST EXPERIENCES IN PARAMETER CALIBRATION

Nonlinear Muskingum Model

Muskingum model is the most extensively used hydrological flood routing model in which inflow hydrograph is routed to a specific point of the river channel and an outflow hydrograph is derived. It is a lumped hydrological model and it takes into account the conservation of mass principle to constitute the storage-discharge relationship as

$$\frac{dS_t}{dt} = I_t - O_t \tag{1}$$

where I_t and O_t are the inflow and outflow observed in the channel at a specific time t, respectively and S_t denotes the channel storage at the related time t. In the Muskingum model, the storage is modeled by the formal properties of the flood wave such that when the rate of inflow exceeds the rate of outflow, the wedge storage occurs. On the contrary, that is, when the outflow rate exceeds the inflow rate the prism storage occurs (the recession process). Considering the wedge and prism storage approaches, the total storage at time t, can be modeled linearly as follows:

$$S_t = K\left[XI_t + (1-X)O_t\right] \tag{2}$$

where K is the storage constant such that it captures the travel time of flood wave through the channel and X is the dimensionless weighting parameter varying between 0 and 0.5 which builts upon the formal property of wedge storage. The related storage model defined in the Equation 2 is called as linear Muskingum model as well. However, the frequently experienced phenomenon in the natural rivers is that the relationship between storage and weighted flows are not linear. Hence, it is more convenient to utilize nonlinear Muskingum flood routing by modeling nonlinearity between storage and river flow in order to derive more precise estimations. The nonlinear Muskingum model mostly used in the literature is defined as follows:

$$S_t = K\left[XI_t + (1-X)O_t\right]^m \tag{3}$$

where m is the additional parameter which reckons nonlinearity between storage and rate of river flow. By Equation 1, 2 and 3 the inflow and outflow in the nonlinear model can be constituted as

$$O_{predicted,t} = \left(\frac{1}{1-X}\right)\left(\frac{S_t}{K}\right)^{1/m} - \left(\frac{X}{1-X}\right)I_t \tag{4}$$

$$\frac{\Delta S_t}{\Delta t} = I_t - \left[\left(\frac{1}{1-X}\right)\left(\frac{S_t}{K}\right)^{1/m} - \left(\frac{X}{1-X}\right)I_t\right] = -\left(\frac{1}{1-X}\right)\left(\frac{S_t}{K}\right)^{1/m} + \left(\frac{1}{1-X}\right)I_t \tag{5}$$

$$S_{t+\Delta t} = S_t + \frac{\Delta S_t}{\Delta t} \cdot \Delta t \tag{6}$$

where $O_{predicted,t}$ is the outflow rate predicted by nonlinear Muskingum model. By integrating Equation 1 with Equation 4, the change in storage volume with respect to time can be calculated by Equation 5 and the successive storage volume can be estimated by Equation 6 until end of the time series. The parameters K, X and m are calibrated in the model and the predicted outflow hydrograph are estimated. Moreover, there are alternative nonlinear Muskingum models constituted for the storage relationship which will not be utilized in the study are as follows:

$$S = K\left(XI_t^m + (1-X)O_t^m\right) \tag{7}$$

$$S = K\left(XI_t^n + (1-X)O_t^m\right) \tag{8}$$

$$S = K\left(XI_t^n + (1-X)O_t^n\right)^m \tag{9}$$

where parameters n and m captures the nonlinearity in the storage volume in Equation 7, 8 and 9. Information about past experiences in calibration of nonlinear Muskingum model was summarized in the next subsection.

Literature Review

The studies about flood routing by nonlinear Muskingum method have been made by several researchers and they generally utilized Wilson's data (Wilson, 1974) in order to compare the diverse approaches in estimation of parameters. The researchers utilized the sum of squared error (SSE) statistical performance criterion to compare the model results which is

$$SSE = \sum_{\forall i \in t} \left(O_{observed,i} - O_{predicted,i} \right)^2 \qquad (10)$$

where $O_{observed,i}$ and $O_{predicted,i}$ denote observed and predicted outflow rate in the river. In the early years, the first precedent study was made by Gill (1978) in which the least squares method was integrated with segmented-curve method (S-LSM) to estimate parameters of nonlinear Muskingum model. Afterwards, Tung (1985) proposed three different algorithms in order to estimate model parameters such that they were mainly built upon the Hooke-Jeeves (HJ) pattern search technique integrated with linear regression (LR) and two different derivative-based methods namely conjugate gradient (CG) and Davidon-Fletcher-Powell (DFP) methods. It was obtained by Tung (1985) that HJ-CG and HJ-DFP methods developed the model performance in comparison to the that of Gill (1978). Yoon and Padmanabhan (1993) utilized nonlinear least squares regression method (NONLR) to in order to estimate model parameters. Mohan (1997) stood out with the study including parameter estimation procedure which was carried out by means of Genetic Algorithm (GA), such that this study was the pioneer study ever made by metaheuristic optimization algorithms until that time. In the related study, it was seen that the model performance is significantly developed when compared to the previous ones. After a little while, Kim et al. (2001), employed another metaheuristic algorithm, that is, Harmony Search (HS) algorithm for parameter estimation of nonlinear Muskingum model and they developed the performance and produced the better results than that of Mohan (1997). Das (2004) utilized a well-known derivative based method, that is, Lagrange multipliers (LM) method and it was obtained significantly lower model performances by this method. Subsequently, Geem (2006) used a parameter estimation procedure based on a type of quasi-Newton method namely Broyden-Fletcher-Goldfarb-Shanno (BFGS) technique. This method generated slightly better results than Kim et al (2001) and showed a significant superiority over another quasi-Newton method, HJ-DFP used in Tung

(1985). Das (2007) utilized Crank-Nicholson method for flood routing and evaluated the flood event with different aspects by chance-constrained model. Chu and Chang (2009) used another popular metaheuristic optimization technique, Particle Swarm Optimization (PSO), and they obtained satisfying results however not as good as the results of Geem (2006). Luo and Xie (2010) used an evolutionary algorithm called as Immune Clonal Selection Algorithm (ICSA) which is performed well and derived close results to the results derived by Geem (2006). In the study prepared by Geem (2011), the Harmony Search method which was employed before in Kim et al (2001), was slightly modified and it showed very good performance when compared to the previous ones, however, it could not develope the model performance when compared to the study made by same researcher (Geem, 2006). A direct search method called as Nelder-Mead Simplex Algorithm (NMSA) was utilized by Barati (2011) and it showed high performances relative to the major quantity of previous studies released for parameter estimation of nonlinear Muskingum model. Xu et al. (2012) performed an algorithm similar to Genetic Algortihm called as Differential Evolution (DE) Algorithm and they derived results similar to Kim et al (2001), Geem (2006), Geem (2011), Barati (2011). Upon evaluating the studies made in recent years, it was distinguished that they derive almost close results relative to each other. Another example to these studies is that of Barati (2013) in which Excel solver based on generalized reduced gradient method (GRG) in order to estimate model parameters. Karahan et al. (2013) used a hybridized optimization method in which Harmony Search and BFGS techniques (HS-BFGS) were integrated and the researchers claim that the method produces the best performance ever had until now. However, the authors feel responsible to state that this claim may not be true in that the results are very close to each other and it is possible that the researchers who has the related claim could not measure the previous model performances precisely such that the result given in the previous studies are not released by long decimal formats. In order to summarize calibration process of nonlinear Muskingum model, the estimated parameters of nonlinear Muskingum model using Wilson's data and the observed inflow and outflow and predicted outflow rates obtained by the above-mentioned researchers were shown in Table 1 and Table 2, respectively.

OPTIMIZATION ALGORITHMS

Due to the fact that conventional derivative-based methods is computationally striving, thus, metaheuristic algorithms are recommended. These algorithms outperform by comparison with the those of traditional methods when the complexity of problem

Table 1. The values of Nonlinear Muskingum model performances by the parameters calibrated by the researchers in a chronological line up

Chronology	Method	*K*	*X*	*m*	SSE
(1)	S-LSM (Gill, 1978)	0.01	0.25	2.347	145.6945
(2)	HJ-LR (Tung, 1985)	0.17	0.24	1.7012	40.1
(3)	HJ-CG (Tung, 1985)	0.0669	0.2685	1.9291	25.2
(4)	HJ-DFP (Tung, 1985)	0.0764	0.2677	1.8978	45.612
(5)	NONLR (Yoon and Padmanabhan, 1993)	0.06	0.27	2.36	156.4399
(6)	GA (Mohan, 1997)	0.1033	0.2813	1.8282	38.2363
(7)	HS (Kim et al., 2001)	0.0883	0.2873	1.863	36.7829
(8)	LM (Das, 2004)	0.0753	0.2769	2.2932	130.4872
(9)	BFGS (Geem, 2006)	0.0863	0.2869	1.8679	36.7679
(10)	PS (Chu and Chang, 2009	0.1824	0.333	2.1458	36.89
(11)	ICSA (Luo and Xie, 2010)	0.0884	0.2862	1.8624	36.8026
(12)	Modified HS (Geem, 2011)	0.0864	0.2869	1.8677	36.768
(13)	NMSA (Barati, 2011)	0.0862	0.2869	1.8681	36.7679
(14)	DE (Xu et al, 2012)	0.5175	0.2869	1.868	36.77
(15)	GRG-based Excel Solver (Barati, 2013)	0.086249	0.286916	1.868087	36.767888
(16)	HS-BFGS (Karahan et al., 2013)	0.086249	0.286917	1.868088	36.767888

Table 2. The observed flow rates and predicted outflow rates obtained by the researchers in the past. The numbers above the each predicted outflow hydrographs denote in which scientific research the time series were generated as similar in Table 1

Time (h)	$I_{observed}$ (m^3/s)	$O_{observed}$ (m^3/s)	$O_{predicted}$ (m							
			(1)	(2)	(3)	(4)	(5)	(6)	(7)	(8)
0	22	22	22	22	22	22	22	22	22	22
6	23	21	22	22	22	22	22	22	22	22
12	35	21	22.8	22.4	22.4	22.4	22.8	22.4	22.4	22.8
18	71	26	29.7	26.3	26.8	26.7	29.3	26.4	26.6	29.6
24	103	34	39.3	35.5	34.9	34.8	37.9	34.2	34.4	38.7
30	111	44	48	47.4	44.5	44.7	45.7	44.2	44.1	47
36	109	55	58.4	60.7	56.7	56.9	56	57	56.8	57.8
42	100	66	67.5	71.7	67.3	67.7	65.5	68.2	68.1	67.6
48	86	75	75.1	79.8	75.9	76.3	73.5	77.2	77.1	75.8
54	71	82	80.7	84.6	81.9	82.2	79.8	83.3	83.3	81.9
60	59	85	83.5	85.6	84.5	84.7	83.2	85.7	85.9	85
66	47	84	83	82.9	83.4	83.5	83.2	84.2	84.5	84.4
72	39	80	80.1	77.9	79.9	79.8	80.9	80.2	80.6	81.4
78	32	73	74.5	70.5	73.6	73.3	75.6	73.3	73.7	75.2
84	28	64	67	62.1	65.8	65.5	68.5	65.1	65.4	67.1
90	24	54	57.8	53.1	56.9	56.5	59.4	55.8	56	57.2
96	22	44	47.6	44.6	47.8	47.5	49.4	46.7	46.7	46.5
102	21	36	37	36.7	38.9	38.7	38.7	38	37.8	35.5
108	20	30	27.7	30.4	31.5	31.4	29.1	30.9	30.5	26.4
114	19	25	21.6	25.8	25.8	25.9	22.3	25.7	25.3	21.1
120	19	22	19	22.5	22	22	19.1	22.2	21.8	19
126	18	19	19	20.6	20.1	20.2	19	20.3	20	19

increases as well (Al-Betar et al., 2018). In this chapter, five metaheuristic optimization algorithms were employed separately for Wilson's data in order to calibrate nonlinear Muskingum model parameters. The optimization algorithms which were utilized in the study are Vortex Search Algorithm (VSA), Gases Brownian Motion Optimization (GBMO), Water Cycle Algorithm (WCA), Flower Pollination Algorithm (FPA) and Colliding Bodies Optimization (CBO). The first two algorithms are single-solution based algorithms, that is, they start with a single solution for parameter estimation and they define a path in accordance with the objective function value as experienced in local search algorithms. Some other popular examples can be given for single-solution based algorithms as Simulated Annealing (Kirkpatrick et al., 1983; Cerny, 1985), Tabu Search (Glover, 1986), Variable Neighborhood Search (Hansen and Mladenovic, 2008) and etc. The rest of the algorithms performed for parameter estimation are population-based algorithms, that is, they initialize the solutions with a group of parameter sets and they are effected by each other in each iteration in order to reach the best solution. They are generally related to Evolutionary Algorithms and Swarm Intellegience. Some of popular population-based algorithms are Particle Swarm Optimization (Kenedy and Eberhart, 1995), Genetic Algortihm (Holland, 1975), Differential Evolution Algorithm (Storn and Price, 1997), Ant Colony Optimization (Dorigo et al., 1996) and etc. (Boussaid, et al., 2013). The algorithms employed in the study include stochastic parameters which direct the solution to find the best parameter set. The values used in the algorithms are inserted as with recommendations of model developers. However, if a model user desires to identify the related parameters with various values, it is recommended to perform sensitivity analysis to determine the optimal ones. The detailed information about optimization algorithms utilized in the study are given in the subsections below:

Vortex Search Algorithm

Dogan and Olmez (2015) proposed a single-solution based method called as Vortex Search Algorithm (VSA). The developed method was inspired from the circular movement of the fluids at the moment of vortex phenomenon. The algorithm begins the solution with a parameter set and envisions a circular pattern in the search space by means of the center and radius parameters which are generated in the initial step as

$$\mu_0 = \frac{L_{upper} - L_{lower}}{2} \tag{11}$$

$$r = \frac{ginv.(L_{upper} - L_{lower})}{2} \tag{12}$$

where L_{upper} and L_{lower} are upper and lower boundaries of the model parameters and μ_0 is the initial center parameter at the beginning of the parameter solution and ginv parameter given in the Equation 12 is inverse incomplete gamma function which uses a parameter for decrement of radius which satisfies the narrower search space when the single particles get close to the optimal solution. The initial value of decrement parameter is defined by user and recommended to be taken as 1 by the model developers and decreases as inversely proportional to the iteration step in which algorithm search for the solution. Thus, the radius parameter decreases in that the inverse incomplete gamma function decreases proportionally to the decrement parameter. The radius parameter is weighted randomly in each iteration and searching process is implemented around the center value by the radius parameter. In each iteration the best solution is converted to the center parameter and these processes continue until the best solution is obtained.

Gases Brownian Motion Optimization

Gases Brownian Motion Optimization method was developed by Abdechiri et al. (2013) and it is built upon Brownian motion of gas molecules under the concept of chemistry-based laws. Although, the method is considered as population-based method at first such that it initializes the solution by parameter sets called as molecules, these parameter groups do not effect each other throughout the solution, hence, the related algorithm turns out a single solution-based algorithm.

The so-called molecules use velocity parameter in order to update the location parameter, which is the parameter value at each iteration step, and this approach corresponds to movement of gas molecules by their kinetic energy. In the method, the kinetic energy, in other words, the velocity parameter is based on several stochastic parameters defined by the model user and the velocity is defined as follows:

$$v_i^d(t+1) = v_i^d(t) + \sqrt{\frac{3kT}{m}} \tag{13}$$

where v_i^d is velocity of d-dimensional ith molecule; k, T and m denote Boltzmann's constant, temperature and mass of molecules, respectively. The Boltzmann's constant and initial temperature value are defined by users and they were taken as 1000 for both in this study. The related velocity term in the method carries out the explorative search at the first iterations and the location parameters are defined as

$$x_i^d(t+1) = x_i^d(t) + v_i^d(t) \tag{14}$$

where x_i^d is location of d-dimensional ith molecule to be estimated. In addition to Brownian motion of the molecules, the developers of the method address turbulent rotational process in order to get precise estimation in the search space by utilizing chaotic sequence generator Circle map and it is defined as

$$x_i^d(t+1) = x_i^d(t) + b - \left(\frac{a}{2\pi}\right)\sin(2\pi x_i^d(t)) \tag{15}$$

In the method, Equation 14 and Equation 15 satisfy explorative and exploitative search in the parameter space, respectively. As the iteration step increases during the running process of the GBMO, the explorative and exploitative searching tasks interchange between Equation 14 and Equation 15 due to the fact that the temperature and mass values decrease while the algorithm is running. The decrement of the temperature and mass values are carried out as follows:

$$T = T - \left(\frac{1}{mean(fitness_i(t))}\right) \tag{16}$$

$$m_i(t) = \frac{fitness_i(t) - worst(t)}{best(t) - worst(t)} \tag{17}$$

where $\text{fitness}_i(t)$ denotes the fitness value of ith molecule at iteration step t and best(t) and worst(t) are the best and worst fitness values obtained until iteration step t. The iterations for each molecule continue until the temperature reaches the value lower or equal to 0.

Water Cycle Algorithm

Eskandar et al. (2012) developed a population-based optimization method mimicking hydrological cycle phenomena called as Water Cycle Algorithm (WCA). They were inspired from the hydrological processes such as precipitation and evaporation and the river network in the real world. The model developers captured the precipitation process by assigning raindrop parameter to each parameter set. In the model, the raindrops are proportionally divided into several parts with regard to their SSE performances in order to join to the river network such that the network consists of stream, river and sea. Sea denotes the best solution and the rest of the raindrop population is allocated to rivers and streams in which the raindrops resulting lower performances are assigned to the latter one. WCA developers addressed the way of how the stream or river flow to the river or sea, respectively as

$$X_{stream}^{i+1} = X_{stream}^{i} + rand.C.\left(X_{river}^{i} - X_{stream}^{i}\right) \tag{18}$$

$$X_{river}^{i+1} = X_{river}^{i} + rand.C.\left(X_{sea}^{i} - X_{river}^{i}\right) \tag{19}$$

Colliding Bodies Optimization

Colliding Bodies Optimization (CBO), developed by Kaveh and Mahdavi (2014), is a method built upon physical laws constituted under the phenomenon of collision of two objects. The population set is called as colliding bodies in CBO and they search the optimal solution in accordance with their mass and velocity values. The mass value of each particle is determined with respect to the fitness value such that they are weighted as directly proportional to their performances. The collision is implemented by the researchers in a way that the parameter set is divided into two as stationary and moving groups. Subsequent to mass assignment to each particle, their initial and updated velocity values are determined by Equation set given as follows:

$$v_i = \begin{cases} 0 & for\, i = 1, 2, \ldots\ldots, \dfrac{n}{2} \\ x_{i-\frac{n}{2}} - x_i & for\, i = \dfrac{n}{2}+1, \dfrac{n}{2}+2, \ldots\ldots\ldots, n \end{cases} \tag{24}$$

$$v_i^{'} = \begin{cases} \dfrac{\left(1+\in\right) m_{i+\frac{n}{2}} v_{i+\frac{n}{2}}}{m_i + m_{i+\frac{n}{2}}} & for\, i = 1, 2, \ldots\ldots, \dfrac{n}{2} \\ \dfrac{\left(m_i - \in m_{i-\frac{n}{2}}\right) v_i}{m_i + m_{i-\frac{n}{2}}} & for\, i = \dfrac{n}{2}+1, \dfrac{n}{2}+2, \ldots\ldots\ldots, n \end{cases} \tag{25}$$

Equation 24 and Equation 25 are used for initialization and collision of the colliding bodies in the optimization process. v_i and $v_i^{'}$ are the velocity parameters at the ith iteration step which are used for determination of updated parameter set by means of mass parameter that is denoted in the equations as m. The ϵ parameter is calculated as $1-(iteration\ /\ iteration_{\max})$, that is, it decreases as the iteration step moves forward and narrows the search while the model gets closer to the optimal solution.

RESULTS

Performance Evaluation of Optimization Algorithms

In order to calibrate nonlinear Muskingum model parameters, five optimization algorithms were utilized such that 40 different simulations with 1000 iterations were performed for each optimization algorithm, separately. The model was employed for simulating outflow component of Wilson data and the objective function of minimization of sum of squares error (SSE) was used so as to enable readers to make comparisons of model performances between the models which performed in the previous studies. The best performance in the simulations (lowest SSE) with parameter sets derived by the each optimization method and the outflow time series of each best solution was shown in the Table 3 and Table 4, respectively. According to Table 3, the best solution obtained among the optimization methods were generated by WCA with lowest SSE value of 36.767888. In the related Table, the SSE values were released with a high amount of significant digits due to the fact that the performance comparisons in the literature were carried out with a high precision before. Moreover, the FPA performed as the second best optimization method at simulating the outflow time series of Wilson data. Generally, the population-based methods were expected to perform better than the single-solution-based methods by their better explorative search capabilities and it was experienced unsurprisingly as anticipated in this study that the population-based WCA and FPA rank the first and second places. However, the single-solution-based VSA method performed better than the population-based CBO and the single-solution-based GBMO and rank the third place in terms of calibration performances. Moreover, VSA method showed better results than the most of the optimization algorithms utilized for estimating Wilson data outflow. GBMO was the worst performing method among

Table 3. The model parameters calibrated by the optimization methods and the values of performances and peak flow errors obtained in the best solutions.

Model	Parameters			SSE	Peak Flow
	K	X	m		Erros
VSA	0.0862491	0.2869166	1.8680871	36.7678884564	0.90065161
GBMO	0.0918227	0.2932169	1.8545336	37.3971874282	1.20589454
WCA	0.0862491	0.2869166	1.8680871	36.7678884564	0.90065146
FPA	0.0862491	0.2869166	1.8680871	36.7678884564	0.90065150
CBO	0.0879285	0.2870235	1.8638159	36.7770374827	0.91475931

the optimization methods utilized in the study. As known, the most important part of the flood prediction studies is to predict the discharge of the peak flow volume and the occurrence time. As seen in Table 3, the algorithms predicted the peak flow rates by the error between 0.9 and 1.2 m^3s^{-1}. When the peak flow errors were compared, it was seen that WCA was best method at this concept and the methods ranked the same places as in the SSE performances. All of the algorithms predicted the occurrence time of the peak flow right. Furthermore, it was evaluated how the optimization methods converge the best results by checking iteration numbers corresponding to the best solutions (Figure 1). Except VSA, all methods converged the near neighborhood (the solutions deriving +0.01 SSE value with respect to the lowest SSE) of the best solution about in between 100th and 200th iterations. VSA reached the region where exploitative capabilities come into prominence about at 700th iteration for the simulation deriving the best results.

Statistical Comparisons of the Optimization Algorithms

Five optimization methods derived better results relative to the several methods employed in the past, however, it was required to make comparisons among the methods utilized in the study. In order to compare the optimization methods, the performances which were obtained at the end of the each forty simulation were evaluated. Upon considering the distributional pattern of the SSE performances obtained in each simulation, it was seen that the optimization algorithms do not have any specific theoretical distribution in terms of their performances. Hence, Kruskal-Wallis nonparametric test was used so as to compare the optimization algorithms. Kruskal-Wallis test is a distribution-free test that is the nonparametric counterpart of one-way ANOVA. The method is used for testing if there is a significant difference between the ranked data of three or more groups. In implementation of the test, the performance data of the optimization algorithms were utilized for the comparison

Table 4. The predicted outflow hydrographs by optimization methods in the best solutions

Time (h)	Optimization method				
	VSA	GBMO	WCA	FPA	CBO
0	22	22	22	22	22
6	22	22	22	22	22
12	22.4222	22.41345	22.4222	22.4222	22.4192
18	26.61135	26.51577	26.61135	26.61135	26.58507
24	34.45625	34.17142	34.45625	34.45625	34.41523
30	44.16709	43.81347	44.16709	44.16709	44.1422
36	56.85496	56.6409	56.85496	56.85496	56.85268
42	68.05868	68.01206	68.05868	68.05868	68.07334
48	77.07135	77.18352	77.07135	77.07135	77.09438
54	83.31787	83.55805	83.31787	83.31787	83.34016
60	85.90065	86.20589	85.90065	85.90065	85.91476
66	84.53631	84.82729	84.53631	84.53631	84.53853
72	80.581	80.82469	80.581	80.581	80.57057
78	73.71064	73.85857	73.71064	73.71064	73.69049
84	65.40672	65.45917	65.40672	65.40672	65.38258
90	55.99733	55.95771	55.99733	55.99733	55.97682
96	46.66744	46.57712	46.66744	46.66744	46.6587
102	37.75378	37.64692	37.75378	37.75378	37.76226
108	30.46871	30.38615	30.46871	30.46871	30.493
114	25.22822	25.19169	25.22822	25.22822	25.26043
120	21.73869	21.74007	21.73869	21.73869	21.76935
126	19.99422	20.00396	19.99422	19.99422	20.01367

Figure 1. The SSE changes at each iteration step for the best solution of the optimization methods

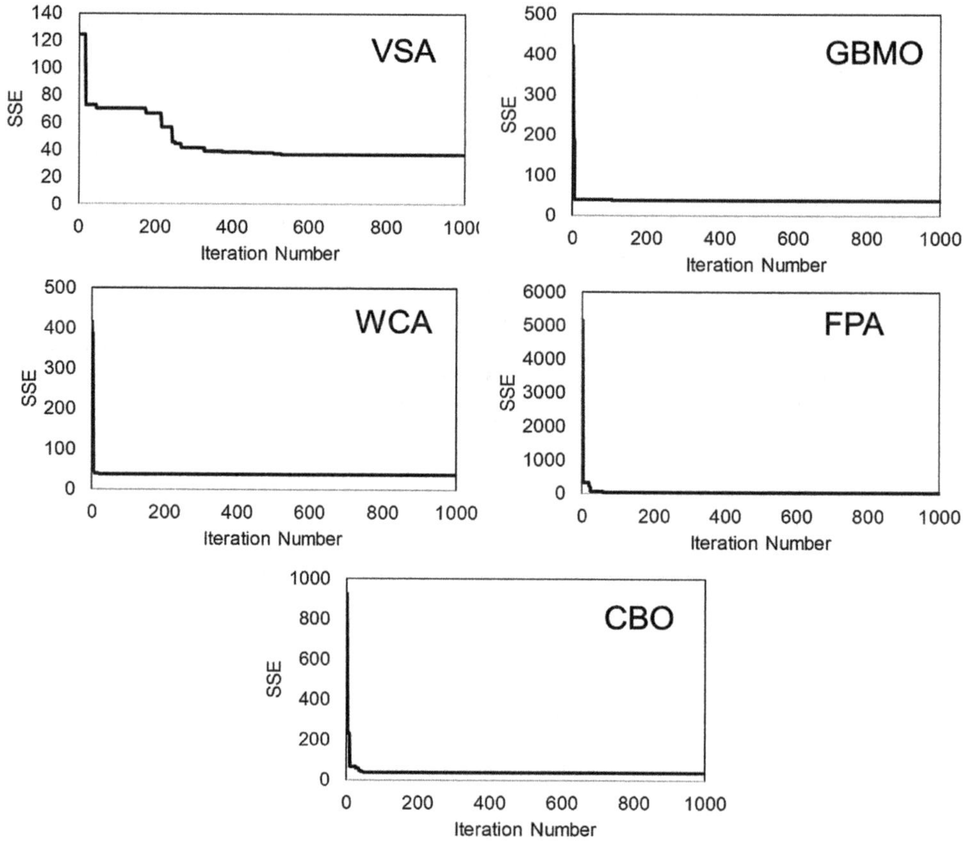

of the methods at α=0.05 significance level which corresponds to a $H_{critical}$ value of 9.45. At the end of the test, an H_{test} statistics with the value of 177.9 was obtained, thus, it was decided that there is a significant difference between the median of the groups due to $H_{test} > H_{critical}$. Subsequently, Dunn's post hoc test was performed for detection of in which model pairs statistical difference occur. The Bonferroni correction was implemented for the adjustment of the significance level to avoid the Type I Error such that it was carried out in terms of the quantity of pairwise comparisons which is equal to 10 for this study. The results of the pairwise hypothesis tests were shown in Table 5. In Table 5, 1 denotes that model pairs are significantly different whereas 0 denotes that they are not significantly different from each other. According to the cross-tabulated results given in the Table 5, there are not significant differences only in between GBMO-CBO and WCA-FPA model pairs. Statistically significant differences were detected in the rest of eight model pairs. Hence, it draws the attention that the statistical similarities are only encountered between the pairs which are constituted from the best and the worst two models (WCA-FPA and GBMO-CBO, respectively). In order to measure the extent of the differences between the model pairs, the z scores obtained in the Dunn's post hoc tests were evaluated. The z scores were calculated in terms of the rank differences of the model performances which were derived in forty simulations. That is to say, the higher a model pair gets z score, the higher level of statistical difference is available within the model pair. The z scores between the model pairs were shown in Figure 2.

Hence, it was distinguished that the lowest difference with z=1.53 was obtained in the hypothesis test in between WCA-FPA which were the best two models employed in the study. The highest difference was obtained in the hypothesis test in between WCA-CBO with the z=11.2. Although the model pair GBMO-WCA was expected to have the highest z score value such that they are the worst and best models, respectively, the highest z score was calculated in testing the statistical differences between WCA-CBO due to the fact that the standard deviation of the model performances within the simulations of CBO was higher than those of all optimization methods.

DISCUSSION AND CONCLUSION

The nonlinear Muskingum model is an efficient tool in order to predict outflow of a flood hydrograph at a river reach, in addition, it has become a benchmark function which is used for examining the performance of optimization algorithms. Upon

Table 5. The pairwise hypothesis test results indicating the availability of significant difference in between model pairs

Optimization Methods					
	VSA	GBMO	WCA	FPA	CBO
VSA	-	-	-	-	-
GBMO	1	-	-	-	-
WCA	1	1	-	-	-
FPA	1	1	0	-	-
CBO	1	0	1	1	-

Figure 2. The z scores obtained in model pairs by Dunn's post hoc tests

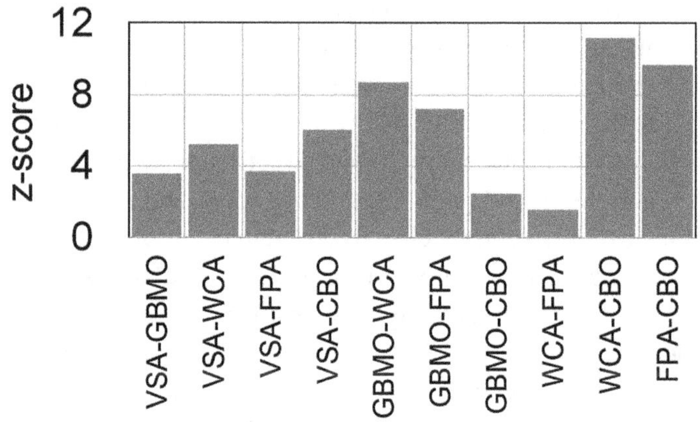

checking the overall results obtained in this chapter, it is seen that employed optimization algorithms estimate satisfactorily outflow hydrograph of given Wilson data when compared to the algorithms which was used by other researchers before. However, especially three of the algorithms namely WCA, FPA and VSA showed significantly better results than the rest of the algorithms that are GBMO and CBO. WCA showed the best SSE performance among five optimization methods. But FPA and VSA derived noticeably good results relative to WCA such that the performance differences between WCA were very small that the quantities of them were 3.6×10^{-13} and 1.14×10^{-12} for FPA and VSA, respectively.

In the forty simulations, the best solution of CBO was also very close to that of WCA, however, the best performances of CBO obtained in each simulation were not as consistent as the rest of the algorithms. Upon the ability of the algorithm in estimating the peak flow of Wilson data, the algorithms showed satisfactory performance together with nonlinear Muskingum method.

In addition, a comprehensive evaluation was carried out so as to go through whether any significant difference was available or not. According to the Kruskal-Wallis test, statistically significant difference was detected among the optimization methods, hence, Dunn's post hoc test were made in order to obtain pairwise comparisons of the methods. In the Dunn's test, z scores were utilized such that they were originated from mean rank of the optimization methods, thus, it made it possible detecting both model pairs in which statistically significant differences were available and the extent to which the differences were among those model pairs. Hence, it was deduced that statistical similarities were encountered only in GBMO-CBO and WCA-FPA model pairs. The highest difference was experienced in WCA-CBO model pair rather than WCA-GBMO model pair such that the fluctuating performance of CBO with relatively higher standard deviations underlay this situation.

In this chapter, a novel research was conducted such that the algorithms employed in this chapter have not been integrated with nonlinear Muskingum method before. The algorithms were considered as up-to-date by the authors due to the fact that each of them were developed by the researchers after 2012. Upon evaluating the SSE performances of WCA, FPA and VSA, one can claim that the algorithms derive the best solutions ever obtained in the literature and WCA is the best among them. However, it cannot be definitely supported this claim because it was not given the precise results in Karahan et al. (2013) with longer decimal formats. That is, the comparisons can be made only with a precision of 10^{-7} and fewer. Hence, it will be more appropriate to state that three of the algorithms are in line for the methods deriving the best results and WCA is the most robust one. Generally, WCA, FPA and VSA can be very useful tools in order to determine the parameters of nonlinear Muskingum model and conceptualize the river characteristics which outline the flood events and it is recommended by the authors that the related methods should be tried to check how they perform in prediction of outflow hydrographs experienced in the another flood events for flood hazard predictions and management as well.

ACKNOWLEDGMENT

The authors feel responsible to thank Prof. Dr. Xin She Yang and grateful for the advices given about Flower Pollination Algorithm. This research received no specific grant from any funding agency in the public, commercial, or not-for-profit sectors.

REFERENCES

Abdechiri, M., Meybodi, M. R., & Bahrami, H. (2013). Gases brownian motion optimization: An algorithm for optimization (GBMO). *Appl. Soft Comput. J.*, *13*(5), 2932–2946. doi:10.1016/j.asoc.2012.03.068

Al-Betar, M. A., Awadallah, M. A., Farris, H., Yang, X. S., Khader, A. T., & Alomari, O. A. (2018). Bat-inspired algorithms natural selection mehanisms for global optimization. *Neurocomputing*, *273*, 448–465. doi:10.1016/j.neucom.2017.07.039

Barati, R. (2011). Parameter Estimation of Nonlinear Muskingum Models Using Nelder-Mead Simplex Algorithm. *J. Hydrol. Eng.*, 946–954. .1943-5584.0000379 doi:10.1061/(ASCE)HE

Barati, R. (2013). Application of Excel Solver for Parameter Estimation of the Nonlinear Muskingum Models. *KSCE Journal of Civil Engineering*, *17*(5), 1139–1148. doi:10.100712205-013-0037-2

Boussaïd, I., Lepagnot, J., & Siarry, P. (2013). A survey on optimization metaheuristics. Information Sciences, 82–117. doi:10.1016/j.ins.2013.02.041

Cerny, V. (1985). Thermodynamical approach to the traveling salesman problem: An efficient simulation algorithm. *Journal of Optimization Theory and Applications*, *45*(1), 41–51. doi:10.1007/BF00940812

Chu, H. J., & Chang, L. C. (2009). Applying Particle Swarm Optimization to Parameter Estimation of the Nonlinear Muskingum Model. *Journal of Hydrologic Engineering*, *14*(9), 1024–1027. doi:10.1061/(ASCE)HE.1943-5584.0000070

Das, A. (2004). Parameter Estimation for Muskingum Models. *Journal of Irrigation and Drainage Engineering*, *130*(2), 140–147. doi:10.1061/(ASCE)0733-9437(2004)130:2(140)

Das, A. (2007). Chance-Constrained optimization-based parameter estimation for Muskingum models. *Journal of Irrigation and Drainage Engineering*, *133*(5), 487–494. doi:10.1061/(ASCE)0733-9437(2007)133:5(487)

Dogan, B., & Olmez, T. (2015). A new metaheuristic for numerical function optimization: Vortex Search Algorithm. *Information Sciences*, *293*, 125–145. doi:10.1016/j.ins.2014.08.053

Dorigo, M., Maniezzo, V., & Colorni, A. (1996). Ant System: Optimization by a colony of cooperating agents. *IEEE Trans. Syst. Man. Cybern. – Part B*, *26*(1), 29–41. doi:10.1109/3477.484436 PMID:18263004

Eskandar, H., Sadollah, A., Bahreininejad, A., & Hamdi, M. (2012). Water cycle algorithm - A novel metaheuristic optimization method for solving constrained engineering optimization problems. *Computers & Structures*, *110–111*, 151–166. doi:10.1016/j.compstruc.2012.07.010

Geem, Z. W. (2006). Parameter Estimation for the Nonlinear Muskingum Model Using the BFGS Technique. *Journal of Irrigation and Drainage Engineering*, *132*(5), 474–478. doi:10.1061/(ASCE)0733-9437(2006)132:5(474)

Geem, Z. W. (2011). Parameter Estimation of the Nonlinear Muskingum Model Using Parameter-Setting-Free Harmony Search. *Journal of Hydrologic Engineering*, *16*(8), 684–688. doi:10.1061/(ASCE)HE.1943-5584.0000352

Glover, F. (1986). Future paths for integer programming and links to artificial intellegience. *Computers & Operations Research*, *13*(5), 533–549. doi:10.1016/0305-0548(86)90048-1

Hansen, P. & Mladenovic, J.A.M. (2008). Variable neighbourhood search: methods and applications. *4OR*, *6*, 319–360.

Holland, J. H. (1975). *Adaptation in Natural and Artificial Systems*. Ann Arbor, MI: University of Michigan Press.

Karaboga, D., & Basturk, B. (2007). A powerful and efficient algorithm for numerical function optimization: Artificial bee colony (ABC) algorithm. *Journal of Global Optimization*, *39*(3), 459–471. doi:10.100710898-007-9149-x

Karahan, H., Gurarslan, G., & Geem, Z. W. (2013). Parameter Estimation of the Nonlinear Muskingum Flood-Routing Model Using a Hybrid Harmony Search Algorithm. *Journal of Hydrologic Engineering*, *18*(3), 352–360. doi:10.1061/(ASCE)HE.1943-5584.0000608

Kaveh, A., & Mahdavi, V. R. (2014). Colliding bodies optimization : A novel meta-heuristic method. *Computers & Structures*, *139*, 18–27. doi:10.1016/j.compstruc.2014.04.005

Kim, J. H., Geem, Z. W., & Kim, E. S. (2001). Parameter estimation of the nonlinear Muskingum model using Harmony Search. *Journal of the American Water Resources Association*, *37*(5), 1131–1138. doi:10.1111/j.1752-1688.2001.tb03627.x

Kirkpatrick, S., Gellat, C.D., Jr., & Vecchi, M.P. (1983). Optimization by simulated annealing. *Science,* (80), 220.

Luo, J. G., & Xie, J. C. (2010). Parameter Estimation for Nonlinear Muskingum Model Based on Immune Clonal Selection Algorithm. *Journal of Hydrologic Engineering*, *15*(10), 844–851. doi:10.1061/(ASCE)HE.1943-5584.0000244

Mantegna, R. N. (1994). Fast, Accurate Algorithm for Numerical Simulation of Lévy Stable Stochastic Process. *Physical Review*, *49*, 4677–4683. PMID:9961762

Mohan, S. (1997). Parameter estimation of nonlinear Muskingum models using genetic algorithm. *Journal of Hydraulic Engineering*, *123*(2), 137–142. doi:10.1061/(ASCE)0733-9429(1997)123:2(137)

Nikoo, M., Ramezani, F., Hadzima-Nyarko, M., Nyarko, E. K., & Nikoo, M. (2016). Flood-routing modeling with neural network optimized by social-based algorithm. *Natural Hazards*, *82*(1), 1–24. doi:10.100711069-016-2176-5

Papaioannou, P., Vasiliades, L., Loukas, A., & Aronica, G. T. (2017). Probabilistic flood inundation mapping at ungauged streams due to roughness coefficient uncertainty in hydraulic modelling. *Advances in Geosciences*, *44*, 23–34. doi:10.5194/adgeo-44-23-2017

Schanze, J. (2006). Flood Risk Management-A Basic Framework. *Flood Risk Management: Hazards, Vulnerability and Mitigation Measures,* 1-20.

Storn, R., & Price, K. (1997). Differential evolution–a simple and efficient heuristic for global optimization over continuous spaces. *Journal of Global Optimization*, *11*(4), 341–359. doi:10.1023/A:1008202821328

Tung, Y. (1985). River flood routing by nonlinear Muskingum method. *Journal of Hydraulic Engineering*, *111*(12), 1447–1460. doi:10.1061/(ASCE)0733-9429(1985)111:12(1447)

Wilson, E. M. (1974). *Engineering Hydrology*. Hampshire, UK: MacMillan. doi:10.1007/978-1-349-02417-9

Xu, D. M., Qiu, L., & Chen, S. Y. (2012). Estimation of Nonlinear Muskingum Model Parameter Using Differential Evolution. *Journal of Hydrologic Engineering*, *17*(2), 348–353. doi:10.1061/(ASCE)HE.1943-5584.0000432

Yang, X. S., Karamanoglu, M., & He, X. (2014). Flower pollination algorithm: A novel approach for multiobjective optimization. *Engineering Optimization*, *46*(9), 1222–1237. doi:10.1080/0305215X.2013.832237

Yoon, B. J., & Padmanabhan, G. (1993). Parameter estimation of linear and nonlinear Muskingum models. *Journal of Water Resources Planning and Management*, *119*(5), 600–610. doi:10.1061/(ASCE)0733-9496(1993)119:5(600)

Chapter 5
Flood Frequency Analysis Using Bayesian Paradigm:
A Case Study From Pakistan

Ishfaq Ahmad
International Islamic University, Pakistan

Alam Zeb Khan
https://orcid.org/0000-0001-6363-2855
International Islamic University, Pakistan

Mirza Barjees Baig
King Saud University, Saudi Arabia

Ibrahim M. Almanjahie
King Khalid University, Saudi Arabia

ABSTRACT

At-site flood frequency analysis (FFA) of extreme hydrological events under Bayesian paradigm has been carried out and compared with frequentist paradigm of maximum likelihood estimation (MLE). The main objective of this chapter is to identify the best approach between Bayesian and frequentist one for at-site FFA. As a case study, the data of only two stations were used, Kotri and Rasul, and Bayesian and MLE approaches were implemented. Most commonly used tests were applied for checking initial assumptions. Goodness of fit (GOF) tests were used to identify the best model, which indicated that the generalized extreme value (GEV) distribution appeared to be best fitted for both stations. Under Bayesian paradigm, quantile estimates are constructed using Markov Chain Monte Carlo (MCMC) simulation method for their respective returned periods and non-exceedance probabilities. For MCMC simulations, as compared to other sampler, the M-H sampling technique was used to generate a large number of parameters. The analysis indicated that the standard

DOI: 10.4018/978-1-5225-9771-1.ch005

errors of the parameters' estimates and ultimately the quantiles' estimates using Bayesian methods remained less as compared to maximum likelihood estimation (MLE), which shows the superiority of Bayesian methods over conventional ones in this study. Further, the safety amendments under two techniques were also calculated, which also show the robustness of Bayesian method over MLE. The outcomes of these analyses can be used in the selection of better design criteria for water resources management, particularly in flood mitigation.

INTRODUCTION

Hydrologists, over the years, have preferred to use best estimation methods and the most suitable probability distribution for extreme events through at-site FFA (Ahmad et al., 2015). Selection of suitable probability distribution not only provides the best fit to the selected site(Ahmad et al., 2013), but also provides efficient and accurate quantiles estimates corresponding to different returned periods (Ahmad et al., 2016). After finding the most suitable distribution for extreme events in FFA the next step is parameters estimation. Different approaches are utilized to estimate parameters of the respective distribution such as Method of Moments (MOM), MLE, and Method of Linear Moments (MLM). Exception of these methods, Bayesian approach based on the theorem by Thomas Bayes in 1763 (Keynes, (1921), is also intensively studied and applied now days in FFA(Coles et al.,1996) and (Gelman et al.,1997). One of the major differences between traditional approaches and Bayesian approach is that the parameters of interest have their own probability distributions and are treated as random variable rather than fixed as in traditional methods. In Bayesian approach, the sample information and prior knowledge are combined to get the posterior knowledge that provides a hypothetically consistent structure of hydrological information in the estimation of uncertainties and flood frequency models. It allows the explicit uncertainties due to flood frequency model and its parameters (Vicens et al., 1975) and (ZhongMin et al., 2011). In this context, Wood and Rodriguez-Iturbe evolved the procedures useful for dealing with the uncertainties resulting from flood frequency competing models and their parameters estimation and also compared these results with Bayesian paradigm (Wood and Rodriguez, 1975). Tang practiced a Bayesian regression analysis to estimate the resulting probability distribution of flood level (Tang et al., 1980). Van Gelder et al. the extension to Tang's work, adopting the procedure of Bayes factor in selection of suitable probability distribution and declared that this procedure provides better results than Tang's work (Van et al., 1999). Kuczera introduced the empirical Bayes technique to deduce hydrological quantities by joining site-specific as well as regional information. A Monte Carlo

Bayesian procedure for finding the expected probability distribution with quantile confidence intervals (Kuczera, 1982) and (Kuczera, 1999). O'Connell et al. presented a comprehensive Bayesian approach to integrate historical information as well as measurement errors in the flood frequency analysis (O'Connell et al., 2002). Reis et al. established a Bayesian methodology for the analysis of a generalized least squares (GLS) regression model for regional analyses of hydrological data (Reis and Stedinger, 2005). Reis and Stedinger presented the Bayesian MCMC method for finding the posterior distributions (Reis et al., 2005). Seidou et al. offered another Bayesian method for merging local and regional knowledge to present the probability distribution of parameters (Seidou et al., 2006). Ribatet et al. presents a regional Bayesian Point Over Threshold (POT) model for FFA in the absence of at-site stream flows (Ribatet et al., 2007). More recently, Micevski and Kuczera offered a Bayesian approach for the GLS regional regression model, which could be utilized with any available at-site information to get design floods estimates with high accuracy (Micevski and Kuczera, 2009). The main purpose of Bayesian analysis in statistical hydrology is to acquire an accurate future prediction of extreme events. These results are helpful to make the design structure in hydrology. In this study, Bayesian MCMC method is utilized for estimation of the posterior distributions. Bayesian MCMC approach offers a computationally and conceptually simple way of suitably integrate (FFA) in the joint distribution of possible errors within rating curves and individual observations (Kuczera, 1999) and (Reis and Stedinger, 2005). Dissimilar sampling approaches may direct to different MCMC algorithms, which is extensively used and applied in hydrological flood frequency analysis. There are many ways to creating these chains; however Metropolish-Hastings algorithm is used in this paper (Metropolis et al., 1953).In general, the basic aim of analysis for flood frequency in extreme hydrology is to estimate the expected values of the extreme flood flows in the future i.e. the future return level or quantile estimates. Bayesian analysis is preferable due to the prediction of return level which is based on quantile estimates can be estimated easily.

Generally, uncertainties in the form of bias and sampling errors are always present in design flood estimation and could not be ignored. These uncertainties are due to many reasons including inefficient parameter estimation like over estimation or under estimation, wrong selection of probability models and incomplete sample information. Usually, empirical methods are considered to estimate the sampling error by adding a safety factor to the design value in order to ensure the engineering safety. Theoretical approaches are seldom used but in this study an alternate to the theoretical approach based on Bayesian Paradigm to estimate quantile sampling errors and distribution for quantiles estimates were used.

BAYESIAN APPROACH FOR PARAMETER ESTIMATION

Bayesian Formula

Bayesian approach has been widely used in extreme flood frequency analysis. In Bayesian approach the sample information and prior knowledge are combined to get the posterior knowledge that provide a hypothetically consistent structure for hydrologic information in estimation of uncertainties and flood frequency models.

The posterior probability density $f(\theta / y)$ of parameter θ with given sample y as follows in equation (1) or in equation (2):

$$f(\theta / y) = \frac{L(\theta / y)h(\theta)}{\int L(\theta / y)h(\theta)d\theta} \tag{1}$$

or

$$f(\theta / y) \propto L(\theta /)h(\theta) \tag{2}$$

Where $h(\theta)$ is probability density of prior for parameter θ, and enlighten that what is identified or known regarding the unidentified parameter prior to the investigation of sample y. Also $L(\theta / y)$ is the likelihood function of sample y, which is sampling distribution of the samples given chosen probability model and parameter (θ).

Prior Distribution

Information about prior or preceding distribution $h(\theta)$ in Bayesian analysis is a necessary step. Generally subjective Bayes priors, non-informative prior and observed Bayesian are frequently recommended rules to fix and join a prior distribution. In this study, non-informative priors were used, because there is not enough knowledge about parameters. Therefore, they also called objective priors and resulting analysis is known as objective Bayesian analysis.

Likelihood Function

After observing data, the likelihood or likelihood function is assembled. The joint probability function of the data (Likelihood function) is however examined as a function of the parameters, delighted the observed data as fixed quantities. Assuming that the data values $y = (y_1, y_2, \ldots\ldots, y_n)$ are attained independently, usually the likelihood function is written as in equation (3):

$$L(\vartheta / y) = f(y_1, y_2, \ldots\ldots\ldots y_n / \vartheta) = \prod_{i=1}^{n} f(y_i / \vartheta) \tag{3}$$

Let $y_1, y_2, \ldots\ldots\ldots\ldots, y_n$ denote the independent annual extreme flood flows observations follow Generalized Extreme Value (GEV) distribution having probability density function in equation (4):

$$f(y \mid \mu, \sigma, \xi) = \frac{1}{\sigma}\left[1 + \xi\left(\frac{y_i - \mu}{\sigma}\right)\right]^{-1-\frac{1}{\xi}} \exp\left\{-\left[\left[1 + \xi\left(\frac{y_i - \mu}{\sigma}\right)\right]^{-\frac{1}{\xi}}\right]\right\} \tag{4}$$

Where μ is location, σ is scale and ξ is the shape parameters respectively with parameter space $-\infty < \mu < +\infty$, $\sigma > 0$ and $-\infty < \xi < +\infty$.

Posterior Distribution

In order to obtain the parameter values, Bayesian Markov Chain Monte Carlo (MCMC) simulation method will be used. Therefore, the likelihood function for $y_1, y_2, \ldots\ldots\ldots\ldots, y_n$ is given in equation (5) as follow:

$$L(\vartheta / y) = L(\mu, \sigma, \xi : y_1, \ldots, y_n) = \prod_{i=1}^{n} f(y_i \mid \mu, \sigma, \xi) \tag{5}$$

Where ϑ is the vector of parameters i.e. $\vartheta = (\mu, \sigma, \xi)$. Thus the density of posterior distribution equation (6) as follow:

$$f(\vartheta / y_1, \ldots, y_n) \propto L(\vartheta / y_1, \ldots, y_n) \times h(\vartheta) \tag{6}$$

or

$$f(\mu, \sigma, \xi / y_1, \ldots, y_n) \propto L(\mu, \sigma, \xi / y_1, \ldots, y_n) \times h(\mu, \sigma, \xi) \tag{7}$$

The above equations (6) and equation (7) can be represented as follow:

$$posterior \propto likelihood \times prior$$

Where $h(\vartheta)$ is prior distribution for GEV parameters.

In this study, the non-informative prior normal distribution is used to indicate that the significant information related to Extreme flood frequency analysis.

Quantile Estimation

In general, the basic aim of analysis for flood frequency in extreme hydrology is to estimate the expected values of the extreme flood flows in the future i.e. the future return level or quantile estimates. Bayesian analysis is preferable due to the prediction of return level which is based on quantile estimates which can be estimated easily. Let" t " denotes the future observation with probability density function equation (8) or equation (9) as follow:

$$h(t \mid y_1,....,y_n) = \iiint f(t \mid \mu,\sigma,\xi) f(\vartheta \mid y_1,....,y_n) d\mu d\sigma d\xi \tag{8}$$

Or

$$h(t \mid y_1,....,y_n) = \iiint f(t \mid \mu,\sigma,\xi) f(\mu,\sigma,\xi \mid y_1,....,y_n) d\mu d\sigma d\xi \tag{9}$$

This can be obtained by using MCMC simulation Bayesian method, given that the posterior density (Equation6) has been estimated by simulation. In other words, the simulated values for the three parameters of best fitted distribution obtained from MCMC simulations will be used to generate the return level or quantile estimates. This procedure will produce a sample $\theta_1,\theta_2,.....,\theta_B$ and the estimate of $m - years$ return level as in equation (10):

$$\Pr(t \leq q_m \mid y_1,.....,y_n) \approx \frac{1}{B}\sum_{i=1}^{B} \Pr\{t \leq q_m \mid \theta_i)\} \tag{10}$$

In this study, return values for 10, 20, 25, 50, 100, 500 and 1000-years are estimated using Bayesian MCMC simulations method.

Markov Chain Mono Carlo (MCMC) Simulations Method

This study prefer a model from some a priori particular range or distribution of the model parameters, and after that suggests a new model, conditional on the existing model, a category of sampling based methods for parameter estimation. MCMC approach is based on a random walk, to construct a chain of models from the model space. In Markov chain the attractive characteristic is that a new model is generated restricted on the previous one. Conversely, this method is independent of how the previous model was arrived at. Dissimilar sampling approaches may direct to different MCMC algorithms, which is extensively used and applied in hydrological flood frequency analysis. There are many ways to creating these chains, however Metropolish-Hastings algorithm is used in this paper.

Metropolis-Hasting (M-H) Algorithm

The M-H algorithm is one of the commonly used and applied technique of MCMC for (FFA). This technique is quite simple for simulation and provides an uncomplicated way as compared to Gibbs sampler, which may not be direct to simulate from these whole conditionals. In multivariate normal distribution the proposal densities with mean ∂^i and appropriate covariance matrix.

The simulation is started from the initial value $\partial_1, \partial^{(i+1)} \dots$ is favored by initial sampling to contender point ∂^* from a proposal density h(∂^* / ∂^i). Mathematically the acceptance probability of ∂^* is follows is equation (11):

$$\alpha i = \left\{ \frac{\Omega\left(\frac{\partial *}{y}\right) h\left(\frac{\partial^i}{\partial^*}\right)}{\Omega\left(\frac{\partial i}{y}\right) h\left(\frac{\partial^*}{\partial^i}\right)}, 1 \right\}, \tag{11}$$

This is the acceptance probability of ∂^*. If the contender point ∂^* is accepted, then the next probability of acceptance becomes $\partial^{(i+1)} = \partial^*$, otherwise.

In Bayesian MCMC technique, convergence of the chain is basically depends on the proposal density as it is selected randomly. The immense advantage of M-H algorithm is that, it only axis on the posterior density in the form of ratios. In this study the M-H algorithm is used.

Figure 1. Flow chart of Diagnostic MCMC Method

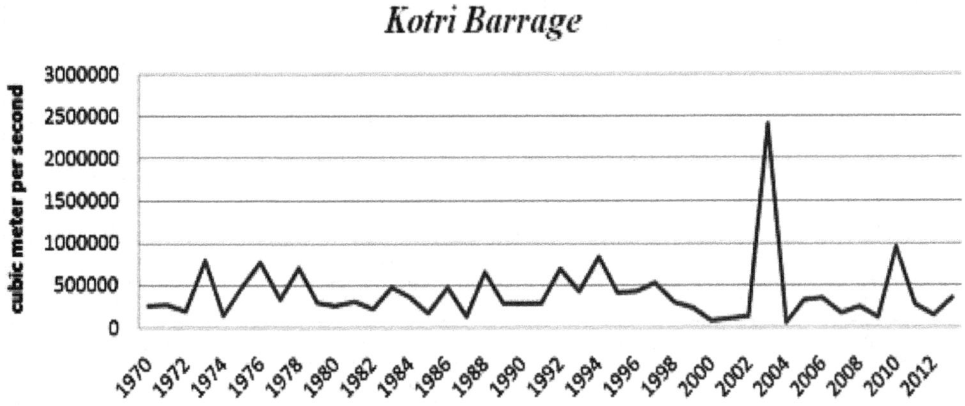


```
Start
"Bur in" periods, generally set y=0
Generate random chain of Proposal Standard Deviation and accept
it
If the Proposal Standard Deviation accepted (by acceptance rate
of MCMC),
then generate proposal parameter value θ~N(μ,δ, ξ)
Compute Parameters values (Parameters of Posterior
Distribution)
End
```

A CASE STUDY

The annual extreme flood flow (in thousands) data of two stations of Pakistan are taken for this study. The data is from 1970 to 2013, providing information about annual extreme flood flow. Sites about which data is available are Kotri and Rasul. The real data series shown in Table 1. Samples were assumed to be from a GEV distribution, which has been extensively used as the population distribution for flood extremes frequency analysis. As for the prior distributions of parameters, the normal prior was used in this study.

*Table 1. Flood peak flow series (discharge 1000*c/f)*

Years	Peak Flow (Kotri)	Years	Peak Flow (Rasul)
1970	25.5495	1970	04.1638
1975	47.6436	1975	12.5597
1980	25.3857	1980	05.8720
1985	15.6489	1985	06.4226
1990	28.0000	1990	11.8484
1995	40.3600	1995	28.5071
2000	06.6500	2000	03.7786
2005	31.0500	2005	09.2246
2010	93.9442	2010	26.3795
2013	34.4866	2013	05.0700

*(Barrages in Pakistan Discharge (1000*cf))

RESULT AND DISCUSSION

Basic Assumption

The time series plots in Figure 2 and Figure 3 present that the data series of two stations have uniform increasing and decreasing trend which indicated the existence of randomness in data and the graph pattern evident to stationarity. Difference Sign Test for Randomness is applied and from the below Table 2, the P-Vale is greater than from 0.05 ($p>0.05$) and hence accept our null hypothesis that "The set of observations

Figure 2. Time series plot for Kotri station

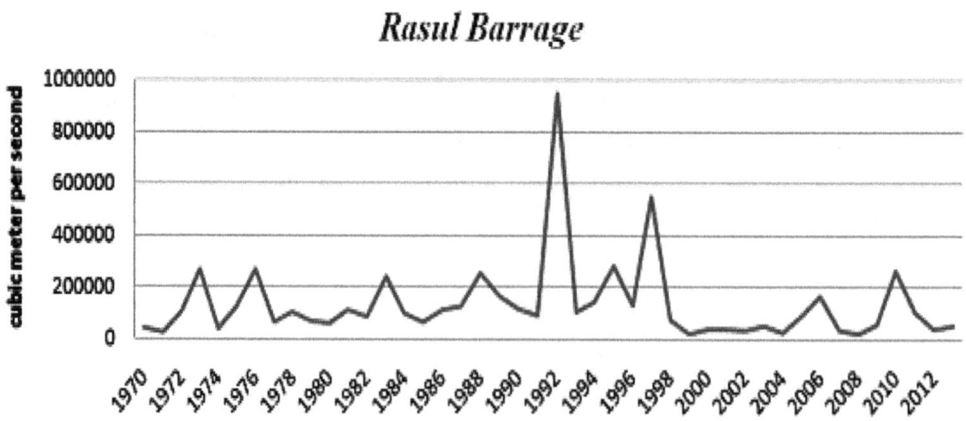

Figure 3. Time series plot for Rasul station

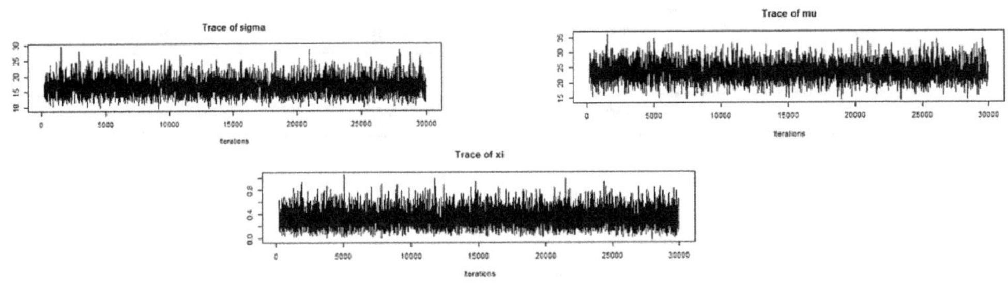

Table 2. Difference Sign Test for Randomness

Site	P-Value	Result	Decision
Kotri	*0.796*	*p>0.05*	*Accept Ho*
Rasul	*0.705*	*p>0.05*	*Accept Ho*

exhibiting randomness". Therefore, the study concludes that there is randomness in data. Ljung-Box Q test for independence is applied and from the below Table 3 the P-Vale is greater than from 0.05 (p>0.05) and hence accept our null hypothesis that "The set of observations are independent". Therefore, the study concludes that there is no dependency in data. The Kruskal-Wallis Test for Homogeneity is applied and from the Table 4, the P-Vale is greater than 0.05 i.e. (p>0.05) and hence accept our null hypothesis that "Two populations come from Identical Distribution." Therefore the study concludes that the data follows the same distribution.

In view of the fact from table 5 and table 6, the goodness of fit test indicates that GEV distribution is best fitted for both sites. Here in (Table 5, Table 6) the study allot each distribution a rank (1 = best fitted model, 2 = Second best fitted model and 3=Third best fitted model). Thus from the below Table 5 and Table 6 for GEV distribution KS, AD and χ^2 goodness of fit test the ranks are 3,1,1 respectively from all other distributions (for Kotri Data) and the ranks are 1,2,1 respectively from all other distributions(for Rasul Data), which showsGEV distribution is the best fitted among all other distributions.

Table 3. Difference Sign Test for Randomness

Site	P-Value	Result	Decision
Kotri	*0.148*	*p>0.05*	*Accept Ho*
Rasul	*0.120*	*p>0.05*	*Accept Ho*

Table 4. Ljung-Box Q test for Independence

Site	P-Value	Result	Decision
Kotri	*0.473*	*p>0.05*	*Accept Ho*
Rasul	*0.428*	*p>0.05*	*Accept Ho*

Table 5. Kruskal-Wallis Test for Homogeneity

S.No#	Distributions	KS-Rank	AD-Rank	χ^2-Rank
1	*GEV*	*3*	*1*	*2*
2	*LP3*	*2*	*5*	*6*
3	*Log Nor*	*6*	*7*	*1*
4	*Norm*	*5*	*2*	*5*
5	*Parto*	*7*	*6*	*7*

Table 6. Probability Distribution of best fitted Kotri data based on AD, KS and Chi Square criterion

S.No#	Distributions	KS-Rank	AD-Rank	χ^2-Rank
1	*GEV*	*1*	*2*	*1*
2	*LP3*	*3*	*1*	*2*
3	*Log Nor*	*6*	*4*	*3*
4	*Norm*	*9*	*6*	*7*
5	*Parto*	*7*	*9*	*8*

Parameter Estimation of GEV Distribution by Using MCMC Bayesian Method

The results of parameters estimates of GEV distribution with their corresponding standard errors for MLE and Bayesian Methods are given in Table 7, Table 8 (for Kotri Station) and in Table 9, Table 10 (for Rasul Station). From the results, the differences between MLE method and Bayesian method the parameters are given and indicated that the values of parameters estimated by Bayesian method are smaller as estimated by MLE method. Furthermore the standard errors for Bayesian method are also smaller than MLE method.

Table 7. Probability Distribution of best fitted Rasul data based on AD, KS and Chi Square criterion

MLE Method	Location (μ)	Scale (δ)	Shape (ξ)
Parameters estimates of GEV distribution	*23.6388189*	*15.6721098*	*0.3074717*
Standard Errors	*2.6971483*	*2.2782627*	*0.1335452*

Table 8. Parameters and Standard Errors of Kotri data by MLE Method

Bayesian Method	Location (μ)	Scale (δ)	Shape (ξ)
Parameters estimates of GEV distribution	*23.2237*	*15.2529*	*0.3102*
Standard Errors	*2.1800945*	*0.2825910*	*0.354536*

Table 9. Parameters and Standard Errors of Kotri data by Bayesian Method

MLE Method	Location (μ)	Scale (δ)	Shape (ξ)
Parameters estimates of GEV distribution	*6.3037397*	*4.6994110*	*0.5837014*
Standard Errors	*1.8538305*	*0.8532623*	*0.1902678*

Table 10. Parameters and Standard Errors of Rasul data by MLE Method

Bayesian Method	Location (μ)	Scale (δ)	Shape (ξ)
Parameters estimates of GEV distribution	*6.1789*	*4.5433*	*0.5844*
Standard Errors	*1.14 7382*	*0.27174309*	*0.4438742*

Trace Plots and Posterior Densities Plots for Parameters of GEV distribution

In this study 30,000 simulations are generated for both stations, which follow the GEV distribution through Bayesian MCMC method, to acquiring the posterior of GEV distribution. The trace plots for 30,000 iterations and posterior densities plots of location (μ), scale (σ) and shape (ξ) parameters of GEV distribution are given in Figure 4 for Kotri and similar trace plots are observed for Rasul station, which shows that all simulated values for the GEV parameters are found to converse within a corresponding region and signifying that no clear tendencies or periodicities in both data. Figure 5 (left side) and Figure 5 (right side) show the densities plots of all three parameters of GEV for Kotri and Rasul station respectively.It can be concluded that the generated series through MCMC simulation is mixing well to the original data series.Furthermore, the posterior densities plots for both stations demonstrate that the estimated densities of three parameters for location (μ), scale (σ) and shape (ξ) of GEV distribution are almost balanced and symmetrical, which indicates that the parameters of the distribution is symmetrical and efficient.

Quantile Estimation and Their Safety Amendments

In engineering of hydrological events, for sake of engineering security, the value of safety amendments (denoted by ΔEp) is estimated for Quantiles estimates or design values. As a rule for calculation of ΔEp the study used the following formula suggested by (Benjiamin and Cornell 1970) is:

$$\Delta Ep = \xi\, \delta_{Ep}$$

Where ξ is a consistency coefficient to measure the reliability of safety improvement, frequently it is taken as $\xi = 0.7$ (in safety engineering), and δ_{Ep} is the Quantiles estimates or design value's standard deviation. The value of δ_{Ep} is more essential for calculation of ΔEp. The calculation of ΔEp is simply product of ξ (0.7) and values of δ_{Ep} (from Table 11 and from Table 12). Results of safety amendments for Quantiles or returned periods for both stations are given below in Table 13 and Table 14. The term $\delta_{Ep,}$ which is known as an estimated term for calculation of the

Figure 4. Trace Plots of Kotri barrage for location (μ),scale(σ) and shape (ξ) parameters of GEV distribution 30,000 iterations.

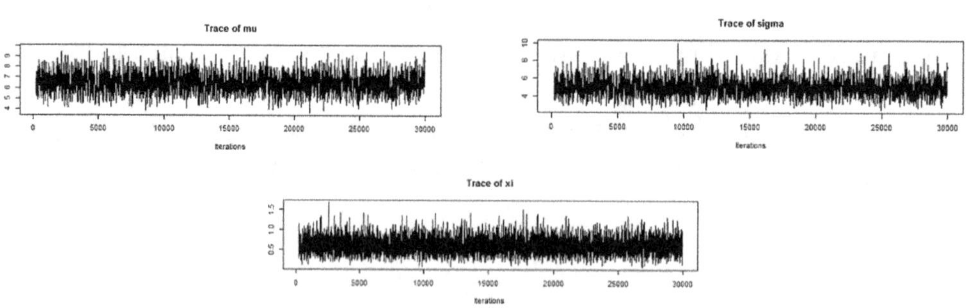

Figure 5. Trace Plots of Rasul barrage for location (μ), scale (σ) and shape (ξ) parameters of GEV distribution 30,000 iterations.

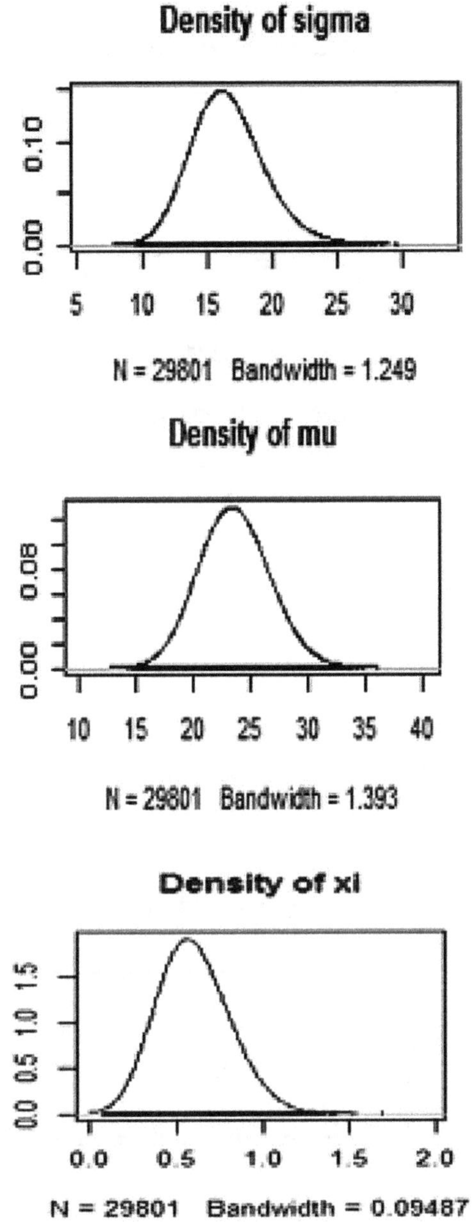

Figure 6. (left) and Figure 5 (right) Posterior Densities Plots of Kotri and Rasul barrage for location (μ), scale (σ) and shape (ξ) parameters of GEV distribution respectively.

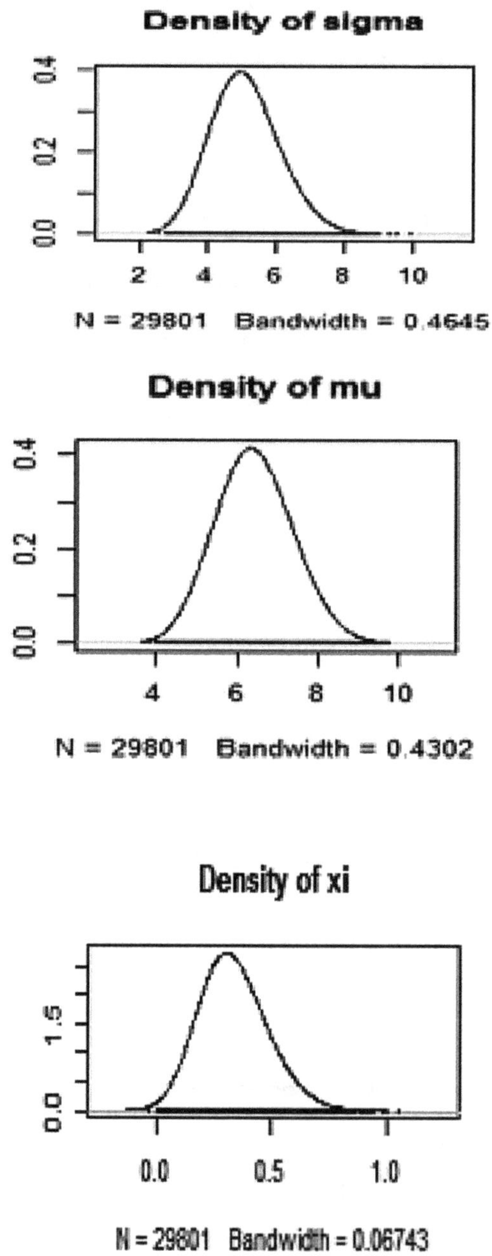

Table 11. Quantiles estimates for Kotri Station

P	T	MLE Method		Bayesian Method	
		Ep	S.D	Ep	S.D
0.1	10	28.19718	1.01001	27.3652	0.9389
0.9	20	43.83456	4.11644	42.5110	2.3588
0.98	50	76.77401	15.9523	74.4337	9.9658
0.99	100	116.2800	24.2166	112.741	12.306
0.999	1000	452.0169	51.3791	438.689	32.143

Table 12. Quantiles estimates for Kotri Station

P	T	MLE Method		Bayesian Method	
		Ep	S.D	Ep	S.D
0.1	10	74.48513	1.34258	72.8796	0.7564
0.9	20	99.70841	3.0962	97.6045	2.8674
0.98	50	141.8525	9.52092	139.010	7.0432
0.99	100	182.3663	17.4292	178.901	13.089
0.999	1000	398.9240	31.5644	393.080	28.039

Table 13. Quantiles estimates for Rasul Station

P	T	MLE Method			Bayesian Method		
		Ep	ΔEp	Ep +ΔEp	Ep	ΔEp	Ep +ΔEp
0.1	10	74.48513	0.939806	75.42494	72.87960	0.529487	73.40909
0.9	20	99.70841	2.16734	101.8758	97.60454	2.007187	99.61173
0.98	50	141.85255	6.664644	148.5172	139.01022	4.930289	143.9405
0.99	100	182.36638	12.20044	194.5668	178.90169	9.16293	188.0646
0.999	1000	398.92403	22.09508	421.0191	393.08071	19.62744	412.7082

quantiles standard deviation. But this is only valid and acceptable if the parameters are estimated by method of moments (MOM). But in case of Bayesian estimation method for estimation of parameters and for sampling distributions of quantiles, it used repetition sampling. Therefore, these values of standard deviation of the design

Table 14. Estimation of Safety amendments for quantiles (Rasul Station)

P	T	MLE Method			Bayesian Method		
		Ep	ΔEp	Ep +ΔEp	Ep	ΔEp	Ep +ΔEp
0.1	10	74.48513	0.939806	75.42494	72.87960	0.529487	73.40909
0.9	20	99.70841	2.16734	101.8758	97.60454	2.007187	99.61173
0.98	50	141.85255	6.664644	148.5172	139.01022	4.930289	143.9405
0.99	100	182.36638	12.20044	194.5668	178.90169	9.16293	188.0646
0.999	1000	398.92403	22.09508	421.0191	393.08071	19.62744	412.7082

valves i.e. $\boldsymbol{\delta}_{Ep}$can be obtained directly in Bayesian estimation method. For results comparison, Bayesian estimation method compared with classical method of MLE in Table 13 (for Kotri station) and Table 14 (for Rasul). The results are presented in the form of the security amendments **ΔEp** from quantiles estimates or design values with non-exceedance probabilities, **p** = 0.1,0.9,0.95,0.98,0.99,0.998,0.99 9 using both estimation methods i.e. Bayesian Estimation Method and MLE. The values acquired through Bayesian Estimation Method for **ΔEp** are all smaller than from MLE method. Therefore, because, of smaller standard deviation and smaller values of **ΔEp**with corresponding non-exceedance probabilities, it is concluded that the Bayesian MCMC method is best for modeling of FFA and safety amendments as compared to MLE method.

SUMMARY AND CONCLUSION

The analyses are based on Annual Maximum Flow (AMF) of two stations as Rasul and Kotriin Pakistan. The data of AMF is retrieved from Federal Flood Commission (FFC) and Indus River System Authority (IRSA). AMF described in cusecs ranging from 1970 to 2013. For estimation of parameters and quantiles, here adopted two well-known methods i.e. Bayesian Estimation method and Maximum Likelihood Estimation (MLE) method. Both methods were applied and analyzed for at-site estimation of extreme hydrologic events. Before the determination of best fitted model, it is pre-requisite in at-site frequency analysis to verify the fundamental assumptions. The data of AMF in this study validated all the fundamentals assumptions. The best fit is selected on the basis of these GoF i.e. Goodness of fit tests like Anderson Darling (AD) test, Chi-Square (χ^2) goodness of fit and Kolmogorov Smirnov (KS) test. GEV distribution appeared as best fit for the data of bothstations in this study. Quantiles estimates, estimated by using MLE and Bayesian MCMC method for the best-fitted probability distribution with respective returned periods and non-exceedance probabilities were also checked. The returned periods and subsequently estimates of quantiles are very significant in the design of water related structures such as culverts, dams, bridges and flood control devices.

The quantiles estimated values established an enormous variation in their magnitudes using MLE method. Furthermore, the analysis indicates that the quantiles estimates or design values based on Bayesian Estimation Method remained less as compared to Maximum Likelihood Estimation (MLE) with same returned period and non-exceedance probabilities. This analysis proves the Superiorly of Bayesian Method over MLE. Therefore, it can be suggested that during the planning phase of flood designs, results obtained from different methods and techniques should be compared.

ACKNOWLEDGMENT

Authors are very grateful to the Deanship of Scientific Research at King Khalid University, Kingdom of Saudi Arabia for their administrative and technical support.

REFERENCES

Ahmad, I., Abbas, A., Saghir, A., & Fawad, M. (2016). Regional Frequency Analysis of Annual Peak Flows in Pakistan Using Linear Combination of Order Statistics. *Polish Journal of Environmental Studies*, *25*(6), 2255–2264. doi:10.15244/pjoes/63782

Ahmad, I., Fawad, M., & Mahmood, I. (2015). At-Site Flood Frequency Analysis of Annual Maximum Stream Flows in Pakistan Using Robust Estimation Methods. *Polish Journal of Environmental Studies*, *24*(6), 2345–2353. doi:10.15244/pjoes/59585

Ahmad, I., Shah, S. F., Mahmood, I., & Ahmad, Z. (2013). Modeling of monsoon rainfall in Pakistan based on Kappa distribution. *Science International Lahore.*, *25*(2), 333–336. doi:10.24949/njes.v10il.1570.g94

Benjiamin, J. R., & Cornell, C. A. (1970). *Probability, Statistics and Decision for Civil Engineers*. New York: McGraw-Hill.

Coles, S. G., & Powell, E. A. (1996). Bayesian methods in extreme value modelling: a review and new developments. *International Statistical Review/Revue Internationale de Statistique*, 119-136, Doi:10.2307/1403426

Gelman, A., Carlin, J. B., Stern, H. S., & Rubin, D. B. (1997). *Bayesian Data Analysis*. New York: Chapman and Hall.

Keynes, J. M. (1921). *A Treatise on Probability*. Macmillan and Co.

Kuczera, G. (1982). Combining site-specific and regional information: An empirical Bayes approach. *Water Resources Research*, *18*(2), 306–314. doi:10.1029/WR018i002p00306

Kuczera, G. (1999). Comprehensive at-site flood frequency analysis using Monte Carlo Bayesian inference. *Water Resources Research*, *35*(5), 1551–1557. doi:10.1029/1999WR900012

Metropolis, N., Rosenbluth, A., Rosenbluth, M., Teller, A., & Teller, E. (1953). Equation of state calcula-tions by fast computing machines. *The Journal of Chemical Physics*, *21*(6), 1087–1092. doi:10.1063/1.1699114

Micevski, T., & Kuczera, G. (2009). Combining site and regional flood informationusing a Bayesian Monte Carlo approach. *Water Resources Research*, *45*(4), W04405. doi:10.1029/2008WR007173

O'Connell, D. R. H., Ostenaa, D. A., Levish, D. R., & Klinger, R. E. (2002). Bayesian flood frequency analysis with paleo hydrologic bound data. *Water Resources Research*, *38*(5), 1058. doi:10.1029/2000WR000028

Reis, D. S. Jr, & Stedinger, J. R. (2005). Bayesian MCMC flood frequency analysis with historical information. *Journal of Hydrology (Amsterdam)*, *313*(1-2), 97–116. doi:10.1016/j.jhydrol.2005.02.028

Reis, D. S. Jr, Stedinger, J. R., & Martins, E. S. (2005). Bayesian generalized least squares regression with application to log Pearson type 3 regional skew estimation. *Water Resources Research*, *41*(10), W10419. doi:10.1029/2004WR003445

Ribatet, M., Sauquet, E., Gresillon, J. M., & Ouarda, T. B. M. J. (2007). A regional Bayesian POT model for flood frequency analysis. *Stochastic Environmental Research and Risk Assessment*, *21*(4), 327–339. doi:10.100700477-006-0068-z

Ribatet, M., Sauquet, E., Gresillon, J. M., & Ouarda, T. B. M. J. (2007). Usefulness of the reversible jump Markov Chain Monte Carlo model in regional flood frequency analysis. *Water Resources Research, 43*(8), W08403. doi:10.1029/2006WR005525

Seidou, O., Ouarda, T. B. M. J., Barbet, M., Bruneau, P., & Bobée, B. (2006). A parametric Bayesian combination of local and regional information in flood frequency analysis. *Water Resources Research, 42*(11), W11408. doi:10.1029/2005WR004397

Tang, W. H., Hydraul, J., & Div-ASCE. (1980). Bayesian frequency analysis. *J Hydraul Div-ASCE, 106*, 1203–1218.

Van, G. P. H. A. J. M., Van, N. J. M., & Duits, M. T. (1999). Selection of probability distribution with a case study on extreme Oder River discharges. *The Tenth European Conference on Safety and Reliability*, 1475–1480. DOI:10.1.1.74.9802&rep:rep1

Vicens, G. J., Rodriguez, I., & Schaake, J. C. Jr. (1975). A Bayesian framework for the use of regional information in hydrology. *Water Resources Research, 11*(3), 405–414. doi:10.1029/WR011i003p00405

Wood, E. F., & Rodriguez, I. (1975). A Bayesian approach to analyzing uncertainty among flood frequency models. *Water Resources Research, 11*(6), 839–843. doi:10.1029/WR011i006p00839

Wood, E. F., & Rodriguez, I. (1975). Bayesian inference and decision making for extreme hydrologic events. *Water Resources Research, 11*(4), 533–542. doi:10.1029/WR011i004p00533

ZhongMin, L., BinQuan, L., ZhongBo, Y., & WenJuan, C. (2011). Application of Bayesian approach to hydrological frequency analysis. *Science China Technological Sciences, 54*(5), 1183-92. Doi:10.100711431-010-4229-4

Chapter 6
Flood Modelling and Mapping:
Case Study on Adyar River Basin, Chennai, India

Brema J.
Karunya Institute of Technology and Sciences, India

ABSTRACT

This chapter presents an overview of the important concepts related to flood hazard assessments and explores the use of remote sensing data from satellites to supplement traditional assessment techniques. The method presented in this chapter can be used in sectoral planning activities and integrated planning studies and for damage assessment. The chapter presents the application of flood modelling to the study area. The study area, Adyar River in Chennai, has experienced major floods in the past decade which is attributed to increased urbanization. The hydrologic model for the Adyar River Basin was set up using HEC geoHMS and was run and calibrated using observed flow in HEC-HMS. The chapter also discusses the results obtained from the IDF analysis and its application in HEC HMS to generate hypothetical storm hydrographs. Furthermore, the chapter goes on to discuss the results obtained from the hydraulic modelling such as the inundation map for the 2005 flood event and the inundation map for hypothetical storms of varying return periods.

DOI: 10.4018/978-1-5225-9771-1.ch006

INTRODUCTION

Flooding is a natural process, which occurs over a river, stream and plains in a recurring manner depending on various factors such as landuse, rainfall, soil etc. As per studies, the mean annual flood occurs every 2.33 years (Leopold et al., 1964). Flooding is due to a heavy or continuous rainfall exceeding the infiltration capacity of soil and the carrying capacity of rivers, streams, and coastal areas. This causes a river or stream to overflow over its banks onto adjacent lands. The frequency of flooding has increased over the past few decades causing havoc to life and property. Though it is impossible to predict when a flood event is going to occur, it is possible to predict the damage which would be caused by a flood, if it ever occurs. This can be done by generating inundation maps for different return periods and performing flood risk modelling. The damage caused by a flood varies from place to place as a variety of factors contribute towards flood risk. The land use and soil type play an important role in estimating how much of the rainfall is converted into runoff. In cities, the rapid urbanization plays an important role in contributing to increased runoff resulting from increased percentage of impervious area. The slope of the flood plain is also an important factor. If the flood plain has very low slope, it takes a longer time for the runoff to flow out and remains stagnant for a longer period. Flood modelling is carried out using spatial data such as satellite imagery, digital elevation model and soil map. In addition to spatial data, historical rainfall data and discharge data is required to generate the IDF curves and to calibrate the hydrologic model.

Accurate modelling of surface has to be done in order to estimate the spatial and temporal distribution of parameters. With the advent of remote sensing and Geographical Information System, it has become easier to manage a large set of spatial and temporal data pertaining to hydrological models.

This chapter deals with detailed procedure of probabilistic assessment of flood using remote sensing data and softwares in addition to the conventional techniques.

The primary objective of application of remote sensing tools is to map flood prone areas with accuracy within a short period. This assessment method will lead to flood prediction which will help the planners to assess the probable impact of disaster due to flood. . The method presented in this chapter can be utilised in regional planning activities and damage assessment due to floods.This section is designed to provide the planner with background information on the nature of floods and the terms and concepts associated with assessing the risks from this natural hazard. The various flood characteristics are shown in Fig.1.

Figure 1. Flood characteristics

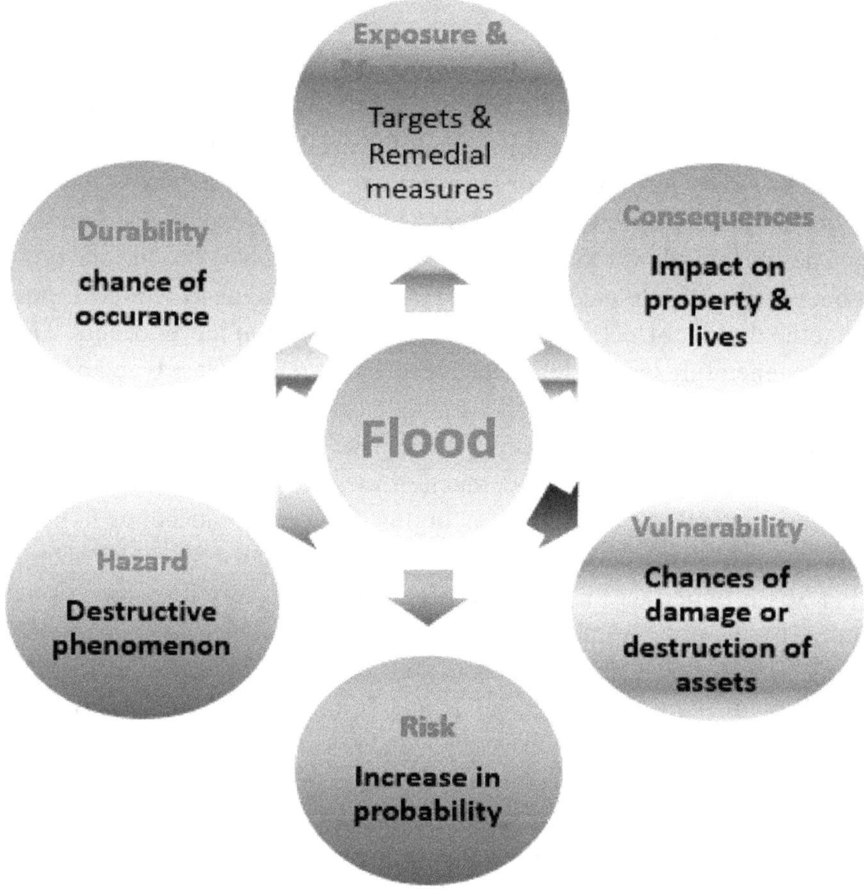

FLOODPLAINS

Floodplains are the regions lying on either banks of the rivers and streams that are subjected to inundation due to excessive flood in the rivers and streams. . Owing to their continually changing nature, floodplains and other flood-prone areas need to be monitored keeping in mind the effects of development. Since floodplains are floodprone, it is developmental activities are to be taken with precautionary measures. In the semi-arid regions, it is a common phenomena to see the widening of rivers and destruction of the floodplain. . If the banks are erosive of nature and have high erosion potential, the phenomenon of change in channel section will take place and the flooding event will cause a large portion of flood waters to be carried in a channel. This type of event normally occurs in arid regions, where high velocity flood waters make drastic changes in the channel configuration during the flooding event. Sediments from these events may be deposited both in the channel and on the floodplains resulting in the change in channel section and thereby increasing or decreasing the capacity of the channel to hold water.

Flood Frequency

The annual flood is usually considered as the single greatest event each year. Generally, the annual floods are used in frequency analysis, the return period of recurrence interval is the reciprocal of probability. Developmental activities, more particularly deforestation and intensive crop production, may drastically alter the runoff conditions, thereby increasing stream flow and increasing the risk of flooding.

A "100-year flood" describes an event or an area subject to a probability of 1% for a certain flood occurring in any given year. Whether it occurs or not in an assessed year, there is still a 1% chance of a similar occurrence in the following year. Assessment of flood prone areas or the boundaries of flood plain for a 100 year flood is to map and then to plan further the flood mitigation programs. Any other statistical frequency of a flood event may be evaluated based on the degree of risk for e.g., 5-year, 20-year, 50-year, 500-year events. Acceptable risk criteria can help in distinguishing between different degrees of risk for different development activities and in evaluating constraints associated with potential investment projects. Depending on the type of development activity, the frequency of a flood event should be selected in a appropriate manner. The flood inundated area depends on the climate, the type of soil, the intensity and duration of rainfall and the stream slope. When substantial rainfall occurs in a particular season, the flood plains get subjected to flooding even in the stretches with streams of larger width. season or each year, the floodplains may be inundated nearly every year, even along wide streams with very small channel slopes.

Flood Hazard Assessment

Flood risk assessment method mainly consist of four steps, including: hazard assessment, exposure assessment, vulnerability assessment and risk assessment (Penning-Rowsell et al.2005b; Foudi et al. 2015). Of all the steps mentioned above, the flood hazard assessment is the one which receives more importance as it aids in disaster preparedness (Koks et al. 2015). Flood hazard determination consists of the following procedure: Determination of the flood characteristics: For a flood event the extent, depth and velocity of flow in the river or stream are to be known. The other information necessary depends on the interpretation of details related to evacuation, property damage etc. If the aim is to secure people and to plan evacuation, therefore the hazard assessment map must include depth and velocity of flow. This analysis is required in detail, to plan evacuation routes and identify shelters. Determination of the probability of occurrence of each flood is required, so that the flood characteristics can be linked to the return period. The level of hazard depends on the interpretation of flood information, type of hazard information and the degree of exposure. A hazard map is not a risk map, a risk analysis includes the potential impact of one or more hazards, taking into account the vulnerability and resilience of the elements at risk. For instance, the flood hazard map does not include information on potential damage to buildings for instance. Regional planning should include the land-surface characteristics such as topography or slope of the land, geomorphology, type and characteristics of soils, hydrological characteristics and the extent of flooding.

Flood Inundation

The extent and duration of flood extent depends on various factors such as stream size, slope of the channel, rainfall and landuse of the catchment etc. In case of small streams, the floods last for few days whereas the duration of flood last for a month or more.The floodwater recedes and normally drains back to the stream. In case of larger floodplains bordered by flood banks or levees, the water recedes slowly, causing inundation in smaller areas which may last for longer periods. These inundated water are disposed by way of downstream drainage, infiltration into the soil and evapotranspiration.

Flood Estimation

Flood estimation is an essential process to be carried out, to design hydraulic structures, disaster mitigation etc. The methods that predict the flood discharge may consider two or more factors. The methods that consider only two parameters are very simple to use however the accuracy will be less. The models which consider more parameters will be laborious but the reliability will be better. None of the methods that are used by the engineers to predict the flood discharges can be considered as reliable and accurate.

SURFACE MODELLING

Surface of earth is continuous and hence one needs an infinite amount of points to accurately model the surface of earth. However the surface of earth can be modeled with a reasonable degree of accuracy by representing the surface as square grids or triangular facets. The method of representing the surface of earth in the form of regular grid of spot heights is known as Digital Elevation Model (DEM) and the method of representing the surface of earth in the form of triangular facets is known as Triangulated Irregular Networks (TIN). During this study, DEM which is the simplest and most common form of digital representation of earth's surface is used for the estimation of flood discharge values.

FLOODPLAIN MAPPING

Flood plain mapping involves many dynamic techniques as listed below: i. continous monitoring of flow changes in rivers or streams, ii.flood estimation based on regression analysis, iii.determination of stream flow and flood characteristics based on return period, estimated from previous hydrologic events/records.-

Flood modeling techniques could become more comfortable with improvements in earth observations even under dense vegetation and densely populated urban areas. Digital Elevation Model (DEM) is one of the most important input parameters for flood modeling. The accuracy of flood modelling and mapping depends on accuracy and resolution of DEM. The simulated flood results should show extent of depth of stagnated water, inundated area and flow velocity. The reliability of the flood plain mapping depends on the resolution and accuracy of the DEM. Therefore, it is inevitable to have high resolution DEM for flood mapping.

Remote sensing technology can be especially useful and desirable when applied during the planning process. Application of this technology eases the planning with recent and repetitive information. Flood inundation and floodplain maps are being prepared from satellite imageries for a long period by hydrologists across the globe. This technique is capable of yielding flood hazard assessment even for a dynamic and long term flood and also facilitates in understanding the study area in a given time. These information are highly useful in the initial stages of integrated planning. A series of observations of flood events, hydrologic data and disaster reports will serve the need for dynamic data. The commonly used Landsat Multispectral Scanner (MSS) data and the high-resolution Landsat Thematic Mapper (TM) and SPOT High

Resolution Visible Range (HRV) data with the potential for larger scale mapping are examples. Also, the small-scale resolution but synoptic regional coverage provided by the NOAA satellite series carrying the Advanced Very High Resolution Radiometer (AVHRR) provides a highly informative aid to planners in determining the extent of flood events. The LIDAR technology also provides accurate elevation data, which can contribute more towards flood modelling. However, LiDAR data comes with huge data size, computational time and limitation of availability and accessibility of the data. LiDAR can penetrate the vegetation cover and provides faster data compared to conventional techniques. Airborne LiDAR data are highly successful in floodplain changes studies. According to Bodoque et al. [19, 2016], Digital Surface Models (DSMs) derived from LiDAR system improved 1D and 2D hydraulic models, provide valuable data for 2D hydrodynamic models, enable grain-scale surface roughness (an essential parameter in flood modeling), and provide very accurate topographic datasets of both the ground and urban landscapes (e.g., buildings, roads, bridges). The SRTM and ASTER DEM have coarser spatial resolution comparatively. In addition, airborne LiDAR can be used to determine water course and water depth of inundated areas [4,5, 2013, 2014]. Despite several opportunities that LiDAR technology provides for flood modeling methodology, there are challenges in using the LiDAR data for flood modeling. According to Baldassarre and Uhlenbrook [15, 2012], successful application of LiDAR dem data for flood modelling not only depends on high resolution of the data but there remains various other challenges also.

The common practice in flood modelling involves hydrologic modelling of rainfall followed by hydraulic modelling of discharge. Rainfall is provided as input to the hydrologic model which transforms the rainfall into runoff. The output of the hydrologic model is given as input to the hydraulic model. The current chapter deals with the procedure and methodology adopted to carry out the flood modelling. This chapter also briefs about the Intensity-Duration-Frequency Analysis of rainfall which is carried out to predict hypothetical storm intensities for different durations and return periods.

INTENSITY-DURATION-FREQUENCY ANALYSIS

The Intensity-Duration-Frequency Analysis of rainfall is carried out to predict the intensity of rainfall corresponding to different return periods and durations. The IDF curve is the one which represents the variations of rainfall intensity with reference to rainfall duration for a specified return period. The intensity values in the IDF table, when multiplied with their respective durations will produce the Depth-Duration-Frequency analysis or the DDF analysis. For generating hypothetical storms using frequency storm in HEC-HMS, the depth values are required instead of the intensity. Sherman (1931) proposed a method of generating the IDF relationship when the rainfall data for the study region is not available. An empirical equation was developed to find the intensity of rainfall depending on the return period and storm duration. The equation contained constants which were dependant on the geographical location.

$$i = \frac{k \times T^a}{\left(t + b\right)^n}.$$

Where, *i* is the intensity of rainfall, cm/hr; T is the return period in years; *t* is the rainfall duration in hours; *k*, *a*, *b* and *n* are the constants. The IDF curves obtained for the model is shown in Fig.2.

Probability Density Function

A Probability Density Function (or PDF) is used to specify the probability that a random variable lies within a specific range. In this case, the random variable is taken as the rainfall depths/intensities. Most variables in the field of hydrology are

Figure 2. IDF Curves

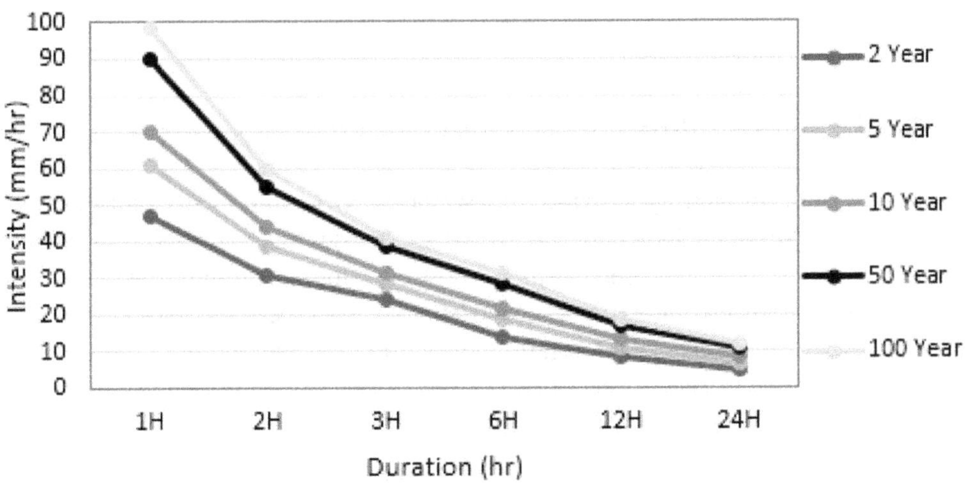

random as they are generated by natural processes. The probability that a random variable lies within a specific range is given by

$$P\left[x_1 < X < x_2\right] = \int_{x_1}^{x_2} f\left(x\right) dx .$$

Where f(x) is a function of the probability distribution and X is the random variable.

Cumulative Density Function

A Cumulative Distribution Function (or CDF) is used to specify the probabily of a random variable being less than a specific value. It is denoted by F(x).

$$F\left(x\right) = P\left[X \leq x\right] = \int_{-\infty}^{x} f\left(x\right) dx .$$

In the field of hydrology, the x is taken as the threshold value above which the event is considered to be a flood. Hence, if we say that F(x) is the probability of non-exceedance, then we may be able to conclude that,

$$F\left(x\right) = 1 - P .$$

Where P is the Exceedance Probability,

$$F\left(x\right) = 1 - \frac{1}{T}$$

Gumbel's Distribution

ere are multiple probability distributions which may be fitted to obtain the IDF curve, but the Gumbel's distribution is one which deals with extreme values. While Normal distribution can be used for monthly streamflow or rainfall, the Gumbel's Distribution is used for the extreme events such as extreme highs like floods and extreme lows likeroughts. The distribution was proposed by Gumbel (1935).

The PDF of Gumbel's distribution is given by

$$f\left(x\right) = \frac{1}{\alpha} e^{\left\{-\frac{(x-\beta}{\alpha} - e^{\left\{-\frac{(x-\beta)}{\alpha}\right\}}\right\}}.$$

The CDF of Gumbel's distribution is given by

$$F\left(x\right) = e^{-e^{-\left\{\frac{x-\beta}{\alpha}\right\}}}.$$

The Gumbel's distribution has two parameters which must be estimated, location parameter, μ and scale parameter β > 0. The random variable x denotes the rainfall depth/intensity.

Parameter Estimation

The method of moments is used to estimate the parameters of Gumbel's distribution (Bowman and Shenton, 1998). In statistics, moment is a concept used to represent the shape of a set of points. In the case of a probability density, the zeroth moment presents total probability, and the first moment is the mean. The equation to calculate the nth moment is

$$\mu_n = \int_{-\infty}^{\infty} \left(x - c\right)^n f\left(x\right) dx.$$

The zeroth and first moments are taken about the origin. Hence the value of *c* is zero. Since we know that the first moment is the mean of the distribution, we can set the values 0 and 1 to *c* and *n* respectively in the equation and equate it to the mean of the distribution.

Using the method of moments, the parameters are estimated to be,

$$\alpha = \frac{\sqrt{6}}{\pi}\left(S_d\right) \text{ and } \beta = \mu - 0.5772\alpha$$

Where S_d = Standard deviation of the distribution; and μ = Mean of the distribution.

Estimation of Frequency Factor

The transformation $y = \frac{x - \beta}{\alpha}$ can be used to simplify the expression of the PDand CDF.

The CDF may be rewritten as,

$$F\left(y\right) = e^{-e^{-y}}$$

The above function is invertible. The inverse function can be obtained by taking natural log twice on either side.

$$y = -\ln\left(-\ln\left(F\left(x\right)\right)\right)$$

$$y = -\ln\left(\ln\left(\frac{T}{T-1}\right)\right)$$

The transform used above can be rewritten as $x = \beta + \alpha y$

$$x = \mu - 0.5772\frac{\sqrt{6}}{\pi}\left(S_d\right) - \frac{\sqrt{6}}{\pi}\left(S_d\right)\ln\left(\ln\left(\frac{T}{T-1}\right)\right).$$

$$x = \mu - \frac{\sqrt{6}}{\pi}\left(0.5772 + \ln\left(\ln\left(\frac{T}{T-1}\right)\right)\right)S_d$$

Chow (1964) proposed that,

$$x = \mu + K_T S_d$$

where, K_T is the frequency factor.

Comparing both the equations, it is concluded that

$$K_T = -\frac{\sqrt{6}}{\pi}\left(0.5772 + \ln\left(\ln\left(\frac{T}{T-1}\right)\right)\right)$$

The frequency factor is dependent on the value T. Hence, the frequency factors corresponding to different return period is obtained.

HYDROLOGIC MODELLING

Many hydrologic models are available which are capable of accurately simulating hydrologic processes by using certain parameters which are specific to the study area. One such model is the Hydrologic Modelling System (HEC-HMS) developed by the Hydrologic Engineering Center; which is part of the United States Army Corps of Engineers (USACE). The HEC-HMS can be used to simulate the precipitation-runoff processes that occur in various basins. It converts the rainfall into stream flow using different routing, transform and loss methods. HEC-HMS is widely used for flood modelling because of its simplicity and advanced simulation capabilities. When properly optimized, the system generates outputs which accurately synchronize with the observed results in the field. In addition to the HEC-HMS, the USACE also provides a GIS extension known as the HEC-geoHMS. This extension is linked with ArcGIS to seamlessly generate the hydrologic model in the ArcGIS itself before opening it in HEC-HMS. This extension efficiently extracts basin characteristics such as curve number, area, stream length. etc. from ArcGIS into the HMS model.

A wide range of datasets are required to generate a hydrologic model of the basin. Spatial data includes satellite imagery, Digital Elevation Model (DEM) and soil map. Non-spatial data includes historical rainfall data, observed river discharge data and curve number tables. The steps involved using these datasets to generate a hydrologic model is discussed in the following sections. As a case study, Adyar river basin located in Chennai, India is considered. The basin has an area of 797.63 sq. km. The river originates from the Malaipattu tank located in the Kancheepuram District of Tamil Nadu. The river drains into the Bay of Bengal via the Adyar estuary. The major flood events which occurred in the Adyar river in the current century are during the years 2005, 2008 and 2015 respectively. The 2015 south Indian flood caused the lives of over 500 people and resulted in damages worth INR 20,000 crores. Rapid urbanization along the coasts of the river acted as a catalyst to the issue. The increased urbanization resulted in increased percentage of impervious areas which in turn increased the percentage of precipitation which gets converted into runoff.

Basin Pre-Processing

The delineated basin based on the DEM can be used to clip the other spatial data such as the Landsat image and the soil map. The next process includes setting up of the HEC-HMS in ArcGIS using the HEC geoHMS extension. The HEC geoHMS helps the user to create hydrologic inputs that can directly be used in HEC-HMS. This extension was used to create the basin model, meteorological and the background map files. The HEC-geoHMS uses spatial data such as basin area, basin length etc. to generate the model and import it to the HEC-HMS software. It also helps in the development of the curve number grid which is essential to obtain the curve number value for each sub basin using the land cover information and the soil information.

Basin Delineation

Basin delineation is the process of defining the outline/ perimeter of the basin. It helps to identify the ridgelines that separate one basin from the other.

The delineation process involves the filling of sinks and generating the flow direction raster. The delineated basin is shown in Fig.3.

The superimposed basin map over the satellite imagery is shown in fig.4

Figure 3. Digital Elevation Model of the Basin

Figure 4. Adyar Basin

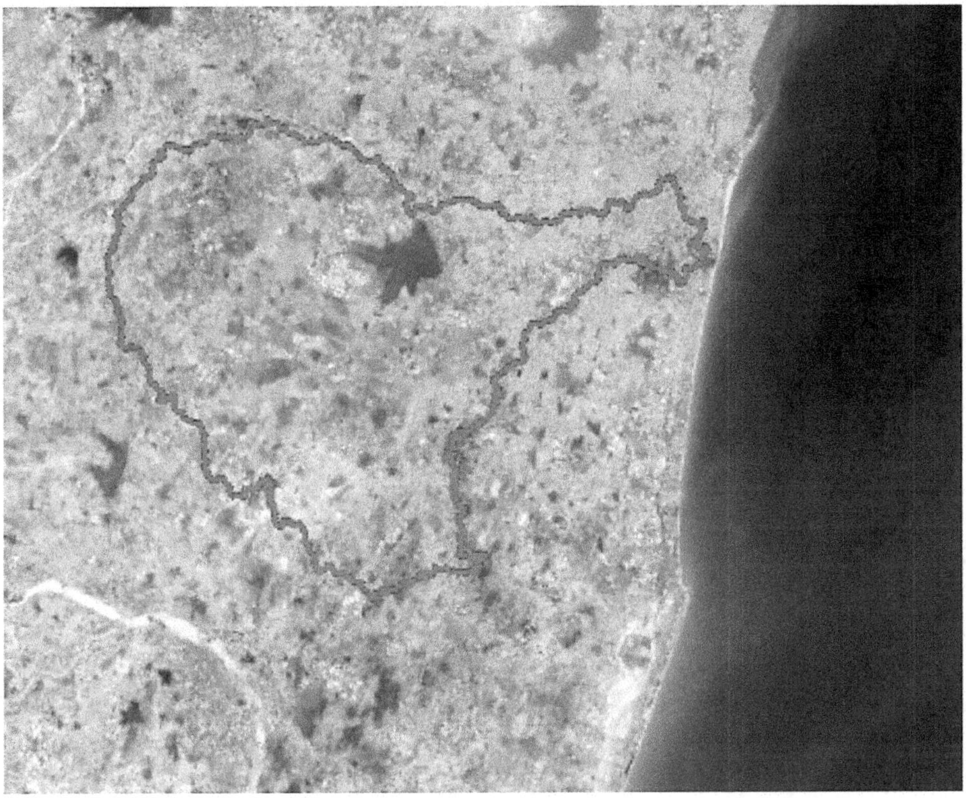

Supervised Image Classification

Image classification is a process which is performed to know the information about the landcover in the study area.

It helps to categorize the land into various categories such as urban, agricultural, forests and barren. In supervised classification, the user selects a few pixels that represent a specific class. These training samples are then used to create the signature file. The supervised classified image from the Landsat -8 imagery is shown in Fig.5. The image was classified into 4 different classes; namely water, urban, forest and barren. The blue cells represent the water bodies and tanks present in the basin. The green represents forests and vegetation and the yellow represents barren land. The black regions signify urban areas.

Figure 5. Landuse map of the Basin

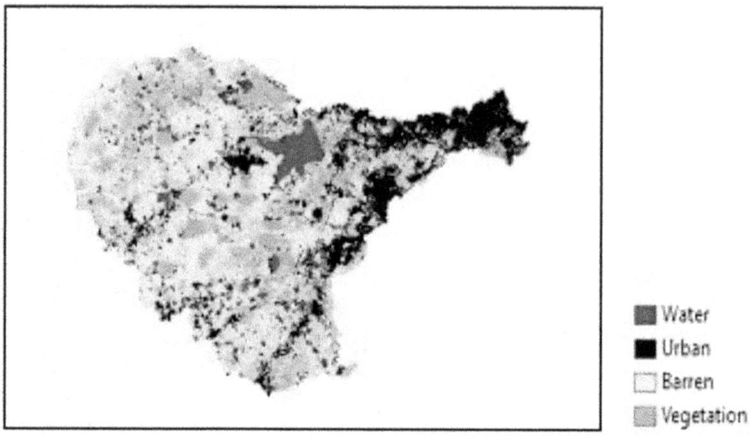

Soil Data Processing

The HWSD is a raster dataset and is clipped using the watershed shapefile to obtain the soil classes pertaining to area under study. Once the soil groups are identified, hydrologic soil group data can be obtained from the database. The soil data was obtained from the Food and Agriculture Organization of the UN which provides a Harmonized World Soil Database (HWSD).

Curve Number Grid Generation

Runoff Curve Number, in the field of hydrology, is an empirical parameter used to predict the direct runoff or infiltration that would occur during excessive rainfall. The curve number is specific to the hydrologic soil group and the land use information of the particular area. Urban Hydrology for Small Watersheds, Technical release 55 (TR55) specifies the various curve numbers corresponding to different hydrologic groups and land cover. The CN is 100 for water bodies and decreases as the perviousness increases. The curve number for the various sub-basins can be estimated by first generating a curve number grid. To prepare a curve number grid, the spatial datasets required are the Land-use raster and the Soil map raster.

HEC-HMS model

The DEM is processed once again using the ArcHydro Tools. Fig 6 shows a schematic representation of the basin preprocessing using the ArcHydro tools.

Figure 6. Schematic representation of basin preprocessing using ArcHydro tools

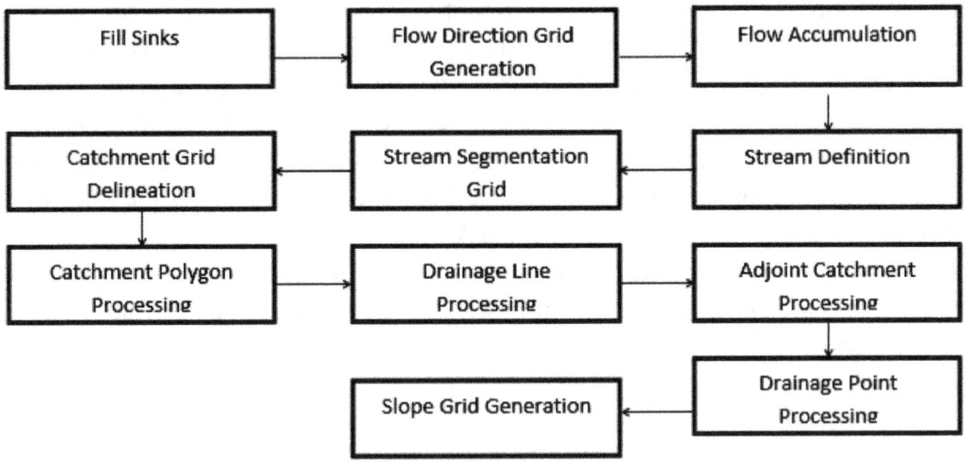

Hydrologic Modelling using HEC-HMS

Once the model is setup in ArcGIS using the HEC geoHMS extension, the model can be run by specifying the rainfall information and other basin related data. The following section deals with the components of the HEC-HMS software and its capabilities.

Components of the HEC-HMS Model

The four basic components of the HEC-HMS model are the basin model, meteorological model, control specifications and times series data.

The Basin Model includes physical information about the basin. It is used to visualize the basin. This includes ridgelines, location of river reaches, location of source and sinks. The basin model further has five different components.

The various components include;

- Subbasin,
- Reach,
- Junction,
- Source, and
- Sink

The Meteorological Model is used to specify the meteorological processes that take place in the basin such as precipitation. The user can specify the method with which rainfall is given as input. The meteorological model is also used to model evapotranspiration and snowmelt. In the present study, precipitation alone is modelled. There are numerous methods used for modelling precipitation such as user specified hyetograph, user gauge weighting, inverse discharge weighting and gridded precipitation. The precipitation method used for the study is the Specified Hyetograph. The precipitation data is given in the Time-Series Data. If numerous Gages are present in the study, the Met model component also allows to user to specify different gages for different sub basins. Control Specification is used to specify the start and end of the simulation. The start time and date and the end time and date along with the time interval for computation can be specified. The current study required both precipitation gages and discharge gages. The time series data manager allows the user to specify the units of the input data and the time interval of the input data. The start and end times and dates can be specified in the time window. There are several methods available in HEC-HMS which can be adopted for estimating the losses. They include; Deficit and Constant Method, Exponential Method, Green and Ampt Method, Gridded Deficit and Constant Method, Gridded Green and Ampt Method, Gridded SCS Curve Number Method, Gridded Soil Moisture Accounting Method, Initial and Constant Method, SCS Curve Number Method, Soil Moisture Accounting Method and the Smith Parlange Method. The loss model adopted for the present study is the SCS Curve Number Loss Method. The SCS-CN loss method takes the precipitation, the soil data, the landcover, and the antecedent moisture content into consideration. The model uses the curve number and the percentage impervious area as the input. The Initial abstraction is not compulsory but can be given. The equation used to calculate the excess precipitation is;

$$P_c = \frac{\left(P - I_a\right)^2}{P - I_a + S}$$

In the above equation, P_c is the accumulated precipitation excess at time t, P is the accumulated rainfall depth for time t, I_a is the Initial Abstraction and S is the Potential Maximum Retention. The time t, is based on the time interval set in the control specifications

The initial abstraction is not a compulsory field which must be filled in. In case the initial abstraction value is not given the system calculates the initial abstraction as

$$I_a = 0.2S$$

This empirical relation was obtained from previous experiments carried out by the SCS for other watersheds.

Using the above two equations, the following can be derived

$$P_c = \frac{\left(P - 0.2S\right)^2}{P + 0.8S}$$

The Maximum Retention, S is completely dependent on one parameter; the Curve Number. The relation is given by the following equation.

$$S = \frac{25400 - 254CN}{CN}$$

The curve number is 100 for water bodies. The lowest value which the curve number can take is 30. The curve number increases as the imperviousness and runoff discharge increases. The curve number for each basin is automatically extracted as the average values of the pixels in the CN grid.

The transform method is used to estimate the direct runoff occurring due to excess precipitation. It is known as the transform method as it is the method used to literally transform the precipitation into runoff. Number of methods can be used perform the transformation and they include; Clark Unit Hydrograph, Kinematic Wave, Modified Clark, SCS Unit Hydrograph, Snyder Unit Hydrograph, User Specified S-Graph and the User Specified Unit Hydrograph. The transform method used in the present study is the SCS Unit Hydrograph method. The SCS unit hydrograph is a single peak unit hydrograph which is dimensionless. The Unit Hydrograph peak discharge is given by the following equation.

$$U_p = C\frac{A}{T_p}.$$

In the above equation, T_p, is the time taken to reach the peak, A is the area of the basin and C is the conversion constant which is taken as 2.08. The time to peak, T_p is dependent on the computational time interval, Δt and the basin lag, t_{lag}. The values are related to each other as,

$$T_p = \frac{\Delta t}{2} + t_{lag}$$

The basin lag (or the lag-time) is the time difference between the center of mass of the rainfall to the peak discharge. It must be noted that the time interval must be taken in such a way that it is less than 29% of the basin lag, else the system will throw an error during computation.

The routing method is used to model channel flow through an open channel, such as a river. The Routing Method helps the system to estimate the downstream hydrograph when the upstream hydrograph is given. HEC-HMS provides numerous methods of defining the flow through an open channel. They include; Kinematic Wave Method, Lag Method, Modified Puls Method, Muskingum Method, Muskingum-Cunge Method and Straddle Stagger Method. The routing method used for the study is the Muskingum method. The following continuity equation can be used,

$$\left(\frac{I_{t-1} + I_t}{2}\right) - \left(\frac{O_{t-1} + O_t}{2}\right) = \left(\frac{S_{t-1} + S_t}{\Delta t}\right).$$

In the above equation, I_t and I_{t-1} are the inflow hydrograph ordinates at times t and t-1 respectively; O_t and O_{t-1} are the outflow hydrograph ordinates at times t and t-1 respectively and; S_t and S_{t-1} is the storage in the reach at times t and t-1 respectively. Δt is the computation time interval for the model. In the above equation, the only variables which are unknown are the outflow and the storage at time t. The storage at time t can be calculated using the equation,

$$S_t = KO_t + KX\left(I_t - O_t\right) = K\left(XI_t + \left(1 - X\right)O_t\right).$$

Where K is the travel time for the water through the given reach and X is dimensionless weight. The Muskingum parameters, K and X and the computational time interval must fall within a specific range to ensure that the model is stable. It is stated that the value of $\Delta t / K$ must fall between 0 and 2 and the value of X must range between 0.0 and 0.5. Once the storage at time t is found, the outflow at t can be estimated. Initially, I_{t-1} and S_{t-1} is taken as zero.

Model Simulation

Apart from simulation of the hydrologic processes, the model evaluation can also be carried out. The HEC-HMS lets the user to calibrate and validate the model using observed discharge data.

Model Calibration

Calibration is the process of using observed hydrometeorological data to estimate the parameters that yield the best fit of the computed results to the observed runoff. This process is known as optimization of the parameters. The parameters which may be estimated are the Muskingum constants X and K and the parameters which may be optimized are the Curve Numbers of each of the basins. The calibration process prompts the program to search for the model parameters that yield the best value of an index, known as the objective function. The objective of calibration is to minimize the value of the objective function. The procedure to calibrate the model is as follows:

Initially, to calibrate the model, the observed rainfall and runoff data is collected. For the system to optimize the parameter, an arbitrary value for the parameters must be initially set. Once the parameters are defined, a simulation run is created and run. The simulated results are compared with the observed results. If the results do not match, an optimization run is created. The parameters which must be optimized is selected and the method used for optimization is specified. The following methods can be used to optimize the parameters; Percentage Error is Simulated Volume, Percentage Error in Simulated Peak Flow, Nash Sutcliffe Efficiency Method and the Coefficient of Determination Method. Under the objective function tab, the method of optimization is selected. The method adopted for the optimization is the Percent Error in Peak. This method measures the goodness of fit of the computed hydrograph peak to the observed peak. The objective function for the Percent Error in Peak method is calculated using the equation

$$Z = 100\left|\frac{q_S(peak) - q_O(peak)}{q_O(peak)}\right|$$

Where $q_s(peak)$ and $q_o(peak)$ are the simulated peak discharge and the observed peak discharge values respectively.

Storm Frequency

The Meteorological Model for precipitation can be set as specified hyetograph when running the model using observed historic rainfall time series data. However, while modelling hypothetical storm events of varying duration and return period, the frequency storm method is adopted. The frequency storm method generates uses the results of the Intensity-Duration-Frequency Analysis to generate a hydrograph corresponding to a different return periods for a given storm duration. When dealing with floods, the concentration time of the basin is taken as the storm duration. Taking any value higher or lesser would not generate an actual peak flow. In the present, the concentration time of the basin was estimated to be approximately 24 hours. Hence in the meteorological model, the storm duration is taken as one day. The intensity position signifies the percentage of water that flows through the channel before the peak has been attained. For the present study, the intensity position is set as 50 percent. The storm area denotes the area of the basin. The annual maximum depth corresponding to different durations can be specified in the table in the meteorological model.

HYDRAULIC MODELLING

Hydraulic modelling is carried out to model the physical property of water (movement) over the terrain. It is used to simulate the flow of water through the river. There is numerous software available to simulate channel flow, the River Analysis System (HEC-RAS) provided by the HEC of the US Army Corps of Engineers has proven to be a suitable option due to its efficiency and simplicity. The model requires the terrain data and the discharge data as inputs. The discharge data is obtained from the HEC-HMS model. The terrain data includes the Digital Elevation Model, River Cross-section Data, Basin Slope and the Manning's Constant. The datasets required depends on the whether the One-Dimensional or Two-Dimensional Model is being carried out. The capabilities of the hydraulic model also include the design of bridges, culverts and channels. In the present study, the HEC-RAS model is used to generate the inundation map for historical flood events and hypothetical storm events. The model can be used to perform one-dimensional and two-dimensional, steady and unsteady flows.

Steady Flow Hydraulics

Steady flow hydraulics is used to obtain the depth of water in the channel resulting from a specific discharge value. It does not require any simulation time interval or time step as the flow is assumed to be constant. Steady Flow is computed using the One-Dimensional Model which uses 1D energy equations. The 1D model requires vector data such as river reaches, the river cross section and banklines as input geometry. The cross-section data for the rivers may be obtained either from the digital elevation model or by carrying out a site survey. The Manning's constants need to be assigned to the river cross-section as the system uses the Manning's equation to estimate the losses that may occur as a result of friction along the river.

Unsteady Flow Hydraulics

Unsteady Flow Hydraulics can be simulated using either the 1D or the 2D Model, however the 2D is widely used for unsteady flow hydraulics. The 2D model requires terrain data such as Digital Elevation Model and an input flow hydrograph to simulate flow through the channel. The simulation time interval need to be specified and the time series flow data must be given as input. The 2D model uses the St. Venant equations for generating the unsteady flow through the channel.

To ensure the efficacy of the model, the 2D model must be stabilized. The model can be stabilized by varying certain simulation variables. To stabilize the model, the Courant-Friedrichs-Lewy condition must be satisfied. According to the condition, the Courant's number must be less than 1. The Courant's number can be estimated using the following equation.

$$C = \frac{u\Delta t}{\Delta x}$$

Where, C is the dimensionless Courant's number. "u" is the magnitude of velocity, which is representative of the velocity of the water flowing through the channel. The unit is length/time. Δt is time interval used for computation and Δx is the length interval used which is representative of the computation mesh size used to define the flow area. Its unit is length. An arbitrary value for velocity of flow must first assumed. The length and time interval must be then assigned such that the condition is satisfied. If the velocity which is assumed is same as the actual flow velocity, then the simulation will run without any errors, else the system will throw an error. In the occurrence of an error, the time and length interval must be varied accordingly until the model is stable.

Geometric Data

The Digital Elevation Model acts as the terrain file. The terrain can be given as input using the RAS mapper found in HEC-RAS. Once the projection and terrain is specified, the 2D flow areas can be drawn in the geometric data editor. The 2D flow area is drawn around the river reach. The computation mesh cell size is assigned, and the mesh is created. The mesh size is assigned in accordance with the Courant-Friedrichs-Lewy condition. A smaller mesh size would mean that the number of cell would be more which may slow down the model. Once the mesh is created, the boundary conditions are specified. The geometric data may also be edited to design levees and embankments by the usage of breaklines. Once drawn, the breakline is enforced into the 2D flow area. The breakline is then converted into an internal SA/2D connection. The elevation of the breakline specifies the elevation of the embankment to be designed and can be specified at different breakline stations. Once the geometric data is edited, the unsteady flow data can be given as input.

Unsteady Flow Data

Different boundary condition types may be specified for different boundary conditions. For the present study, the upstream boundary condition type is flow hydrograph and the downstream boundary condition type is normal depth. In the case of the upstream boundary condition, the time series data obtained as output in HEC-HMS must be given as input. In the case of the downstream boundary condition, the basin slope is given as input.

Unsteady Flow Analysis

The geometry file and the unsteady flow file are used for performing the unsteady flow simulation as specified. The programs to be run include the geometric preprocessor, unsteady flow simulation, post processor and flood plain mapping. The simulation time window is used to specify the start and end dates and times. The computation settings allow the user to specify the computational time interval, hydrograph output interval, mapping output interval and detailed output interval. Similar to the cell size, the computation time interval must also be set so as to satisfy the Courant-Friedrichs-Lewy condition. Should there be any error in the model, the system will throw an error during computation and will not generate the outputs. The simulation which is generated can be viewed in the RAS mapper.

The HEC geoHMS is used for specifying the nodes, ridge lines and for generating the curve number grid. Once the model is set, it is run in HEC-HMS. The HEC-HMS output is calibrated using the observed peak value. Once calibrated, the IDF analysis is carried out and the model is also used to obtain the flood hydrograph for different hypothetical storms using the frequency storm option in HEC-HMS. The discharge time series data obtained from HEC-HMS is used as the input for the hydraulic model. The HEC-RAS model is used to generate the inundation map for floods of different return periods. The various datasets required include Digital Elevation Model, satellite imagery, soil map, curve number table, rainfall data and discharge data.

The Dataset required for the study includes both Spatial and Non-Spatial Data. Spatial datasets include Digital Elevation Model (DEM), Landsat-8 Image and Soil Map. Non-Spatial datasets include the historic rainfall data, curve number values and discharge data.

Hypothetical Storm Modelling

To generate the hypothetical storm for different return periods, the IDF analysis should be carried out to obtain the intensity values for various return periods and durations. Once the intensity values are estimated, the rainfall depths corresponding to the durations can be calculated and given as input in the frequency storm method.

IDF Analysis

The results from the IDF analysis are compared with the results obtained from the empirical equation proposed by Sherman (1931). The IDF relationships can also be represented as a graph. This graphical representation of the variation of rainfall intensity with respect to rainfall duration for various return intervals is known as the Intensity –Duration –Frequency Curve. Fig 2 shows the IDF curve obtained using the Gumbel's equation and Fig 2 shows the IDF curve obtained using the empirical equation.

Soil Map Generation

The Harmonized World Soil Database provides the soil map of the entire world. To obtain the Soil Map for the study region, the HWSD is imported to Arcmap and it is clipped using the basin polygon feature shapefile. Fig 7 shows the 3 different types

Figure 7. Soil Map of the Basin

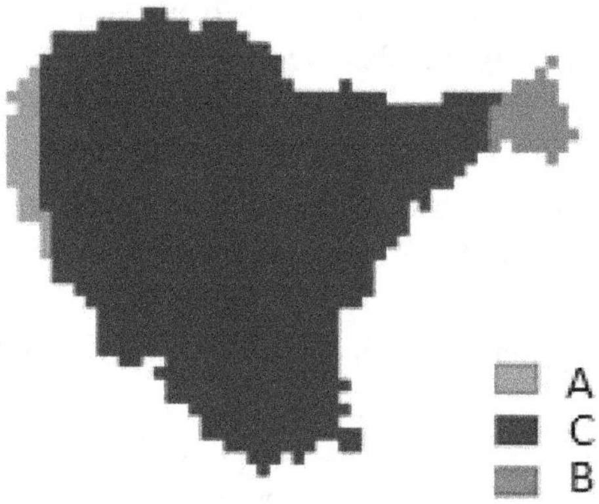

Table 1. Curve Number Look Up Table

Sl.No	Description	Hydrologic Soil Groups			
		A	B	C	D
1	Water	100	100	100	100
2	Barren	77	86	91	93
3	Urban	71	80	85	88
4	Forest	26	40	58	61

of soils present in the study region. The data was interpreted using the additional tools legend provided along with the HWSD by the FAO and were grouped based on their hydrologic soil groups. The soil towards the coast falls in the hydrologic soil group B whereas the soil upstream falls in the category of C.

Curve Number Grid Generation

The values were obtained from (Kumar et al. 1991) who developed the CN values for Indian soil conditions.

Figure 8 represents the Curve Number grid which is generated. The Curve Number Grid is a raster with each cell representing its curve number. As seen in Fig. 8, the lowest curve number value found in the basin is 40. This number is assigned to regions with high infiltration and decreased runoff. The highest number of 100 is assigned to waterbodies such as lakes and tanks.

Figure 8. Curve Number Grid of Adyar Basin

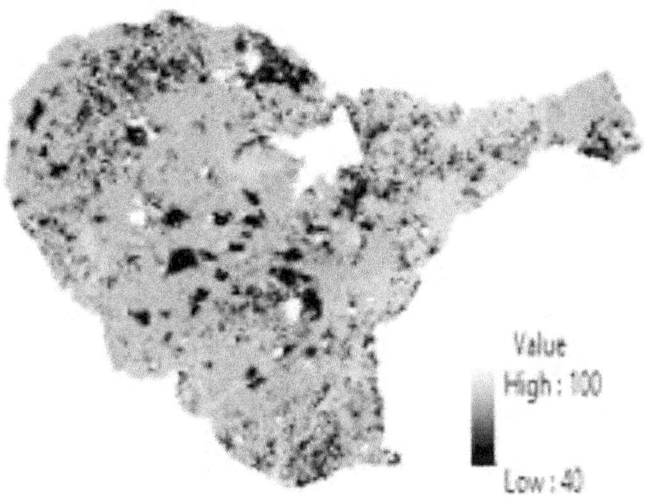

HEC-HMS Model Setup

In Fig 9 c the different sub basins are separated by the dark edged lines. In Fig. 9 d the blue lines signify the river reaches and the light brown lines signifies the longest flowpath of each subbasin. The green lines signify the reaches which meet together at a junction and the red lines represent the reaches that meet connects two junctions. Each subbasin and reach have a unique code to identify them with reaches having the alphabet 'R' and a number and sub-basins having the alphabet 'W' and a number.

Simulating Flows Using Observed Rainfall

In the meteorological model, type of precipitation input is specified as specified hyetograph. The required time series data such as rainfall data and reservoir release data are given in the time series data manager. Fig. 10 shows the input hyetograph for the period 1-5 Dec 2005 which is required to run the model. Fig. 11 shows the output discharge graph obtained at the outlet.

Under observed flow data at the outlet, the observed discharge data is given. This data is used to calibrate the model. Since the calibration only involves the matching of the peaks, the observed peak flow is given as input.

Figure 9.

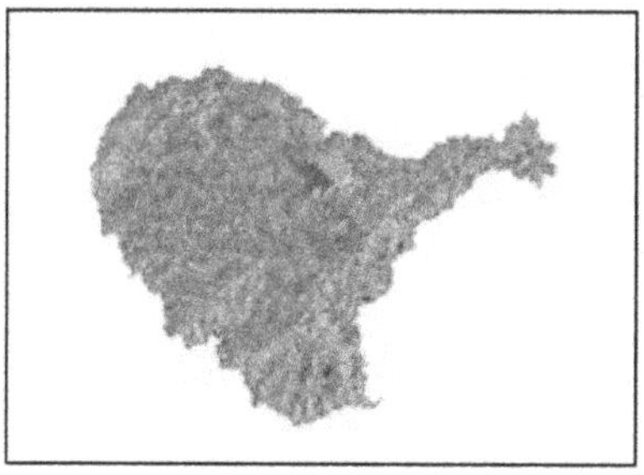

a. Flow Direction Raster

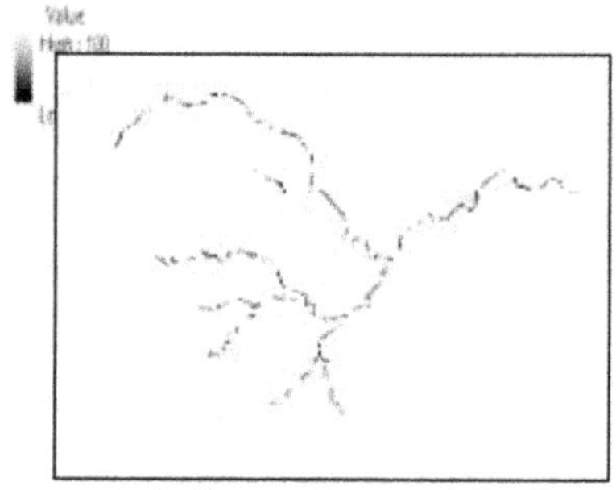

b. Stream Segmentation Raster

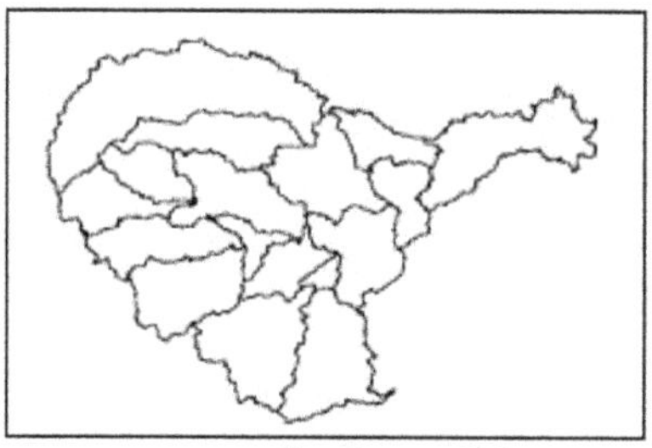

c. Catchment Polygon

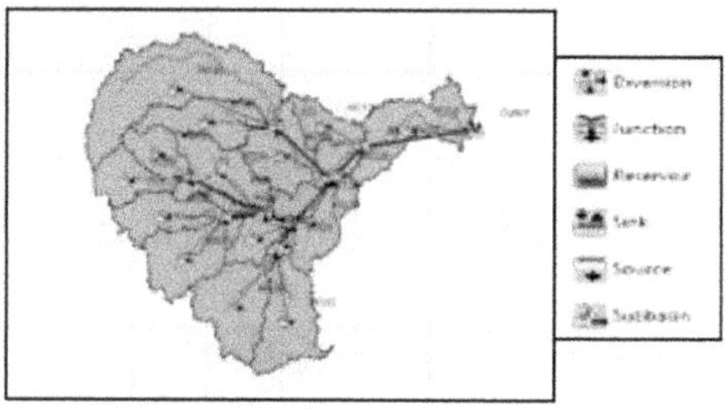

d. Graphical representation of the HEC-HMS basin model for the Adyar Basin

Figure 10. Hyetograph for the 2005 December Flood Event

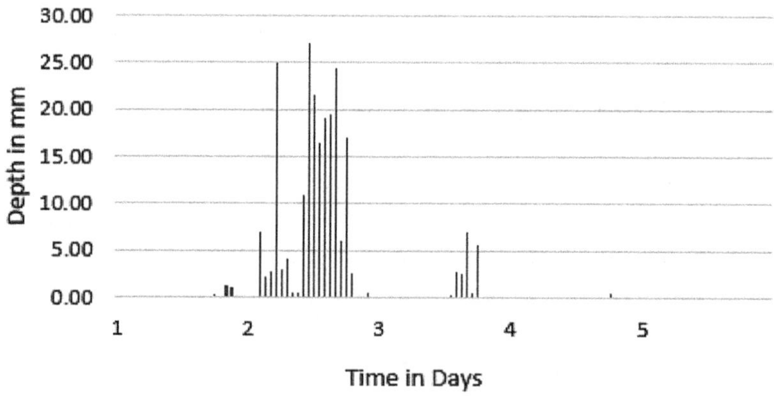

Figure 11. Simulated Flow Hydrograph

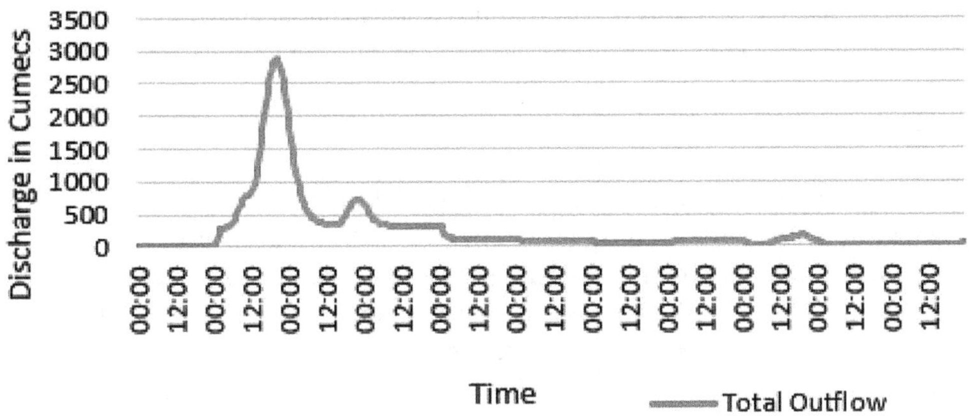

Figure 12. Simulated Hydrograph after calibration

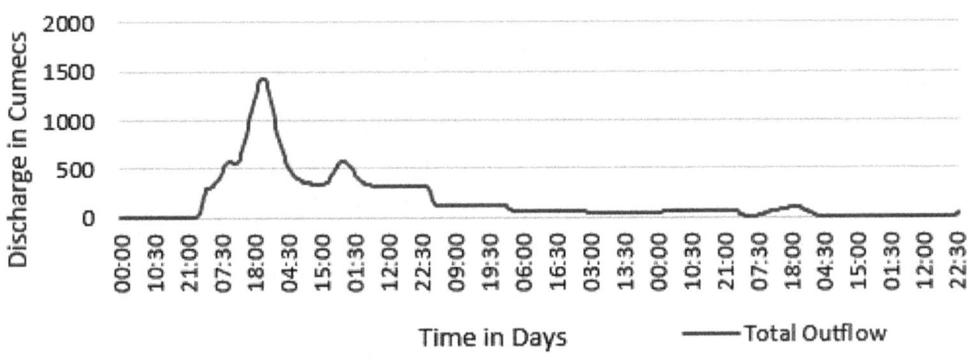

Calibrating the HEC-HMS Model

The observed maximum runoff for the 2005 flood event was 1400 cubic meters per second. The observed peak does not match the simulated peak. Hence an optimization trial was run and the curve number values for all the sub-basin along with the Muskingum constants is calibrated. It was found that the curve number must be multiplied with a factor of 0.44.i.e. the curve number are too high hence resulting in a higher runoff. The reason that the curve numbers were so high is due to the presence of an error in the image classification process where the scrublands were classified as barren land. Scrublands have a very low curve number compared to barren land. The curve Numbers were optimized, and the simulation was run a second time with the same control specifications and the following results were obtained.

Frequency Storm Analysis

Using the frequency storm method, the hypothetical storm may be specified to generate the hydrographs for different return periods. While generating hypothetical storm events, the duration of the storm must be set equal to the concentration time of the basin. If the storm duration is less, then the storm will end before all the water reaches the confluence point. If the duration is too high, then the intensity of the rainfall will be decreased, and the discharge will be less.

HYDRAULIC MODELLING

In the geometry editor, the DEM is given as input and a polygon is drawn around the course of the river to specify the flow area. Once the flow area is drawn, the flow area mesh must be generated by specifying a suitable grid size. The grid size is specified using the Courants' number condition as specified in the methodology. The cell size was taken as 50m. Once the mesh is generated, the upstream end and the downstream end is specified in the flow area. This tells the system the starting end and the confluence point of the river. Fig 13 shows the geometric data with the mesh and the boundary conditions.

The unsteady flow data is given as a flow hydrograph in the upstream end of the river. The hydrograph given as input is obtained from the output generated by the HEC-HMS model. Once the geometric data is generated and the unsteady flow data is given,the model is run and the inundation maps for different historic flood events and hypothetical flood events of different return periods are generated. Fig. 14 shows the inundation maps generated for the 2005 chennai flood as well as that of different hypothetical storms.The water depths go up to 15m in the rivers during flood s with higher return period.

SUMMARY

The chapter presents the application of flood modelling to the study area. The study area, Adyar River in Chennai, has experienced major floods in the past decade which is attributed to increased urbanization. The hydrologic model for the Adyar river

Figure 13. Geometric Data to generate 2D flow

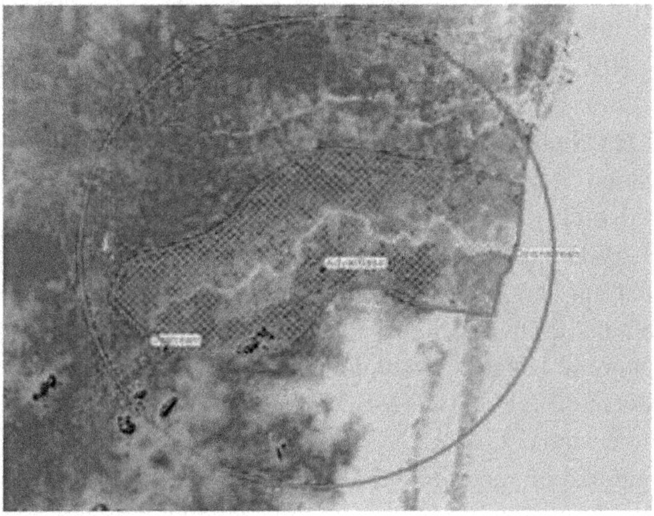

Figure 14. Inundation Map of the Adyar River for the December 2005 Flood event

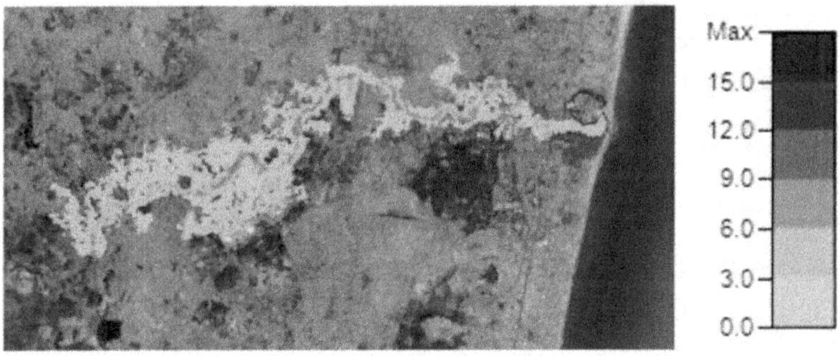

basin was set up using HEC geoHMSand the model was run and calibrated using observed flow in HEC-HMS. The chapter also discusses the results obtained from the IDF analysis and its application in HEC HMS to generate hypothetical storm hydrographs. Furthermore, the chapter goes on to discuss the results obtained from the hydraulic modelling such as the inundation map for the 2005 flood event and the inundation map for hypothetical storms of varying return periods.

CONCLUSION

Based on the results obtained from the hydrologic and hydraulic modelling using the HEC HMS, HEC RAS and ArcGIS tools, the following conclusions are drawn.

- The HEC HMS proved to be an efficient method for finding the discharge for different flood events. The model works to a good extent even with the lack of a good dataset.
- The Intensity-Duration-Frequency Analysis of rainfall for the site in Chennai was carried out and the results obtained was found to synchronize with the results obtained from the empirical equation. Hence, it can be concluded that the IDF curve obtained is accurate to some extent, even though the number of years of data availability is less.
- The 2-Dimensional Unsteady Flow Hydraulics proved to be a simple yet efficient method of generating the inundation maps for different events. The 30m Digital Elevation Model proved to be sufficient to an extent.

REFERENCES

Abdullah, A., Rahman, A., & Vojinovic, Z. (2009). LiDAR filtering algorithms for urban flood application: Review on current algorithms and filters test. *Laser Scanning, 9*(38), 30-36.

Abdullah, A. F., Vojinovic, Z., Price, R. K., & Aziz, N. A. A. (2012). Improved methodology for processing raw LiDAR data to support urban flood modelling–accounting for elevated roads and bridges. *Journal of Hydroinformatics, 14*(2), 253–269. doi:10.2166/hydro.2011.009

Alho, P., Hyyppä, H. & Hyyppä, J. (2009). Consequence of DTM precision for flood hazard mapping: A case study in SW Finland. *Nordic Journal of Surveying and Real Estate Research, 6*, 21-39.

Arora, K., & Singh, V. P. (1989). A Comparative Evaluation of the Estimators of the Log Pearcon Type (LP) 3 Distribution. *Journal of Hydrology (Amsterdam), 105*(1-2), 19–37. doi:10.1016/0022-1694(89)90094-2

Avanzia, A., Franka, E., Righetto, M., & Fattorellia, S. (2013). LIDAR Data Resolution Versus Hydro-Morphological Models for Flood Risk Assessment. ISPRS-International Archives of the Photogrammetry. *Remote Sensing and Spatial Information Sciences, 1*, 119–124.

Babu, R., Tejwani, K. K., Agrawal, M. C., & Bhusan, L. S. (1979). Rainfall Intensity-Duration-Return Period Equations and Nomographs of India. CSWCRTI, ICAR, Dehradun, India.

Bater, C. W., & Coops, N. C. (2009). Evaluating error associated with lidar-derived DEM interpolation. *Computers & Geosciences*, *35*(2), 289–300. doi:10.1016/j.cageo.2008.09.001

Bobee, B. (1975). The Log Pearson Type 3 Distribution and Its Application in Hydrology. *Water Resources Research*, *11*(5), 681–689. doi:10.1029/WR011i005p00681

Bodoque, J. M., Guardiola-Albert, C., Aroca-Jimenez, E., Eguibar, M. A., & Martinez-Chenoll, M. L. (2016). Flood Damage Analysis: First Floor Elevation Uncertainty Resulting from LiDAR-Derived Digital Surface Models. *Remote Sensing*, *8*(7), 604. doi:10.3390/rs8070604

Brown, K. M., Hambidge, C. H., & Brownett, J. M. (2016). Progress in operational flood mapping using satellite synthetic aperture radar (SAR) and airborne light detection and ranging (LiDAR) data. *Progress in Physical Geography*, *40*(2), 196–214. doi:10.1177/0309133316633570

Candela, L., Tamoh, K., Olivares, G., & Gomez, M. (2012). Modelling Impacts of Climate Change on Water Resources in Ungauged and Data-Scarce Watersheds. Application to the Siurana Catchment (NE Spain). *The Science of the Total Environment*, *440*, 253–260. doi:10.1016/j.scitotenv.2012.06.062 PMID:22795259

Cook, A., & Merwade, V. (2009). Effect of topographic data, geometric configuration and modeling approach on flood inundation mapping. *Journal of Hydrology (Amsterdam)*, *377*(1-2), 131–142. doi:10.1016/j.jhydrol.2009.08.015

Costabile, P., Macchione, F., Natale, L., & Petaccia, G. (2015). Flood mapping using LIDAR DEM. Limitations of the 1-D modeling highlighted by the 2-D approach. *Natural Hazards*, *77*(1), 181–204. doi:10.100711069-015-1606-0

CRED. (2015). *Centre for Research on the Epidemiology of Disasters Emergency Database*. CRED.

Dell, V. K., Arthur, S., & Pender, G. (2008). *The use of a LiDAR DEM in modelling sewerage derived urban flood wave routing. 11th International Conference on Urban Drainage*, Edinburgh, UK.

Demir & Kisi. (2016). Flood Hazard Mapping by Using Geographic Information System and Hydraulic Model: Mert River, Samsun, Turkey Hindawi Publishing Corporation. *Advances in Meteorology*, *2016*, 9. doi:10.1155/2016/4891015

Di Baldassarre, G., & Uhlenbrook, S. (2012). Is the current flood of data enough? A treatise on research needs for the improvement of flood modelling. *Hydrological Processes*, *26*(1), 153–158. doi:10.1002/hyp.8226

Dupont, B. S., & Allen, D. L. (2000). *Revision of the Rainfall Intensity Duration Curves for the Commonwealth of Kentucky*. Kentucky Transportation Research Center Research Report, 292.

Elsebaie, I. H. (2012). Developing Rainfall Intensity-Duration-Frequency Relationship for Two Regions in Saudi Arabia. *Journal of King Saud University. Engineering and Science*, *24*, 131–140.

Flood, M. (2001). Laser Altimetry: From Science to Commercial Lidar Mapping. *Photogrammetric Engineering and Remote Sensing*, 67.

Ghimire, S. (2013). Application of a 2D hydrodynamic model for assessing flood risk from extreme storm events. *Climate (Basel)*, *1*(3), 148–162. doi:10.3390/cli1030148

Guhathakurta, P., Sreejith, O. P., & Menon, P. A. (2011). Impact of Climate Change on Extreme Rainfall Events and Flood Risk in India. *Journal of Earth System Science*, *120*(3), 359–373. doi:10.100712040-011-0082-5

Gumbel, E. J. (1941). The return period of flood flows. *Annals of Mathematical Statistics*, *12*(2), 163–190. doi:10.1214/aoms/1177731747

Gumbel, E. J. (1954). *Statistical Theory Of Extreme Values and Some Practical Applications.* U.S. Department of Commerce, National Bureau of Standards.

Haile, A. T., & Rientjes, T. H. M. (2005). Effects of LiDAR DEM resolution in flood modelling: A model sensitivity study for the city of Tegucigalpa, Honduras. *Isprs*, *3*, 12–14.

Ho, L. T. K., Umitsu, M., & Yamaguchi, Y. (2010). Flood Hazard Mapping by Satellite Images and SRTM DEM in the Vu Gia- Thu Bon Alluvial Plain, Central Vietnam. *The International Archives of the Photogrammetry, Remote Sensing and Spatial Information Sciences*, *38*, 8.

Hohenthal, J., Alho, P., Hyyppä, J., & Hyyppä, H. (2011). Laser scanning applications in fluvial studies. *Progress in Physical Geography*, *35*(6), 782–809. doi:10.1177/0309133311414605

Kenbl, M. R., Yang, Z. L., Hutchison, K., & Maidment, D. R. (2005). Regional Scale Flood Modelling Using NEXRAD Rainfall, GIS and HEC-HMS/RAS: A Case Study for the San Antonio River Basin Summer 2002 Storm Event. *Journal of Environmental Management*, *75*(4), 325–336. doi:10.1016/j.jenvman.2004.11.024 PMID:15854726

Klemas, V. (2014). Remote sensing of floods and flood-prone areas: An overview. *Journal of Coastal Research*, *31*, 1005–1013.

Kumar, P., Tiwart, K. N., & Pal, D. K. (1991). Establishing SCS Runoff Curve Number from IRS Digital Data Base. *Photonirvachak (Dehra Dun)*, *19*(4), 245–252. doi:10.1007/BF03023971

Li, J., & Wong, D. W. (2010). Effects of DEM sources on hydrologic applications. *Computers, Environment and Urban Systems*, *34*(3), 251–261. doi:10.1016/j.compenvurbsys.2009.11.002

Liu, X. (2008). Airborne LiDAR for DEM generation: Some critical issues. *Progress in Physical Geography*, *32*(1), 31–49. doi:10.1177/0309133308089496

Ma, R. (2005). DEM generation and building detection from lidar data. *Photogrammetric Engineering and Remote Sensing*, *71*(7), 847–854. doi:10.14358/PERS.71.7.847

Mason, D. C., Cobby, D. M., Horritt, M. S., & Bates, P. D. (2003). Floodplain friction parameterization in two-dimensional river flood models using vegetation heights derived from airborne scanning laser altimetry. *Hydrological Processes*, *17*(9), 1711–1732. doi:10.1002/hyp.1270

Miller, J. D., & Hess, T. (2017). Urbanization Impact on Storm Runoff Along a Rural-Urban Gradient. *Journal of Hydrology (Amsterdam)*, *552*, 474–489. doi:10.1016/j.jhydrol.2017.06.025

Mujere, N. (2011). Flood Frequency Analysis Using the Gumbel Distribution. *International Journal on Computer Science and Engineering*, *3*, 2774–2778.

Ohl, C. A., & Tapsell, S. (2000). Flooding and human health: The dangers posed are not always obvious. *BMJ: British Medical Journal, 321*(7270), 1167–1168. doi:10.1136/bmj.321.7270.1167 PMID:11073492

Phine, H. N., & Hira, M. A. (1983). Log Pearson Type-3 Distribution: Parameter Estimation. *Journal of Hydrology (Amsterdam), 64*(1-4), 25–37. doi:10.1016/0022-1694(83)90058-6

Sanyal, J., & Lu, X. X. (2004). Application of remote sensing in flood management with special reference to monsoon Asia: A review. *Natural Hazards, 33*(2), 283–301. doi:10.1023/B:NHAZ.0000037035.65105.95

Schumann, G., Matgen, P., Cutler, M. E. J., Black, A., Hoffmann, L., & Pfister, L. (2008). Comparison of remotely sensed water stages from LiDAR, topographic contours and SRTM. *ISPRS Journal of Photogrammetry and Remote Sensing, 63*(3), 283–296. doi:10.1016/j.isprsjprs.2007.09.004

Sherman, C. W. (1931). Frequency and intensity of excessive rainfall at Boston, Massachusetts. *Transactions of the American Society of Civil Engineers, 95*, 951–960.

Suriya, S., & Mudgal, B. V. (2011). Impact of Urbanization on Flooding: The Thirusoolam Sub Watershed- A Case Study. *Journal of Hydrology (Amsterdam), 412*, 210–219.

Miller & Hutchins. (2017). The impacts of urbanisation and climate change on urban flooding and urban water quality: A review of the evidence concerning the United Kingdom. *Journal of Hydrology: Regional Studies, 12*, 345–362.

Turner, A. B., Colby, J. D., Csontos, R. M., & Batten, M. (2013). Flood modeling using a synthesis of multi-platform LiDAR data. *Water (Basel), 5*(4), 1533–1560. doi:10.3390/w5041533

Wang, Y. (2002). Mapping extent of floods: What we have learned and how we can do better. *Natural Hazards Review, 3*(2), 68–73. doi:10.1061/(ASCE)1527-6988(2002)3:2(68)

Wedajo, G. K. (2017). LiDAR DEM Data for Flood Mapping and Assessment; Opportunities and Challenges: A Review. *J Remote Sensing & GIS*, *6*(04), 211. doi:10.4172/2469-4134.1000211

Zhang, K., Chen, S. C., Whitman, D., Shyu, M. L., & Yan, J. (2003). A progressive morphological filter for removing nonground measurements from airborne LIDAR data. *IEEE Transactions on Geoscience and Remote Sensing*, *41*(4), 872–882. doi:10.1109/TGRS.2003.810682

Zope, P. E., Eldho, T. I., & Jothiprakash, V. (2016). Impact of Land Use-Land Cover Change and Urbanization on Flooding: A Case Study of Oshiwara River Basin in Mumbai, India. *Catena*, *145*, 142–154. doi:10.1016/j.catena.2016.06.009

Chapter 7
Quantification and Evaluation of Water Erosion:
Application of the Model SDR - InVEST in the Ziz Basin in South-East Morocco

Souad Ben Salem
Cadi Ayyad University, Morocco

Abdelkrim Ben Salem
https://orcid.org/0000-0002-2283-5928
Cadi Ayyad University, Morocco

Ahmed Karmaoui
https://orcid.org/0000-0003-3881-4029
Southern Center for Culture and Sciences (SCCS), Morocco

Mohammed Khebiza Yacoubi
Cadi Ayyad University, Morocco

Mohammed Messouli
Cadi Ayyad University, Morocco

ABSTRACT

The Ziz Watershed is located in the arid zones of South-Eastern Morocco and belongs to the large basin of Ziz-Rheris. In this basin, floods are related to natural factors and mainly to the occupation of the hydraulic public domain and the human intervention on the courses of the rivers. Increases in sediment yield are observed in many places in the Ziz, dramatically affecting water quality and reservoir management. In order to map overland sediment generation and delivery to the stream (studying the service of sediment retention), the InVEST sediment delivery ratio (SDR) model was applied. The sedimentation analysis in the Hassan Dakhil Dam, located in this watershed, shows that there is a very important erosion rate. The proof is the rapid filling of the dam. This is due to the transport of sediments in the rivers. If this situation continues at the current rate, the dam will no longer be fully operational for irrigation by 2050.

DOI: 10.4018/978-1-5225-9771-1.ch007

INTRODUCTION

Soil erosion and soil loss are natural resource degradation processes affecting, with varying intensities a large part of the national territory of Morocco. (Tahiri et al., 2017). Thus, a total area of watersheds close to 20 million ha, the risk areas represent 75% of the 15 million hectares studied until now. The cumulative annual soil loss is estimated at some 100 million tons and storage capacity lost to siltation was valued at 50 million cubic meters per year (*National Plan for Watershed Management*).

The Moroccan experience of erosion control and its effects is rich and diversified. Historically, the anti-erosion schemes initiated by the farmers have had the purpose of agricultural development and protection against flood damage. The collective dimension was limited to the size of the group that participated directly or indirectly in the development work.

The Ziz watershed located in south-eastern Morocco; in the upstream basin found the Imilchil region located to the peaks of the High Atlas Oriental, water split point in four watersheds: Umm Errabia, Ziz and Malouiya Ghris is an area known for its land very rugged, geology, its harsh climatic conditions and the deterioration of its vegetation cover due to anthropogenic pressure, is a typical case for the phenomenon of water erosion. Indeed, several studies (Lahlaoi et al., 2015; Ouallali et al., 2016) made in this direction in Morocco, revealed that water erosion affects 23 million hectares, causing a specific degradation from 500 t / km^2 / year to more than 5000 t / km^2 / year with soil loss over 10 t / ha / year in the High Atlas. These figures far exceed what the soil formation can occur to offset these losses. The consequences of such degradation are reflected in lower crop yields, a reduction in the area of already poor agricultural soils and an acceleration of siltation of reservoirs, and downstream water infrastructure. In addition to this phenomenon, which threatens the lives of vulnerable and needy populations, the situation is aggravated by a high rate of poverty and illiteracy that exceeds 80% for women and 65.8% according to official figures, which endangers the development of this territory (Imilchil., 2015).

To determine the rate of water erosion on agricultural soils and to evaluate the efficiency of appropriate response practices, as in other countries and regions the InVEST Sediment Delivery Ratio (SDR) model is used.

The objective of the Sediment Delivery Ratio (SDR) model is to map the production and distribution of surface sediments in the watercourse and to study the sediment retention service in a watershed.

SOILS EVOLUTION

Soil formation (pedogenesis) is dependent on natural factors such as geology (bedrock), topography, climate and vegetation, as well as anthropogenic factors. For millennia man has profoundly transformed soils, by clearing, planting, cultivating, by arranging terraces, benches, hydraulic structures (Plan Bleu, 2003).

Climate Aggressiveness

The main characteristic of the Mediterranean climate is the existence of dry months in summer. During these months, potential evapotranspiration (ETP) is well below the rainfall contribution. Moreover, rainfall variability is very large from one year to another. The rains, which occur especially in the cold season (over 90% of annual rainfall occurs between the months of September and March), can be very violent, and reach very high instantaneous intensities (100 millimetres per hour, and even more) (Roose, 1991). They then cause considerable runoff, which can cause two types of erosion: erosion diffuse tablecloths, and concentrated erosion in gullies, streams or wadis. In practice, the vulnerability of a soil is highly dependent on its vegetation cover, and its exposure to the sun, drying winds and showers (Roose, 1991).

Favourable Topography Erosion

In Morocco, tectonics has created a landscape with rugged topography. The Alpine folding has given rise to mountain ranges sometimes up to 3000 or even 4000 meters, very folded and fractured, and frequently dug deep canyons (Plan Bleu, 2003). The slopes are often so strong. Yet it is commonly accepted that the slope is, especially in areas where the vegetation cover has been disturbed, one factor influencing most soil loss by water erosion. Exposure may also have a role; Indeed, it is observed that south-facing slopes, because of their canopyless dense, are more vulnerable to the erosive action of water (Plan Bleu, 2003)

Climate Change and Floods

With a semi-arid climate, Morocco has not escaped the effects of floods directly related to climate change. Periods of long and frequent droughts and succeed and sometimes alternating with wet and rainy weather to the point of flooding (Rhazi et al., 2017)

Indeed, many floods have occurred. The flood that devastated the city of Sefrou September 25, 1950, and which caused damage to human and material(El Fellah Idrissi et al., 2017). Also, the one that devastated the Ziz Valley November 5, 1965, leaving 25,000 people homeless and who imposed the dam Hassan Adakhil upstream, north of the city of Errachidia (Rhazi A et al., 2017).

CHARACTERISTICS OF ZIZ CATCHMENT

Geographic Setting

The Ziz Basin located in the Errachidia province in south-eastern Morocco (Figure 1) extends from a higher altitude 3423 m in the High Atlas Mountains to the plain of Tafilalet, where the altitude is equal to593 m. Itis divided into three major physiographic sub-basins according to the topography (Fig. 2): (1) the Upper Ziz Basin (UZB), starting from the southern flank of high Atlas Mountains south to Hassan Addakhil Dam; (2) the Middle Ziz Basin (MZB) between the dam and the city of Erfoud; and (3) the Lower Ziz Basin (LZB) south of Erfoud until to the southern bound of the basin in Ouzina. Within the Ziz Basin there are a series of oases, along the main stem of the Ziz Wadi, the most important of which are Mdaghra, Rteb, Tizimi and Tafilalet. Within each of the oases, the population depends primarily on irrigation, using both surface water released from the dam and local groundwater to support agriculture, particularly date production. Due to the arid nature of the majority of the basin, agricultural production without irrigation is only marginal and performed on only small areas in the upper basins (Lgourna et al., 2014).

Climate and Surface Water

The climate of the Ziz basin is arid, it is generally colder and wetter in the north, warmer and drier in the south, the average annual rainfall ranges from 250 mm in the southern flank of the High Atlas to 130 mm in Errachidia region and less than 75 mm in the Tafilalet plain. The annual average of temperature ranges from 15.2 °C in the upstream part of the basin to 21.5 °C in the downstream part. The maximum temperature rises above 42 °C in summer and drops to as low as 0.5 °C during the winter. The potential evaporation increases from 2,700 mm/year in Errachidia to 4,500 mm/year in Taouz in the south (MIDGCL, 2015).

During the last decade the water table has been drawn down because the lack of groundwater replenishment under these arid conditions.

Figure 1. Location map of the study area, showing the Ziz Basin

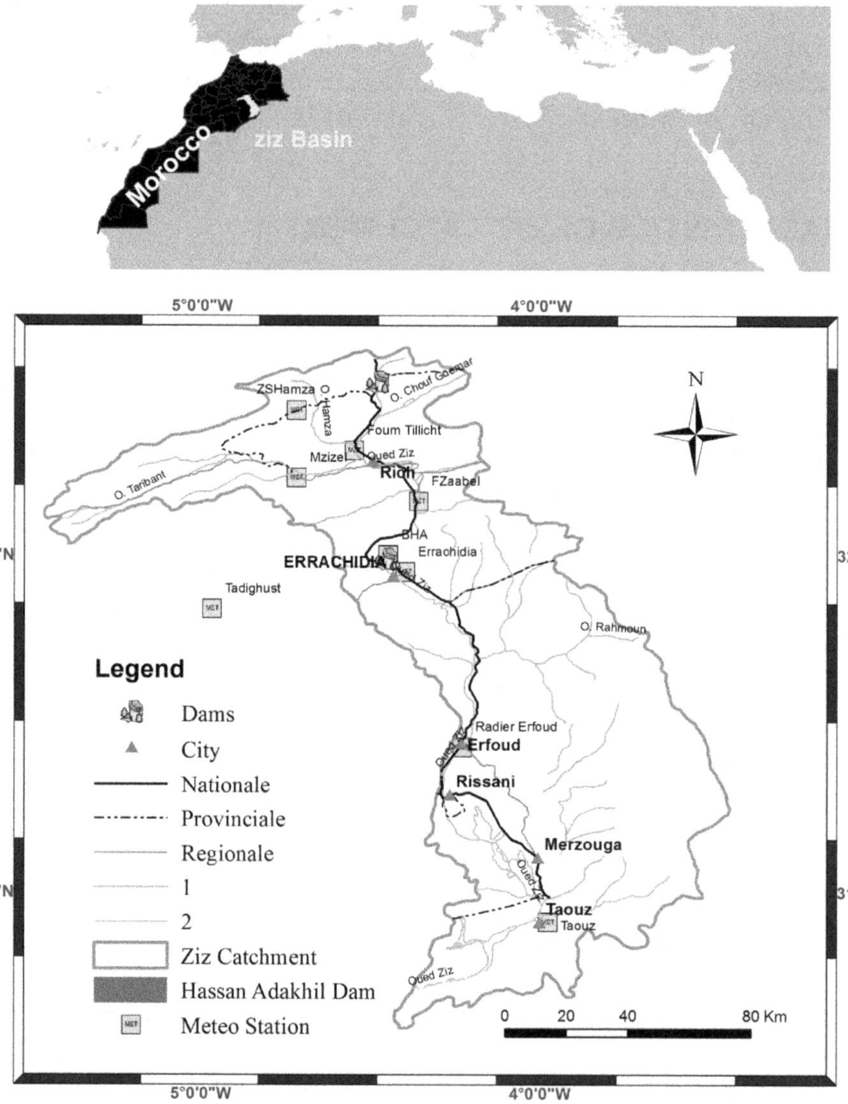

Ziz Wadi is the main surface water system in the basin; it flows from the north to the south and its flow rate is maintained by several tributaries derived from High Atlas Mountains as well as by Meski spring in the central portion of the basin (Figure 1). In order to control and regulate the surface water flow, the Hassan Addakhil Dam was built in 1971 with a capacity of 380 Million cubic meter (Mm^3). Since the beginning of operation of the dam in 1971, the drainage of the natural ecosystem has changed completely.

Based on inflows to the reservoir, water is released through canals only three to four times per year, typically during the summer. Each release occurs for a period of about 20 days. The annual outflow from the reservoir is about 105 Mm^3, from which 85% is used for irrigation while15% is lost due to evaporation and leakage(DRH-GRZ 2008). In the MZV and Tafilalet plain, the overall potential of surface water is about 154 Mm^3, which includes the reservoir input with 84 Mm^3, the contribution of the intermediate basin (including springs) with 55 Mm^3, and an external source of 15 Mm^3 transferred via a small diversion dam from the neighboring basin to the west, the Rheris basin. The diverted portion from the Rheris basin, compensates the evaporation losses at the reservoir (DRH-GRZ 2008).

Geology and Hydrogeology of the Study Area

Several studies have discussed the geology of the region (Choubert and Faure-Muret, 1962; Margat, 1962; Michard, 1976; Ruhard, 1977; El Ouali, 1999). The geology varies substantially between the three geographical sub-basins of the Ziz (Figure 2): (1) the UZB is composed of the Jurassic limestone of High Atlas Mountains extending from the southern flank of the Mountains to the major South Atlas Fault in the piedmont; (2) the MZB is composed of Cretaceous and Cenozoic formations from the South Atlas fault to Jbel Erfoud; and (3) the LZB exposes Paleozoic formations outcropping to the southern part of the studied basin. The Jurassic series outcropping in the UZB is formed mainly by limestone and marl formations (Margat, 1962). The Triassic formations are composed of dolerite, basalts and evaporitic marls and clays, which play an important role as an impermeable substratum for the Jurassic series. Within the MZB, the Cretaceous formations correspond to an asymmetric and tabular syncline and are composed of a series of alternating sedimentary layers from the Paleozoic to the Tertiary range. The Upper Jurassic and the Lower Cretaceous formation, called "Infra-Cenomanian or Continental intercalaire" (100–500 m thick) is composed of sandstones at the bottom, followed by the Cenomanian clays and marls (50 m depth) and Turonian limestone (40–100 m depth).

These limestone formations plunge from the tabular outcrops in the south toward the north where they show a subvertical dip. In the northern part, they are sometimes reversed, affected by faults and are in contact with the Jurassic limestone of the High Atlas (Margat, 1962; DuDresnay, 1979; CAG, 1988; Jossen and Filali, 1989). The top of the Turonian displays an anticline ridge at the north, where the thickness of the Turonian is 100 m. In some parts of the Senonian formation, gypsum occurs in sandstones and clayey sands (50–120 m depth).

Figure 2. The scheme of the methodological steps
Source:(Bouguerra & Jebari, 2017)

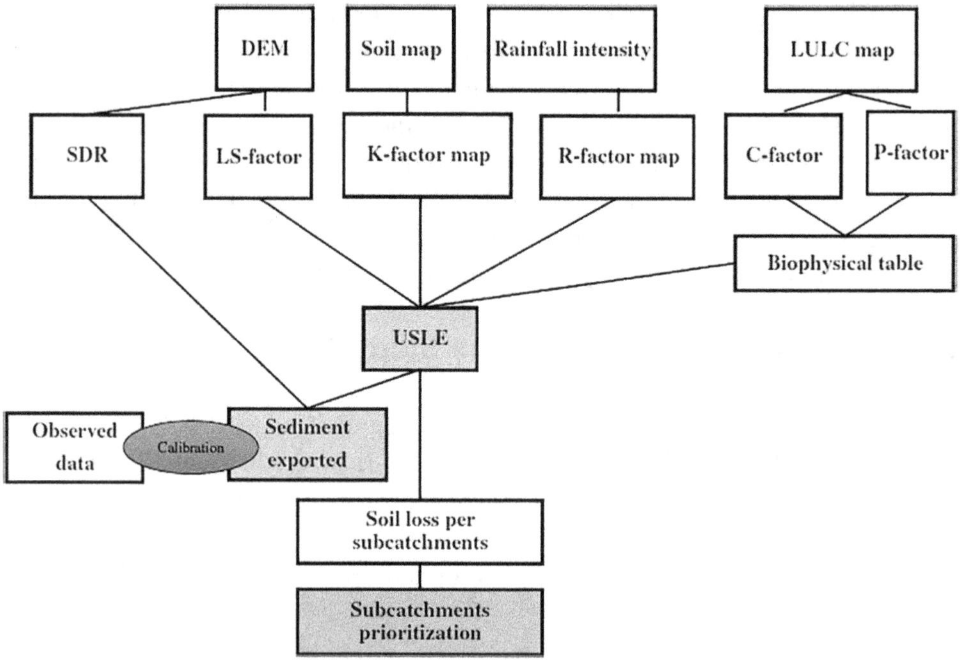

MODEL DESCRIPTION

In many countries, various efforts have been made to reduce soil loss and conserve soil and water quality. Some of these attempts involve the development of models for predicting the amount of soil loss or sediment yield such as; Universal Soil Loss Equation Model (Renard et al., 1997; Wischmeier and Smith, 1978), Water Erosion Prediction Project model (Haregeweyn et al., 2013), Water and Tillage Erosion Model/Sediment Delivery Model (Van Oost et al., 2000; Van Rompaey et al., 2001), Korean Soil Loss Estimation Model (Park, 2017) and so on. USLE (Universal Soil Loss Equation) (eq:1) and RUSLE (Revised Universal Soil Loss Equation) are well-known empirical models for estimating the long-term global average annual soil loss (Renard et al., 1997; Wischmeier and Smith, 1978). Over the last 40 years, the (R)USLE model has been widely used for predicting the average rate of soil erosion from arable land. Pandey et al., (2016) showed that 21 of 50 soil erosion and sediment yield models were adopted input parameters from the USLE model.

Sediment Delivery Ratio (SDR)

The sediment delivery module is a spatially explicit model working at the spatial resolution of the input digital elevation model (DEM) raster (Figure 3).

For each pixel, the model first computes the amount of annual soil loss from that pixel, and then computes the sediment delivery ratio (SDR), which is the proportion of soil loss actually reaching the stream. Once sediment reaches the stream, we assume that it ends up at the catchment outlet, thus no in-stream processes are modeled. This approach was proposed by Borselli et al. (2008) and has received increasing interest in recent years (Cavalli et al., 2013; López-vicente et al., 2013; Sougnez et al., 2011).

Differences Between the Invest SDR Model and the Approach Developed by Borselli et al., (2008)

The InVEST SDR model is based on the concept of hydrological connectivity, as parameterized by Borselli et al., (2012). This approach was selected since it requires a minimal number of parameters, uses globally available data, and is spatially explicit. In a comparative study, Vigiak et al., (2012) suggested that the approach provides: "(i) large improvement in predicting specific sediment yields, (ii) ease of implementation, (iii) scale-independency; and (iv) a formulation capable of accounting for landscape variables and topology in line with sedimentological

Figure 3. Conceptual approach used in the model. The sediment delivery ratio (SDR) for each pixel is a function of the upslope area and downslope flow path.
Source: Sediment delivery ratio of InVEST (Sharp et al., 2016, 149).

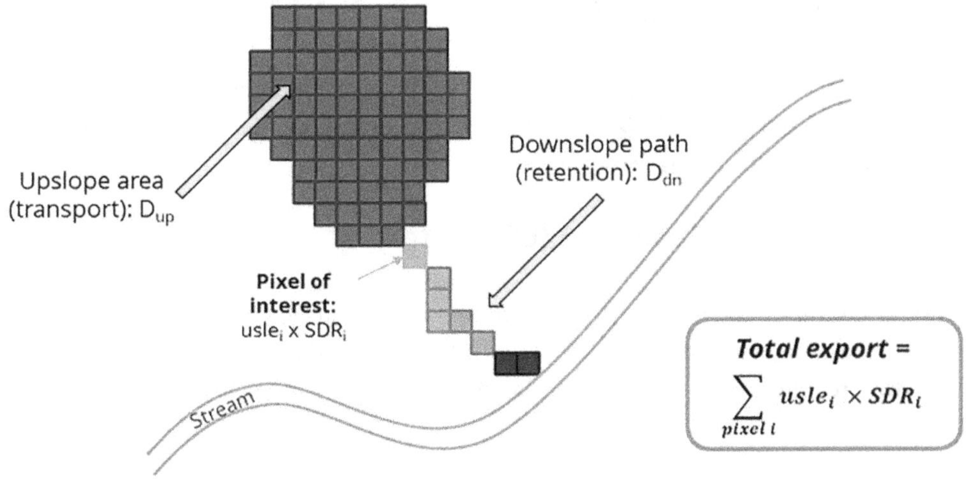

connectivity concepts". The approach has also been used to predict the effect of land use change (Jamshidi et al., 2013).

The following points summarize the differences between InVEST and the Borselli model:

- The weighting factor is directly implemented as the USLE C factor (other researchers have used a different formulation, e.g. roughness index based on a high-resolution DEM (Cavalli et al., 2013),
- The *SDRmax* parameter used by Borselli et al. is set to 0.8 by default to reduce the number of parameters. Vigiak et al. (2012) propose to define *SDRmax* as the fraction of topsoil particles finer than coarse sand (<1 mm).

The average amount of annual soil loss is calculated by the universal soil loss equation:

USLEi =(R×K×LS×C×P)i (eq:1)

Where, R is the rainfall erosivity (MJ mm (h.hr)−1), K is the soil erodibility, LS the slope length gradient factor, C is the cover management factor, and P is the support practice factor (Renard et al., 1997).

The sediment load from a given pixel i, E (ton ha−1 year−1) is given by (eq.2):

Ei=usleixSDRi (eq:2)

The sediment delivery ratio for a pixel i is then directly derived from the conductivity index IC using a sigmoid function (Vigiak et al., 2012).

Digital Elevation Model (DEM)

This is a digital representation of elevations for the study area as raster format with a resolution of 30 meters. DEM for Ziz basin was downloaded from the site: https://dwtkns.com/srtm30m/ (Figure 4).

The Topographic Factor LS (LS-Factor)

Topography plays an essential role as an engine of water erosion. Indeed, the soil losses rates climbs increasingly that the slopes and the slopes length amount. The Most recent studies of water erosion by RUSLE model, using the equation

Figure 4. Digital elevation model (DEM) of ziz Basin
Source: Authors

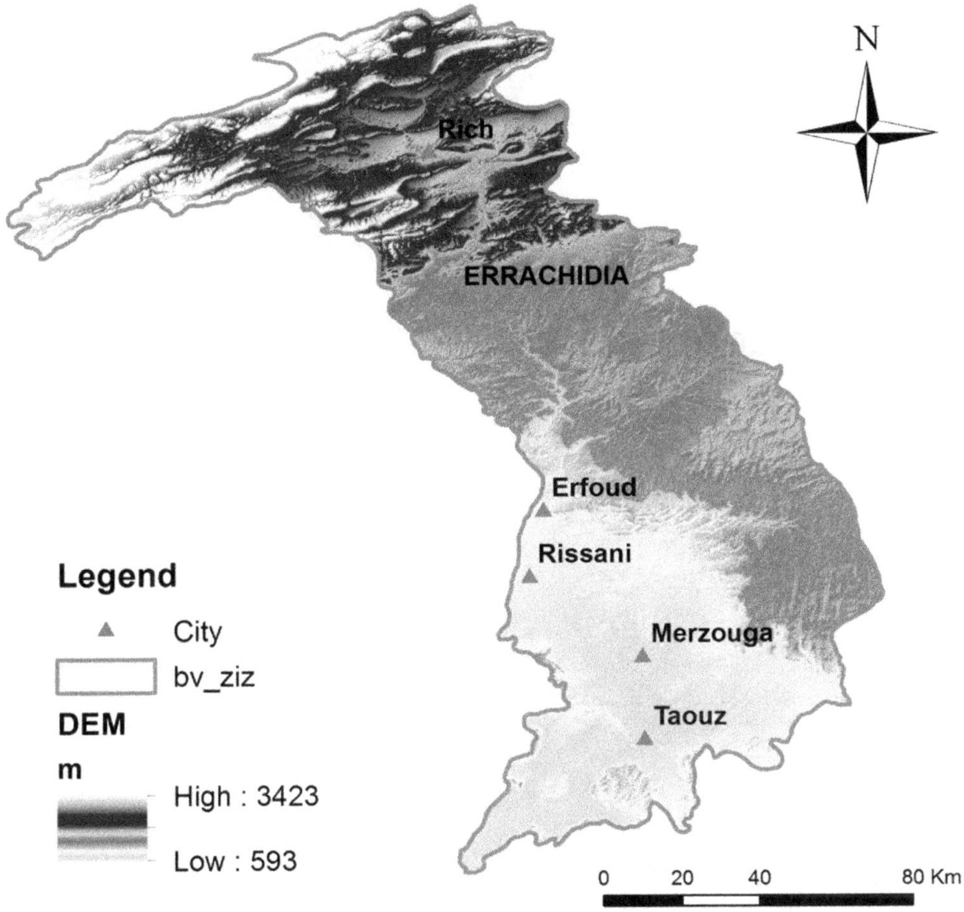

of Mitasova et al., (1996), hat depends on the slope gradient, flow direction and flow accumulation. The LSi factor used in this study was given from the method developed by Desmet and Govers, (1996).

Calculations of the LS factor of the Ziz watershed are carried out directly by the model (SDR) (Figure 5). It uses the digital elevation model in LS calculations and other parameters (slope, orientation, commutative slope length, etc.) by the method developed by Desmet and Govers (1996) for the surface of two dimensions.

Figure 5. The topographic factor LS (LS-Factor)
(http://data.naturalcapitalproject.org/nightlybuild/invest-users-guide/html/sdr.html)

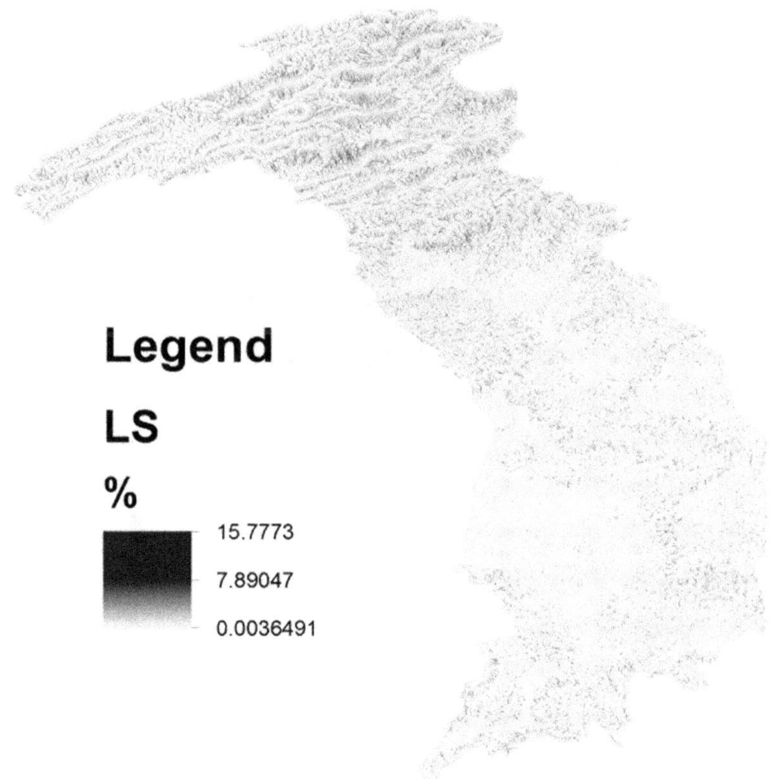

Soil Map / K-Factor Map

Soil erodibility, K, is a measure of the susceptibility of soil particles to detachment and transport by rainfall and runoff. This factor depends on structures and soil texture and organic matter percentage. The resistance of a soil to the disintegration and transport of its constituent elements depends on its structure and its content of organic matter and calcium. Soils rich in humus and clay offer good resistance to erosion because of their structural stability. Coarse-textured soils limit erosion by high permeabilities, but when the intensity of the rain exceeds the soil's adsorption capacity, these soils become vulnerable because they are poorly structured and weakly cohesive (Boukheir, 2001).

For the Ziz basin, it was calculated applying the equation developed by Wischmeier & Smith, (1978) (eq.3).

100K=2.1M1.14 .10-4(12-OM)+3.25(S-2)+2.5(P-3)(eq.3)

Where, M is calculated by the formula:

M=(% fine sand+silt)×(100-clay)(eq.4)

OM is the percentage of organic matter,
S is the permeability index, P is the structure index
The erodibility of soils represents their resistance to degradation by the kinetic energy of raindrops or runoff. Estimated K values for lithology groups range from 0.016 to 0.034 t.ha.h / ha MJ mm. The erodibility map (Figure 6) shows that the most erodible terrain is in the UZB part, upstream of the basin and some area of the LZB, where the land is dominated by the limestone, sandstone, and clayey formations tending to erosion. In the MZN and LZB parts of the basin, the vast majority of the area has a relatively low erodibility coefficient.

Rainfall Erosivity Factor (R)

The original equation of (R) uses the kinetic energy of the rain and requires measurements of rainfall intensity (Wischmeier & Smith, 1978). This direct method of Wischmeier and Smith can only be applied in areas equipped with autographic recorders.

An alternative formula developed by Rango & Arnoldus (1987)involves only annual and monthly precipitation to determine the R factor.

The Rango & Arnoldus (1987) formula (eq.5) was applied to seven stations in the ZIZ catchment.

logR=1.74×logPi2P+1.29 (eq.5)

The annual and monthly precipitation was recovered from 7 weather stations for 30 years. R values were calculated for the selected stations (Table 1). They were interpolated over the whole watershed using geostatistic model (kriging).

Table 1. Distribution of upstream and downstream costs of erosion in the Ziz catchment

Basin /Dam	Basin Area	Dam Retaining Capacity	Annual Silting	Downstream Losses	Areas at Risk	Specific Degradation	Upstream Losses	Total Losses
	Km2	*Mm3*	*Mm3/an*	*MDH*	*% total*	*t/ha/an*	*MDH*	*MDH*
Ziz / Hassan Eddakhil	4400	343	1,17	34	85%	3,99	79	113

Figure 6. K factor values

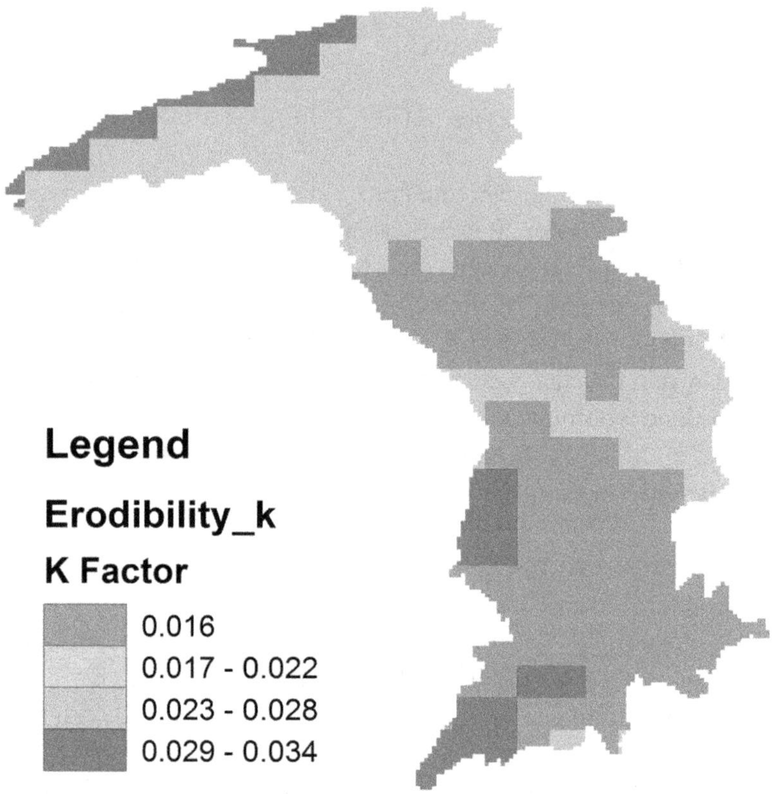

The calculation and mapping of the spatial distribution of the rain erosivity, factor R, (Figure 7) shows that the Ziz catchment is subject to significant rainfall aggression through varied decrease distribution of rainfall erosivity from the downstream to the upstream. Thus, the values of R vary between 54.97 and 293.23 Ml.mm/ha.yr, with an average established at 116.22 MJ.mm/ha.h.yr.

LULC Map

The land cover map was extracted using space remote sensing technology, as it was established through the processing of Landsat OLI8 satellite imagery. The classification conducted by the method of maximum likelihood established on the image processing tool ENVI (4.4).Map has been downloaded from this links: http://glovis.usgs.gov/.

Figure 7. Erodibility factor (R).

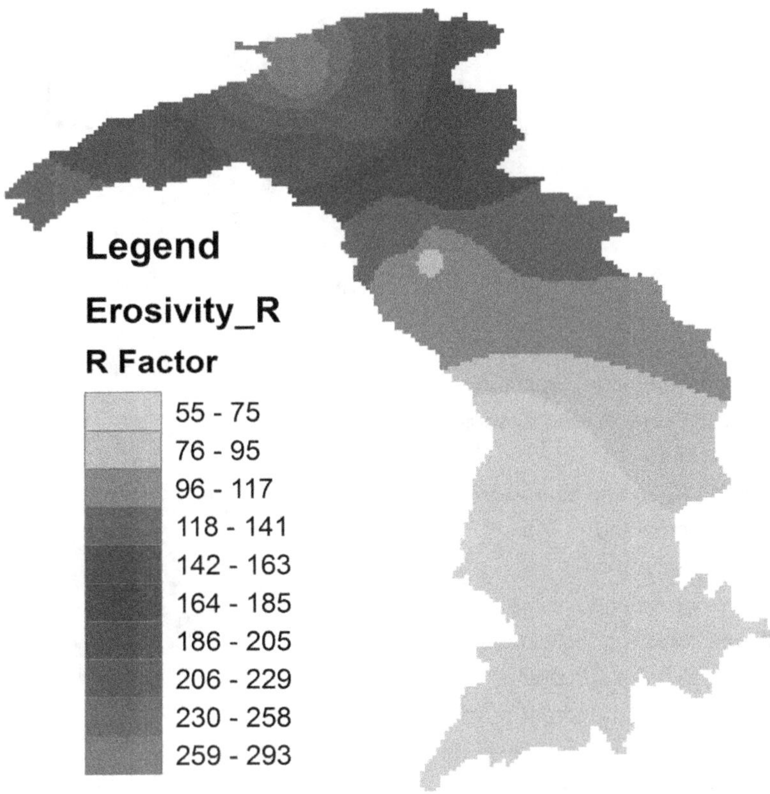

The land cover map (Fig. 8) is developed to assign each soil component a Plant Cover Factor (C) value and an Anti-erosive practices (P) factor value, in the biophysical table (data used by the model) containing model information corresponding to each of the land use classes. Each row is a land use/land cover class and columns should be named and defined (Table 2).

C- Factor

The C-factor represents the effect of soil-disturbing activities, plants, crop sequence and productivity level, soil cover and subsurface biomass on soil erosion (Wischmeier and Smith, 1978). It is defined as the ratio of soil loss from land cropped under specific conditions to the correspondingloss from clean-tilled, continuous fallow

Figure 8. The land cover classes map

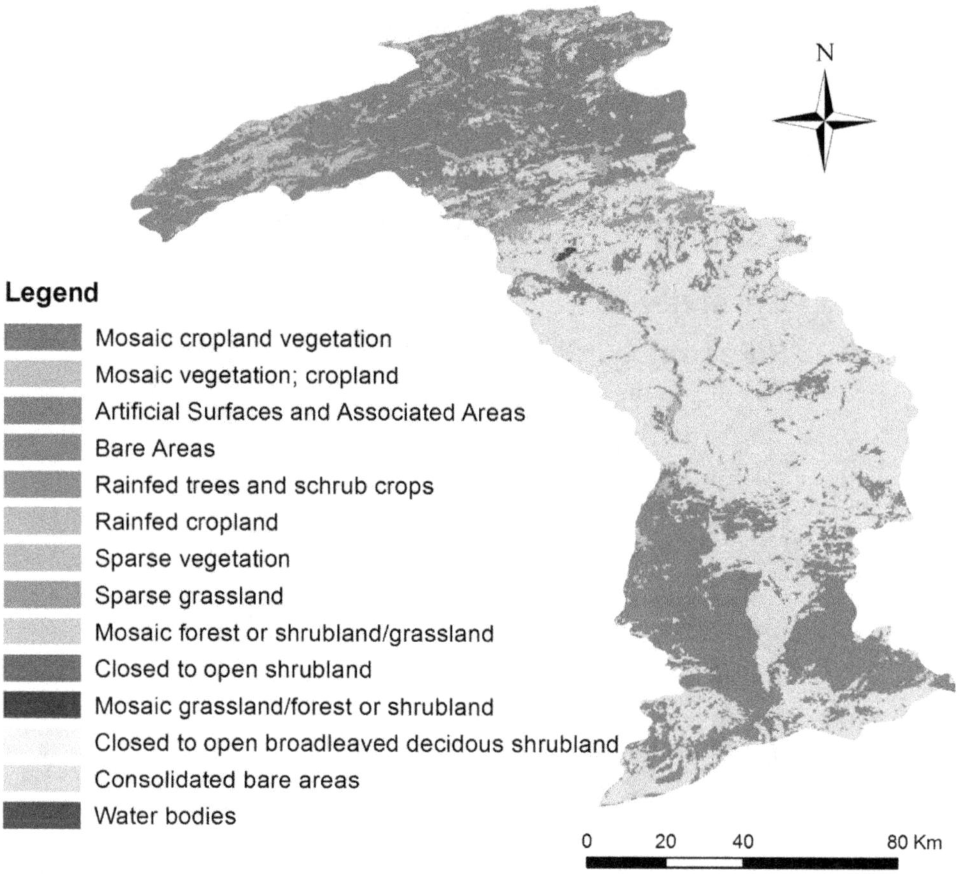

Table 2. Physical and morphological characteristics of sub-watersheds of Ziz

Catchment	Station	Oued	X (m)	Y (m)	S(Km2)	Perimeter (Km	Thalweg (Km	K Compacity	Average Altitude (m	Average Slope (%)
Ziz	Zaouia	Sidi Hamza	564 412	204 407	102	45	17	2.93	2 238	18.68
Ziz	Tillicht	Sidi Hamza	579 979	192 269	1309	199	55	2.46	1 899	10.76
Ziz	M'zizel	Ziz	560 057	185 134	1190	201	92	7.21	2 112	14.71
Ziz	Zaâbel	Ziz	597 400	174 478	4012	404	152	5.86	1 839	11.07
Ziz	Barrage HA	Ziz	615 284	103 868	4417	560	183	7.6	1 808	11.12
Ziz	R.Erfoud	Ziz	615 284	103 869	8106	662	262	8.48	1 482	7.55
Ziz	Taouz	Ziz	632 334	36 133	13480	939	365	10.42	1 247	5.48

Source: (http://www.abhgzr.ma/)

(Wischmeier and Smith, 1978). Currently, due to the variety of land cover patterns with spatial and temporal variations, satellite remote sensing data sets were used for the assessment of C-factor (Karydas et al., 2009; Tian et al., 2009). The Normalized Difference Vegetation Index (NDVI), an indicator of the vegetation vigor and health is used along with the following formula (eq. 6) to generate the C-factor value image for the study area (Zhou et al., 2008; Kouli et al., 2009).

$$C = \exp\left[-\alpha \frac{NDVI}{\left(\beta - NDVI\right)}\right] \quad (6)$$

Where, a and b are unitless parameters that determine the shape of the curve relating to NDVI and the C-factor. Van der Knijff et al., (2000) found that this scaling approach gave better results than assuming a linear relationship and the values of 2 and 1 wereselected for the parameters a and b, respectively. This equation was successfully applied for assessing the C-factor of areas with similar terrain and climatic conditions (Prasannakumar et al., 2011a, b). The values assigned to each land use are recorded in Table 2; it is ranging between 0.69 and 0.75.

P- Factor

Land use/land cover map was used as a tool in the allocation of C and P factors for each land use/land cover class. The P factor depends on the type of conservation measures implemented in slope classes. Contour ridges dominate conservation measures. Lands with no conservation measures are assigned a P factor value of 1 (Table 3).

Average Annual Soil Loss of Ziz Basin (USLE)

The multiplication of the different factors using the InVEST SDR model translates the universal soil loss equation in its revised form (RUSLE). This procedure provides the amount of eroded soil in each pixel (30m× 30m) over the entire catchment area and on each sub-catchment initially introduced considering that, only a part of the eroded particles finds their way to the outlet and constitute the stream sediments (Bouguerra & Jebari., 2017).

Table 3. C and P factors values for each land cover class

LU_DESCP	LUCODE	USLE_C	USLE_P
Rainfed_croplands	1	0.706386	0.25
Rainfed_shrub	2	0.71026671	0.25
Mosaic_cropland	3	0.71766498	0.5
Mosaic_vegetation	4	0.73231917	0.8
Mosaic_forest	5	0.72732177	0.9
Closedto_open	6	0.68542919	0.92
Sparse_vegetation	7	0.75110116	1
Sparse_ grassland	8	0.72043154	1
Artificial_surfaces	9	0.6966307	1
Bare_areas	10	0.77054218	1
Consolidated_bare_areas	11	0.72709653	1
Water	12	0.7303	0.001

The spatial maps, the biophysical table, and the InVEST calibration parameter were integrated in InVEST SDR model. The model was calibrated using the K-parameter equal to 2. Results were analyzed in Arc GIS module. Figures9 and 10 present annual erosion map of the study area. It helps in the identification of vulnerable areas. Table 4indicates the spatial distribution of different erosion classes.

The mean annual soil loss estimated for the entire watershed is 10.32 ton/ha/year. Soil loss presents the amount of eroded soil. Generally, soil losses vary between 0 and values above 23.73 t / ha / year spread over the entire study area.

The values of potential erosion have been grouped into 5 classes of values:(1) Very High (between 13.66 and 23.73 t / ha / year), (2) High (between 6.5166 and 10.1065 t / ha / year), (3) Medium (between 3.0255 and 5.3346 t / ha / year), (4) Low (between 1.207 and 1.75 t / ha / year) and (5) very low (less than0.5283 t / ha / year).

DISCUSSION

Water induced soil erosion is a severe problem in the Ziz Basin region as demonstrated by the analysis of sedimentation in the reservoir Hassan Dakhil, located in the middle of the catchment (MZB) (Messouli et al., 2008).

Figure 9. Soil loss maps in ton/ha/year

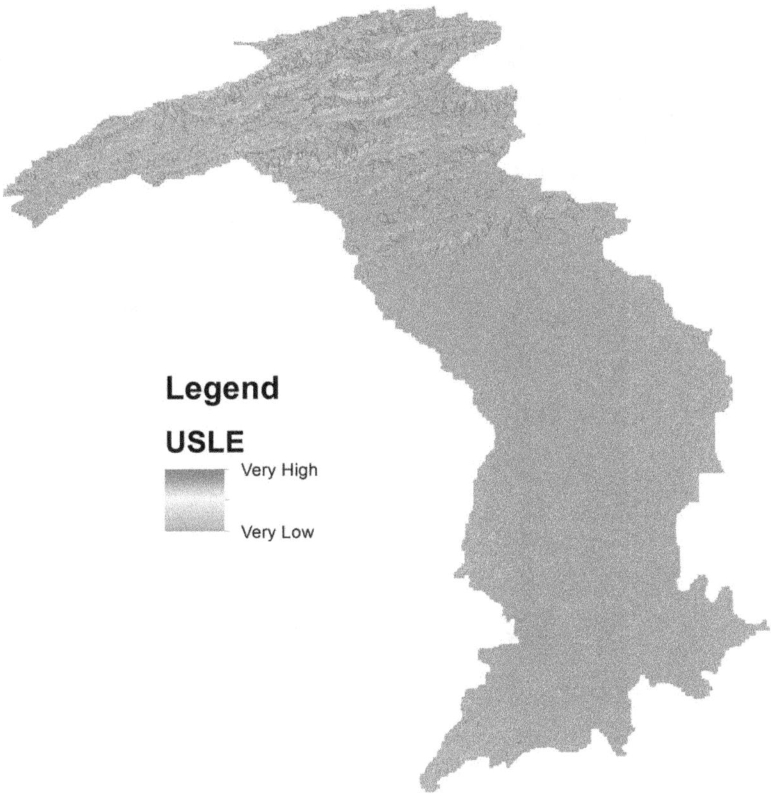

If the rapid dam infill due to high sediment transport into the rivers continues at the present rate, the dam will no longer be fully operational for irrigation by the year 2050 Ben Salem et al., 2011).

Increase in sediment clogging and in clogging intensity is currently observed, related to human activities. Xerophytes are widely exploited as a source of domestic energy near the grouped or scattered houses, thus contributing to the expansion of desertification (Messouli et al., 2008). Livestock farming is dominated by goats and sheep which have increased along with the population. The resulting increase in grazing pressure has caused a decrease in the vegetation density and some changes have occurred in the vegetation diversity in both the lowlands and the highlands (Messouli et al., 2008).

The role of vegetation is crucial. The average C value in the watershed is 0.5. This reflects the modes of space use characterized by both a dominance of agriculture that less protects the soil as well as the scarcity of the forest domain which well protects the soil both by its root system and its canopy.

Figure 10. Soil loss maps class /sub-watershed

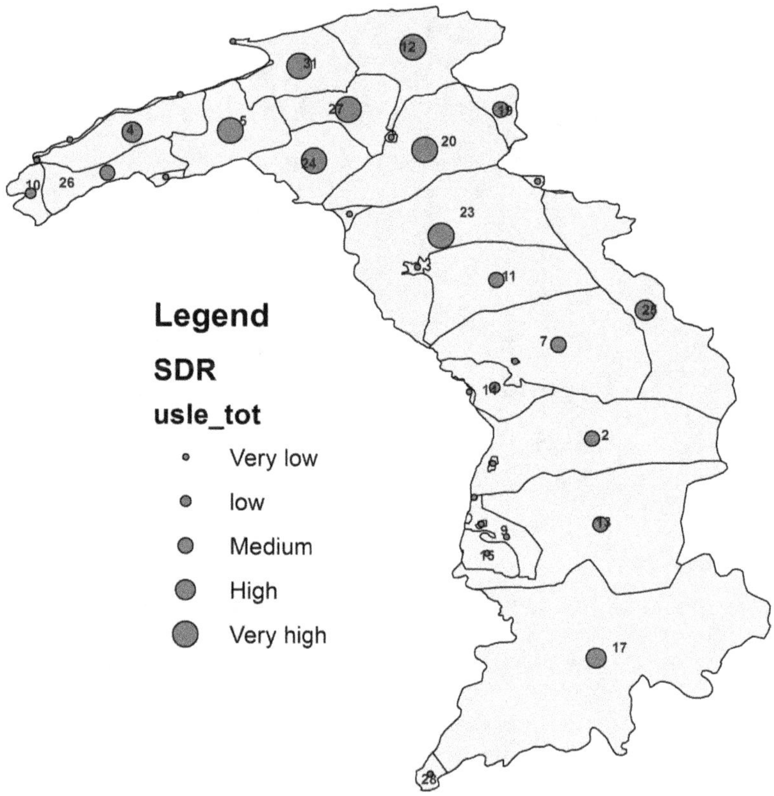

The rate of erosion is generally considered moderate in most of the watershed with spatial variability. Moreover, it appears from the analysis of the potential erosion map that the majority of produced sediment comes from UZB and MZB sub-catchment whilst the minimum sediment produced is generated in LZB sub-catchment.

In general, USLE is used to estimate average annual soil loss, the use of remote sensing and GIS allows us to spatialize the potential risk of erosion.

Agricultural systems require a higher resilience to cope with climate change, both in case of excess water (due to heavy rainfall) and lack of water (due to prolonged droughts).

The key element to overcoming these two problems is soil organic matter, which improves and stabilizes the soil structure so that it can absorb larger amounts of water without causing surface runoff, which could lead to soil erosion.

Soil organic matter also improves the water absorption capacity of the soil for during extended drought (Ben Salem et al., 2019).

Table 4. Soil loss maps in ton/ha/year and in ton/sub-watershed and Rank of priority

ws_id	Priority	Communes	Sub Catchment	usle_tot t/ha/ year/	Loss of Soil
12	1	CR EN-NZALA	UZB	23.7321	Very High
31	2	CR ZAOUIAT SIDI HAMZA	UZB	22.0287	
20	3	CR GUERS TIAALLALINE	UZB	17.834	
23	4	CR LKHENG	MZB	17.1963	
5	5	CR AMOUGUER	UZB	14.9323	
27	6	CR SIDI AAYAD	UZB	14.9221	
24	7	CR M ZIZEL	UZB	13.663	
4	8	CR AIT YAHYA	UZB	10.1065	High
17	9	CR ET-TAOUS	LZB	8.3391	
25	10	CR OUED NAAM	LZB	6.5166	
26	11	CR OUTERBAT	UZB	5.3346	Medium
7	12	CR AOUFOUS	LZB	4.6763	
2	13	CR AARAB SEBBAH ZIZ	MZB	4.6131	
11	14	CR CHORFA M DAGHRA	MZB	3.5288	
13	15	CR ER-RISSANI	LZB	3.2082	
19	16	CR GOURRAMA	UZB	3.0255	
10	17	CR BOU AZMOU	UZB	1.7523	Low
14	18	CR ER-RTEB	MZB	1.207	
8	19	CR ASSOUL	UZB	0.5283	very Low
29	20	CR TADIGHOUST	MZB	0.3324	
9	21	CR BNI M HAMED SIJELMA	LZB	0.3147	
15	22	CR ES-SFALAT	LZB	0.2879	
28	23	CR SIDI ALI	UZB	0.2265	
21	24	CR GUIR	MZB	0.1313	
33	25	MU ER-RICH	UZB	0.1049	
22	26	CR IMILCHIL	UZB	0.0457	
34	27	MU ERRACHIDIA	MZB	0.0433	
18	28	CR FEZNA	MZB	0.029	
32	29	MU ARFOUD	MZB	0.0103	
35	30	MU MOULAY ALI CHERIF	LZB	0.0094	
1	31	AC AOUFOUS	MZB	0.0047	
16	32	CR ES-SIFA	LZB	0.0016	
6	33	CR ANEMZI	UZB	0.0009	
3	34	CR AGOUDIM	UZB	0.0007	
30	35	CR TOUNFITE	UZB	0	

CONCLUSION

This study represents a first attempt at applying and testing the InVEST-SDR model as a means to address water and soil erosion in the Ziz basin.

The watershed of Ziz suffered severe climatic aggressiveness with an average of 116.22 $MJ{\bullet}mm{\bullet}ha^{-1}{\bullet}h^{-1}{\bullet}yr^{-1}$. The weighted average value of k for the watershed is 0.023, confirming the susceptibility of soils to erosion in this watershed.

Increases in sediment yield are observed in many places in the upper and middle Ziz, dramatically affecting water quality and reservoir management. The InVEST Sediment Delivery Ratio (SDR) model and its application proved useful for catchment and sub-catchment scale at arid and semiarid areas. Results confirm that soil erosion is a severe problem in the Ziz basin region as demonstrated by the analysis of sedimentation in the reservoir Hassan Dakhil, located in the middle of the catchment found by Messouli et al., (2008).

Critical sub-catchments were identified and prioritized based on average annual soil loss. These areas need conservation measures for sustainable use of lands and BHD dam.

REFERENCES

Lahlaoi, H., Rhinane, H., & Hilali, A. (2015). Potentialerosion risk calculation using Remote Sensing and GISin Oued El Maleh Watershed, Morocco. *Journal of Geographic Information Systems*, *7*(2), 128–139.

Ministère de l'Intérieur Direction Générale des Collectivités Locale (MIDGCL). (2015). *La région de Drâa Tafilalet, Monographie générale*. Author.

Pandey, A., Himanshu, S. K., Mishra, S., & Singh, V. P. (2016). Physically based soil erosion and sediment yield models revisited. *Catena*, *147*, 595–620. doi:10.1016/j.catena.2016.08.002

Park, Y. S. (2017). Development of Korean soil loss estimation model. *Geoenviron. Eng.*, *18*, 14–17.

Plan Bleu. (2003). *Les menaces sur les sols dans les pays méditerranéens Etude bibliographique*. Author.

Renard, K. G., Foster, G. A., Weesies, G. A., McCool, D. K., & Yoder, D. C. (1997). *Predicting Soil Erosion by Water: A Guide to Conservation Planning with the Revised Universal Soil Loss Equation (RUSLE)*. U.S. Department of Agriculture, Agricultural Research Service.

Rhazi, Essahlaou El Hmaidi, & El Ouali. (2017). Etude De La Vulnérabilité Aux Risques D'inondations Dans La Ville De Meknès. Apport Des SIG, Du MNT et des Modèles Empiriques. *European Scientific Journal, 13*. doi:10.19044/esj.2017.v13n36p102

Roose, E-O. (1991). Conservation des sols en zones méditerranéennes. *Synthèse et proposition d'une nouvelle stratégie de lutte antiérosive: la GCES-Cahiers ORSTOM, Série Pédologie, 26*(2), 145-181.

Tahiri, M., Tabyaoui, H., & El Hammichi, F. (2017). *Quantification of water erosion and sedimentation using empirical models in the Tahaddart watershed*. Northwestern Rif.

Van Oost, K., Govers, G., & Desmet, P. (2000). Evaluating the effects of changes in landscape structure on soil erosion by water and tillage. *Landscape Ecology, 15*(6), 577–589. doi:10.1023/A:1008198215674

Van Rompaey, A. J., Verstraeten, G., Van Oost, K., Govers, G., & Poesen, J. (2001). Modelling mean annual sediment yield using a distributed approach. *Earth Surface Processes and Landforms*, *26*(11), 1221–1236. doi:10.1002/esp.275

Wischmeier, W. H., & Smith, D. D. (1978). *Predicting Rainfall Erosion Losses - A Guide to Conservation Planning*. Hyattsville, MD: USDA, Science and Education Administration.

Chapter 8
Hydrologic Modeling Using SWAT:
Test the Capacity of SWAT Model to Simulate the Hydrological Behavior of Watershed in Semi-Arid Climate

Zineb Moumen
University of Sidi Mohamed Ben Abdellah of Fez, Morocco

Soumaya Nabih
University of Sidi Mohamed Ben Abdellah of Fez, Morocco

Ismail Elhassnaoui
University Mohammed V, Morocco

Abderrahim Lahrach
University of Sidi Mohammed Ben Abdellah of Fez, Morocco

ABSTRACT

The Innaoune Watershed represents an important hydric potential of the oriental part of Morocco. However, the basin exhibits a set of hydrologic drawbacks, such as floods, erosion, and pollution. This chapter is focused on flood forecast study. In order to help managers and decision makers to adopt the appropriate land management strategies for protecting the population from flood damages, the study of the hydrological behavior and quantification of water yield are paramount. According to this perspective, the main goal of this chapter is to test the ability of the SWAT model to simulate and reproduce the hydrological behavior of the upstream of Innaouene Watershed. The output of the model could be used to map, delineate, and forecast the floods expansion for a particular rainfall event. SWAT was performed on a daily time step from 2004 to 2012 for calibration and 2012 to 2014 for validation. The model accuracy was evaluated by measuring the Nash-Sutcliffe coefficient and R^2.

DOI: 10.4018/978-1-5225-9771-1.ch008

INTRODUCTION

Over the history of Earth development climate has known many changes, the end of the ice age 7000 years ago, marks the beginning of a new climate cycle (IPCC, 2007). Population growth and pollution combined withclimate change, are some of the main issues that decision makers face when managing the most important resource, which is water, especially in developing countries like Morocco, indeed the country has a complex climate variability characterized by a long period of drought, extreme evapotranspiration, and early or late rains (FalhiS.& al., 2016). The main sequels of these changes are; changes in precipitation patterns, excessive drought and heat waves, the hurricanes will become more devastating, and the sea level will rise 1-4 feet by 2100 (IPCC, 2013). Similar deviations will have a harmful impact on ecosystem balance generally, and human life specifically. According to the fifth report of the Intergovernmental Panel of Climate Changes (IPCC), a decrease of up to 20% in rainfall is predicted by the end of this century, and the increase in temperature is expected to reach 2.5 °C to 5.5 °C under the same scenarios (IPCC, 2015).

Not to mention the availability of water has decreased from 3500m^3/person/the year 1960 to 1000m^3/person/year in 2000, a drop of 420 m^3/person/year in 2020 is predicted, and therefore the water demand will increase in the future (IPCC, 2014).

The three major sources of water pollution in Morocco are; municipal wastewater discharge, industrial effluents, and agricultural activities, urban area generate around 500 m^3/year of wastewater, and it may reach 900 m^3 in 2020. The high salinity and nitrate concentration affect the underground water, streams are generally contaminated by phosphorus, ammonia, organic matters, and high coliform counts. This polluted water is directly discharged to a natural water body, for example, Sebou watershed which constitutes 29% of Moroccan water resources is heavily polluted by nitrates, phosphorus and pesticide residues (EnnabliM. &al.,2017).

Even though Innaouene watershed presents an important hydric potential of the oriental part of Morocco, because of its geomorphological situation and its karstic reservoir, the basin faced a set of problems such as floods, drought and water quality.

In the last decay, a devastating flood caused important humans and materials damages, in 1973, a hydroelectric reservoir was built in order to manage water and mitigate the floods (OECD,2017). However, the area still suffers from several problems, which relay its socio-economic development.

In the other hand, Innaouene River's tributariesareexposed to sewage and margins release. The sequels of such actions are not only limited in the river level, but they alsospread and may affect the groundwater, the reservoir (which is essentially used for irrigation), the neighboring area, and on sanitary and environmental field.

Moreover, the basin did not evade the impact of global climate change, this later present semi-arid climate that has two major phases; dry and humid, therefore, the hydrological cycle does not assure a complete water renewal since evaporation and evapotranspiration are limiting factors.

In order to help manager and decision makers to adopt the appropriate land management techniques, the study of the hydrological behavior and quantification of water resources is paramount. Hydrological models are important tools for managing and planning a sustainable use of water resource to meet various demands. According to this perspective, this chapter remains part of hydrologic simulation using the SWAT model, the study is axed as below;

The first part of the chapter is dedicated to the study backgrounds, the inventories of drawbacks and the potential solution to deal with it, the second part concerns hydrologic modeling using SWAT (Soil and Water Assessment Tools), which is divided into two subpart, preparation of soil attribute table using ISRIC database (ISRIC, world and soil information, 2013) and the calibration and validation part, the last part concerns the result's interpretation as well as the recommendations.

BACKGROUND

Water Is a Fundamental Resource

Water is the most important natural resources on Earth, 97% of water supply is salty, the reminding 3% is frozen in icesor glaciers, or stored in the underground, that leaves only 1% of surface water which is useful to meet everyone needs (Peter al.,1993). From cellular metabolism to a big industry, water is the basic element of human life, indeed, on microscopic level water molecule is required to carry on nutrients, and oxygen to body cells, and transform the food into energy, water is also required to maintain a good life quality and for economic and industrial development (Alaoui M., 2013).

The Hydrologic Cycle

Hydrology treats the circulation, distribution, dynamic and proprieties of water through the study of water cycle components; precipitation, evaporation, transpiration, interception, infiltration, runoff, dissolved substances transport(PERRIN, 2000).

Precipitation is the process by which water falls directly into land surface, it may take many forms, liquid (rain), frozen (snow or hail) or vapor (mist and fog), the rain that falls into the oceans become a part of them, the part that falls into rivers and streams constitute the stream flow and the part that falls on the overland and vegetation forms the intercepted rain (PERRIN, 2000), this later may be defined as the temporary storage of water by vegetation (ward, 1990), in this section three categories are distinguished;the throughfall which is the water directly evaporated or dripped by vegetation, steam flow is water that flows along twigs, and three branches and interception loss is the precipitation that is retained by plant absorbed, and evapotranspiration (ward, 1990).The spatial distribution of precipitation varies according to the soil properties, land use, and land cover, during the precipitation events, water flows rivers and land surfaces, if the soil is permeable (sandy or gravelly) the water infiltrate, the gravity force pulls the water down, and through cracks, fractures and pores, it goes further to the deep underground aquifer, the water continue moving sideways and laterally toward rivers, lakes, and oceans. (USGS, 2016).

Evaporation is the transformation of water liquid to water vapor, the process is powered by the heat of sun energy, latent heat of evaporation of water rise with water temperature, and break the intermolecular links (Likens,1981). Water evaporates from wet areas; plant roots draw up groundwater and release it through evapotranspiration (USGS, 2016), the density of water vapor is low, mixing with the suspended particles in the atmosphere, the water condenses into clouds, and after reaching the dew point temperature the water vapor falls again into the Earth surface.

Water Disparity

The global water amount has been estimated by $1386x10^9$ Km^3, lakes and rivers form 225 000 Km^3 of the total amount (table1).

Around 30 percent of the total mass of Earth is covered by land, then the continental area receive just 1/3 of the total precipitation. Due to climatic and geographic characteristics, the total income water is unequally distributed over the world, some regions have a water excess when others suffer from water scarcity. According to the world water development fourth report, Saharan and sub-Saharanregions are more

Table 1. World Water Distribution

Water	Global Water Volume Km3)
Oceans	$1348x10^6$
Polar ice, glaciers and icebergs	$2782x10^4$
Groundwater, soil moisture	$8062x10^3$
Lakes and Rivers	$2250x10^2$

Sources: Baumgartner and E. Reichel, the World Water Balance (Munich: Elsevier, 1975), P.179

exposed to water scarcity stress, 700 million people in 43 different countries suffer today from water shortage. For example, in the in south-western America, there are years in which they may receive no rain, and when it rains the precipitation total amount doesn't exceed 508 millimeters (Miller et al., 1992).

Watersheds, Hydrologic Response, and Flood Hydrographs

Watershed is the topographic area drained by the principal stream network and its tributaries, from the upstream to downstream, all the hydrologic process take places in this section, those later are impacted by the climate, vegetation, topography, pedology, and geology (Yonaba, 2009), watershed's sizes may vary from some fraction of meter square to kilometer squares, its limits are defined from topographic crest line, its hydrogeologic limits are determinate by the expansion of water in deep aquifer.The hydrologic response is the way watersheds behave regarding solicitations,during a precipitation event, before reaching the land surface, the majority of this water is lost to evaporation, transpiration, an interception. The remained portion constitute runoff when the soil moisture reaches a maximum value, the water or infiltrate into the deep aquifer or flows to the outlet.

The flow of a river at determinate point is defined as the variation of the volume of water per time unit, this change is measured in m3/s, the change of this discharge in time unit define the flood hydrograph, they take the shape of a dissymmetric bell, and they have three main components (Maria A, & all, 2015):

- Surface Runoff: Water flowing over the ground surfaces.
- Intermediate Runoff or Interflow: Water which moves laterally beneath the ground surface in the unsaturated zone, and returns directly to the ground surface or stream banks,
- Baseflow: Water drained from saturated zone to the river flow.

Figure 1 shows a hydrograph for a flood event with hyetograph, which is the intensity of precipitation over time, presented in the top of the graph.

t0:The rain started, the water loses are firstly extracted from the total amount,

t0-t1: The rest of rainfall is converted to direct runoff; the discharge rises progressively (rising limb)until the peak flow, which is the maximum streamflow of a rain event. The lag time is the difference between the peak of rainfall and the peak of flow.

t1-t2: The decrease of flow discharge is known as a falling limb and the water return to the baseflow conditions (Maria A, & all, 2015).

Figure 1. Flood hydrograph during a single precipitation event
Source: Maria et al., (2015)

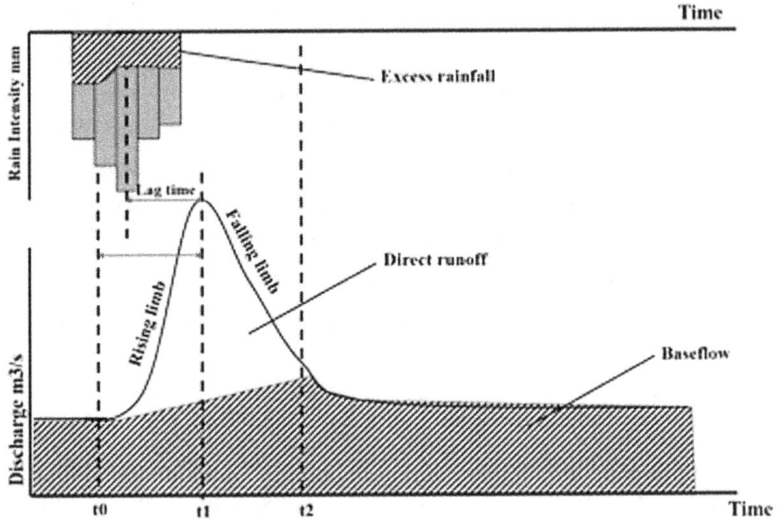

The inundation happens when water covers an area that is normally dry (Australian government), it's amplified by human activities; deforestation and urbanization. In deforested areas, during heavy precipitation, water is not intercepted either absorbed by vegetation, the flow is rapidly vehiculated to the downstream, through the drainage way it erodes soils. The paved roads and the urban areas prevent water from infiltration.The water may overflow rivers also when the rainfall intensity exceeds the infiltration capacity of the ground (Horton & Robert,1933).

Flood Risks

The water availability is the pylon of development, the distribution of this source over the world goes hand by hand with the rate of countries' development. For example, Egyptian civilization was developed near the Nile river, which floods and fertilize the lands. However, the water excess may destroy humanity, if it exceeds the needed rate, a heavily unexpected rain may cause human and economic damages. For instance, in 1993 the Mississippi River and its tributaries overflew and inundated more than 9.3 million of hectares, 50 were dead and the materials damages were estimated by 12 million dollars (Raven & all, 1992). According to Munich Re's natural catastrophe database (NatCatService), in Europe during the last 15 years, flash floods and storms have caused economic damages between 1 and 18 billion US dollars per incident. The Elbe and Danube floods in Germany and other parts of Eastern Europe in 2002 and 2013, caused more than 10 billion US dollars of damages (Munich, 2014), and according to a recent estimate by (Jia et al., 2016), the risk of inundation in China is eight times higher than the global average.The economic damagescan be split into tangible and intangible damage (Smith &Ward,1998),the difference whether the two types is that the first one could be estimated in monetary terms, when the second is not measurable, for example, environmental impacts, health impacts,heritage cultural impact (Meyer & Schwarze, 2019). Such damages place the vulnerable area for inundation under a critical risk, in order to protect humankind, and the environment the foreseeing of floods is mandatory.

Flood Forecast

Forecasting is the estimation of the future conditions at a certain period of time, in order to warn the extreme events such as floods, there are twotypes of flood forecasting methods the short term and the long term (David R.1992).The first one is based on channel routing and rainfall-runoff modeling.

The Channel routing forecast method is a one-dimension simulation of the river represented by St. Venant equation, the models that use this equation, computes the water surface expansion into rivers, and stream from the data flow. HEC RAS model is an example of a dynamic wave model, it was created by the American army in order to simulate and map the overflows, this model was applied in various study for forecasting floods, sediment, and nutrient yield (Khaddor, 2016)(Pierre, & Moussa, 2009) (Base Modelisation et al., 2013)(Abdulrazzak et al., 2019)(Berterreche et al., 2008). Therainfall-runoff modeling or hydrologicmodelssimulate the flow in watershed'sdownstream,usingrain and soilscharacteristics as an input.

Long term forecasting method groups three methods; index-variable, storage accounting and conceptual simulation, the first one relate the forecast period runoff to one or more variable index that is likely to affect stream flow, like soil moisture, the second estimate the quantity of water stored in the catchment which may contribute in the future in the surface run-off, the latest one uses a rain-flow simulation gathered from observed data, from which the forecasting period variable is estimated, which is the study case presented in this chapter. (David, 1992)

MAIN FOCUS OF THE CHAPTER

Issues, Controversies, Problems

As a developing county, Morocco is more exposed to floods risk, the flash floods have affected the different regions of the kingdom with significant economic damageand large loss of life. Namely: The Ourika 1995 flash flood, which caused 732 of people mortality and caused 80 million dirhams of economic damage, Casablanca - Mohammedia 1996, Tetouan 2000, Mohammedia 2002, Errachidia 2006, Al Gharb, Al Haouz, Le Sous 2009 and 2010 and Casablanca 2010 (MEM, 2014). Indeed it is predictedthat the flash inundation maycause 21.000 loss of life per year by 2030 (DARA, 2012).

Each year, the damage impacting essential facilities, government buildings, and infrastructure is estimated by1.4 billion dirhams or 0.6% of the central government's annual budget. For example, floods with a 100 year, could inflict damage to public property beyond 4.4% of the central government's annual budget (world bank, 2012).

Out of a total population of 30 million, 18 million currently live in urban areas, and it is estimated that the urban population will increase by 4 million by 2025. Global climate change should lead to an increase of the intensity and frequency of droughts and floods; this will expose up to 25 million urban dwellers to floods in the Middle East and North Africa and, will result in a 30-50% reduction in water availability, thus exacerbating a significant water shortage (World Bank, 2012).

Given its geographical location, combined with the context of global climate change, the Kingdom of Morocco is facing a climatic trend, characterized by short but intense rain events.

All the above combined with population growth, land use and uncontrolled urban development in inappropriate areas, made hazardous events such as floods become more and more frequent, and the damages generated by these catastrophes are costly. The probabilistic risk assessments for the whole country indicate that floods, earthquakes, and tsunamis cost on average 5.0 billion DH per year, of which floods constitute the largest part (World Bank, 2012).

Through history, Morocco experienced this extreme event in multiple areas, we can site:

- The 25/9/1950 Sefrou, the city was flooded with a water slide of 6m high making a hundred victims,
- On 5/11/1965 Ziz, leaving 25000 inhabitants homeless,

- Moulouya occurred on May 23, 1963, and was so violent that it took away the left bank of the dam Mohammed V (the flood had a peak flow of 7200 m3 / s and a volume of 570 million m3 either the equivalent of the capacity of the restraint),
- Ourika in 1995, El Hajeb in 1997, Settat and Mohammedia in 2002 and Tan Tan, Nador, Al Hoceima and Khénifra in 2003, Sefrou 2008.... and the list goes on.

The determination of the areas likely to be affected by floods is essential for better management of this risk. In Morocco, 391 sites have been listed as flood sites (National Plan of Protection against Floods: PNPI). The recurrence of floods is linked to a return period of flood, which involves frequency calculations on a historical series of flows or on a synthetic series, reconstructed from precipitation measurements via an integrated hydrological model, which is considered a priority for countries to conserve and protect their water resources.

Hydrologic Modeling

The introduction and development of digital tools and approaches, combined with geographical information system advance, allowed to progress in hydrological/ hydraulic models and made the process of understanding the hydrological system better.

Model is a representation of a physical phenomenon in order to understand the mechanisms which control the system, this simulation may be physic, analogic or mathematics.

Hydrologic model is the total or partial representation of the hydrologic cycle (Marchandise, 2007), it describes the system and allows to predict how it may behave in the future if one of its parameters (rain, temperature, pollutants…) is modified. This representation is realized by a set of the mathematic equation which reproduces the system, in general, hydrologic models, presents six types of variables (Boudahiraa,2007).

- **Input Variable**: The set of parameters required by the models, which depend on the time and space (evaporation, precipitation, hydrodynamic characteristics of the study area).
- **Output Variable**: Variables by which the model responds (flow, pollutants concentration…).

- **Stats Variable**: Characterize the state of the simulated systems and could vary according to the time (soil depth, rate of soil saturation, slopes, relative humidity…).
- **Mathematics Equations**: Links Input variables, Output variables, and State variables in order to represent the process implied in the watershed dynamic.
- **Parameters**: Related to the model's equation, serve to adapt the relation which regulates the model to the real observed functionality
- **Calibration Parameters**: Certain parameters could not be directly measurable from the field, the calibration has as scoop the determination of the value of this variable, in order to find the best fit between the observed and the simulated results.

Figure 2 is a simplification of the hydrologic simulation process, this later is subdivided to two major part the transfer and the production;

- The production is the estimation of water runoff quantity from the total incoming water,
- The transfer is the function which assures the transformation of water runoff already determined by the production function to flood hydrograph.

Simulation Approaches and Model Typology

The literature outlines two simulation approaches (Grayson, 2000):

- **Ascendants approach**: The diversity of the whole system is considered in this approach. The rain-flow relation is determined by physics laws, a part of the parameters derived from these laws are not directly measurable, then, in order to estimate their value, the model should be calibrated,
- **Descendant approach**: The watershed is considered as a functional unit, the system's outputs are defined by its inputs (rain and state variables), the parameters of rain-flow relation do not have a physical significance.

Classification

The equivocation of interaction between the components of the hydrologic cycle and the complexity of the studied systems, lead to the elaboration of a widerange of hydrologic models (SINGH,1995), (BEVEN, 1989), (ASCE, 2000), (AMBROISE B., 1998) suggest different classification.

Figure 2. Schematic representation of hydrologic models
Source: the representation was reproduced from (PERRIN, SAMI, & HENDRICKX, 2009) Scheme

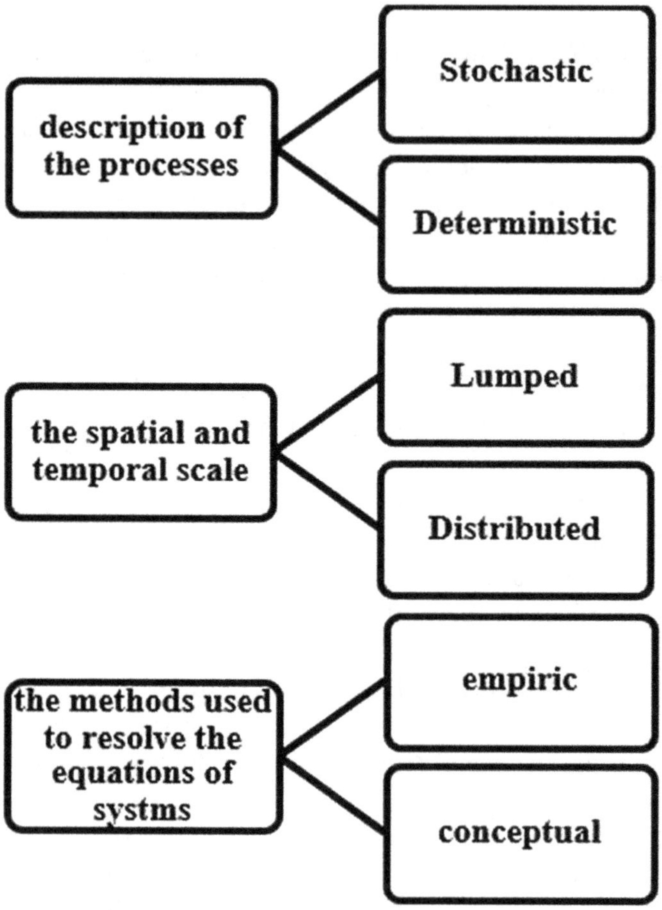

REFSGAAED, (1997) categorize hydrologic models according to the description of physical processes, as a conceptual or physic, and according to spatial discretization as a lumped or distributed. According to Singh, a model is characterized by five elements; the geometry of the systems, Inputs, formalization laws of processes, the initial state and boundary condition of the system and then Outputs. In order to class hydrological models Sigh proposes three criteria, the processes description, spatial and temporal scales and the methods used to resolve the equations of systems (Figure 3).

Figure 3. Hydrologic models' classification.
Source: The scheme was derived from the hydrologic model classification proposed by Singh, 1995

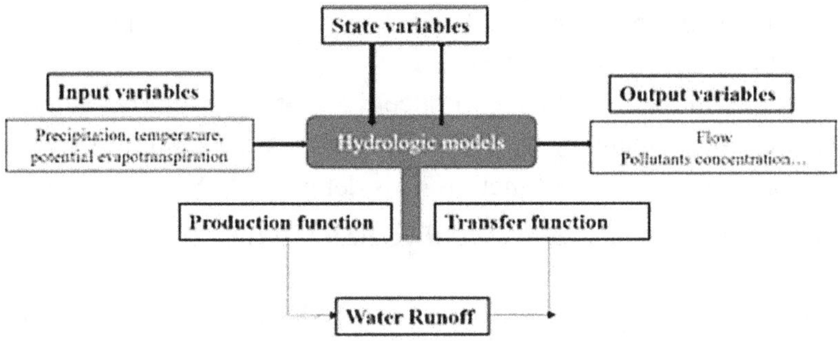

Generally, in the literature, hydrologic models are classified:
According to the abstraction degree:

- **Physics Models**: The representation of the systems in reduced scale, the equation used is derived from physics laws (mass conservation, the quantity of energy…), the final simplified form has a physical meaning.
- Mathematics Models: Describe the hydrologic process with mathematical equations which link the Input variables to the Output variables

According to the type of variables:

- Deterministic Models: the relation between Inputs and outputs is phenomenological, a determinate input produce always the same output.
- Stochastic or probabilistic models: these models generate randomly the output using statistic distribution laws.

According to spatial discretization:

- **Lumped Models**: this model considers the whole watershed as one homogeneous unit, all the characteristics parameters of the basin are described by their mean value (MAISON, 2000).
- **Distributed Models**: take into account the spatial variability and the heterogeneity of watershed, if the system component is not completely spatialized the models is semi-distributed.

According to the description of the processes involved in the hydrologic cycle (precipitation, infiltration, interception)

- **Conceptual Models**: are vertical interconnection between humidity reservoirs, every reservoir presents a level of water storage, this model is based on the determination of physical phenomena which impact Inputs and then Outputs (AMBROISE, 1991).
- **Empirical Models**: used to reproduce the global system behaviorwithout a description of elementary processes, they are called black box models when the $Flow = f\left(rain\right)$, the function f is determinate by statistical regression from the rain-flow data series (AMBROISE, 1991).

There are numerous typologiesof these models that are widely used (Philip et al., 2012), such as single-event models that are used to simulate or calculate the resulting storm hydrograph, from a well-defined watershed area, for a given pattern of rainfall intensity (ex; HEC-HMS), continuous models (ex; SWMM and SWAT, presented in this chapter) that take in consideration soil moisture storage, evapotranspiration and antecedent rainfall over long time periods, statistical models that generates time series of rainfall, and finally distributed models (ex; MIKE) that can handle input, output and data manipulation at the watershed level (Figure 4).

The soil and water assessment tool (SWAT) which is semi-distributed physical based hydrologic model was used in this research, it is a widely applied model in the world for the study of the watershed hydrologic behaviors, the impact of land use, urbanization and climate change on water resource (Kannan et al., 2019), (Zhou et al., 2013), (Wang et al., 2014), (Luzio et al., 2003), and in Morocco in particular, itis even considered better performing hydrological model in Moroccan context (Brirhet, 2016), its popularity is attributed to the fact that the model is open source and relies on GIS-based interfaces to collect, create, visualize, analyze and manipulate its inputs and provides sensitivity, calibration, and uncertainty analysis tools (Srinivasan and Arnold, 1994).Thus not only hydrological modelers can use it by modifying its source code but also hydrological model users.

The study presented inthis chapter was carried on the upstream part of Innaouene watershed.The pilot zone is facingplenty of environmental drawbacks, related to pollution, erosion and flood risks (Jamal et al., 2012). Those problems are mainly a consequence of geology and climatic factors. Different models have been used to forecast the floods runoff, like HEC RAS for hydraulic modeling and HEC HMS for hydrological modeling. Swat model was applied on Bab Marzouka watershed (Bouslihim et al., 2016). However,to date no actualized research has been reported, furthermore the last cited study used a climatic data from the global weather website, and a soil database from the FAO, which subdivide to basin to only three pedologic types.

Figure 4. The history of the development of hydrological Model
Source: Zemzami, 2013. Applied hydrology.

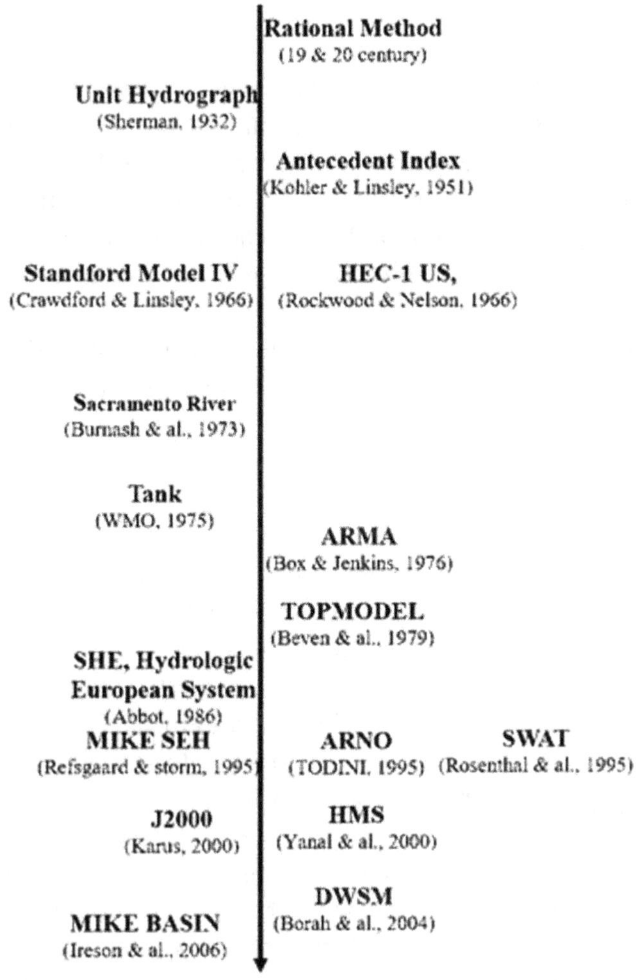

This study begins to address issues that are surrounding the flood forecasting in Innaouene watershed, more particularly test the ability of swat model to reproduce flow using actualized data series. It proposes also a methodology of building soil map, by merging data from ISRIC soil database and data collected from water and forest directory, using the geographic information system. The results of the model could be used to estimate the flow for different return period and introduced in a dynamic wave model to map and delineate the overflow expansion.

Materials and Methods

Study Area Overview

The present study concerns the upstream of Innaouene watershed (Figure5), limited by the meridians (33.78W; 4.91W) and the parallels (33.84N; 34.58N) and located between Fez and Tazatowns(geographically known as the Fes-Taza corridor), it covers an area of 2690Km^2, (Figure 5). The annual average of precipitation is around 598 mm, the annual temperature average is 22.1 °C, the maxima are reached in June with a value of 40 °C, the minimum temperature is reached in January with an average of 7.8 °C (ABHS, 2014).

Innaouene watershed is submitted to the influence of both climate type; Mediterranean and semi-arid, this fact is traduced by an unbalanced water supply during the year.

Description of the SWAT Model

SWAT is a basin-scale, continuous-time model that operates on a daily time step and is designed to predict the impact of management on water, sediment, and agricultural chemical yields, in ungauged watersheds. Major model components include weather, hydrology, soil temperature and properties, plant growth, nutrients, pesticides, bacteria and pathogens, and land management (Gassman et al., 2007). The catchment is divided, on a topographic basis, into spatially linked sub-basins and this later issegregated into unique hydrological response units (HRUs), by integrating the combination of land use, soil type, and slope to describe the system physical heterogeneity. The modeled hydrological components include surface runoff, percolation, lateral flow, groundwater flow, evapotranspiration (ET), and transmission losses. The simulation of watershed hydrology with SWAT is split into the land phase and the channel or routing phase of the hydrologic cycle, which controls the amount of water, sediment, and nutrients into the main channel in each sub-basin and through the channel network to the watershed outlet (Neitsch et al., 2011).

Figure 5. Geographic situationof Innaouene watershed
Sources:1: Map data© 2019 Google
2: hydraulic agency of Sebou (Fez),3: DEM of Sebou ASTER Global Digital Elevation Model (ASTER GDEM), 4: The stream network was extracted from the topographic maps of Innaouene watershed;
AKNOUL: 1/50000, paperNI-30XV-3a. Published in 1956 by the direction of land conservation and topographic work (Rabat)
FES EST: 1/10000, paperNI-30XIV-I. Updated in 1971 and published by the direction of land conservation and topographic work, Rabat.
MATMATA:1/50000, paperNI-30-XI-V-1b. Published in 1950 by the direction of land conservation and topographic work, Rabat.
MSOUN: 1/50000, paper NI-30-XV-Id. Published in 1945 by the administration of land conservation of cadaster and topography, Rabat.
RIBAT EL KHAYER: 1/50000, paper NI-30-VIII-4c. Published in 1966 by the direction of land conservation and topographic work, Rabat.
TISSA: 1/50000, paper NI-30-XIV-Id. Published in 1956 by the division of maps, the direction of land conservation and cadaster and topography
TAZA: 1/50000, Paper NI-30-XIV-2b. Published in 1987 by the division of cartography
Map 2, 3 and 4 were edited by Arc Gis software

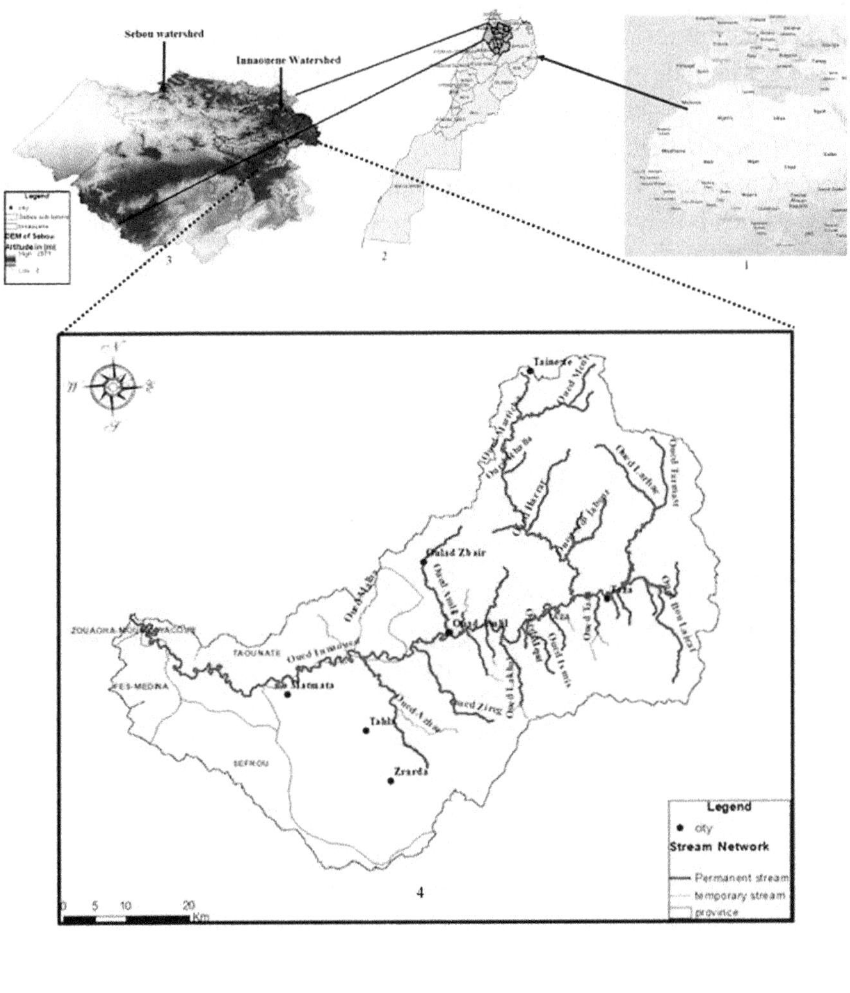

Water Balance

SWAT estimates water balance using this formula

$$SW_t = SW_0 + \sum_{i=1}^{t}(R_i - Q_i - ET_i - P_i - QR_i)$$

Where *SWt* is the soil water content, SW0 the initial soil water content *i* is the time in days for the simulation period *t*, and *R, Q, ET, P, and QR* respectively are the daily precipitation, runoff, evapotranspiration, percolation, and return flow (Neitsch et al., 2011).Water enters the SWAT model's watershed system boundary predominantly in the form of precipitation. It is partitioned into different water pathways depending on system characteristics. The water balance of each HRU in the watershed contains four storage volumes: snow, the soil profile (0-2 m), the shallow aquifer (2-20 m), and the deep aquifer (>20 m). The soil profile can contain several layers. The soil-water processes include infiltration, percolation, evaporation, plant uptake, and lateral flow. Surface runoff is estimated using the SCS curve number or the Green-Ampt infiltration equation. Percolation is modeled with a layered storage routing technique combined with a crack flow model. Potential evaporation can be calculated using Hargreaves, Priestly-Taylor or Penman-Monteith method (Arnold et al.,1998)

Figure 6. the needed data for SWAT simulation
Source: The scheme is reproduced from SWAT model user Manuel

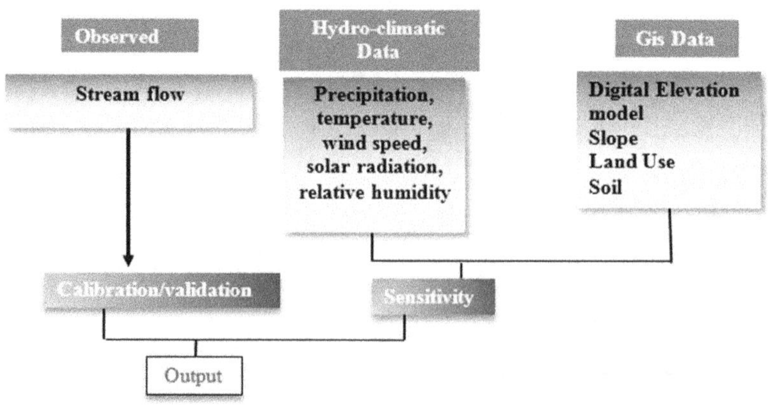

Table 2. DEM characteristics

Acquisition Date	Sensor	Spatial Resolution	Radiometric Resolution
17/10/2017	ASTER Global DEM	30m	16 bits

(Source: http://earthexplorer.usgs.gov/)

Data Collection

SWAT is a highly parametrized model, it requires a large platform of spatial and hydro-climatic data (Figure 6), the problem that its application in developing country face is the scarcity of data (Gassman et al., 2010).To overcome this issue local data gathered from the hydraulic agency and local administrations, combinedwith the global data provided by global weather database (Saha et al., 2010)0

GIS Data

The first step of SWAT model simulation is watershed delineation, the basin is partitioned into a number of sub-basins, each sub-basin is subdivided into a homogenous area called HRU (hydrologic response units), this later result fromthe overlay of land use, soil map, slopeand have the same hydrologic behavior.

- **Digital elevation model**

The Digital elevation model (DEM) was extracted from the ASTER Global Digital Elevation Model (ASTER GDEM) which has the following characteristics.

The DEM is used to delineate the watershed into sub-basin and calculate their topographic parameters such as terrain slope, reach slope, longest flow path andreach length.

- **Land Use**

The Land Use map was extracted from Global cover map which is a European Space Agency project (ESA) (Bicher et al., 2008), it began in 2005 in partnership with JRC, EEA, FAO, UNEP, GOFC-GOLD, and IGBP, the goal of this project is to develop a service of delivering land cover maps. ESA makes available the land cover maps, which cover 2 periods: December 2004 - June 2006 and January - December 2009 from which the map used in this study was obtained (Bicher et al., 2008).

Table 3. The required soil characteristics for SWAT simulation

Characteristics	Signification
SOL_Z	Depth from the soil surface to the bottom of the layer (mm)
SOL_ZMX	Maximum rooting depth of soil profile (mm)
SOL_BD	Bulk density (Mg/m3 or g/cm3) the values should fall between 1.1 & 1.9
SOL_AWC	Available Water Capacity of the soil layer (mmH_2O/mm soil)
SOL_K	Saturated hydraulic conductivity (mm/hr)
SOL_CBN	Organic Carbon Content (%soil weight)
SOL_CLAY	Clay content (%soil wright)
SOL_SILT	Silt content (%soil weight)
SOL_SAND	Rock sand content (%sand weight)
SOL_ROCK	Rock fragment content (%soil weight)
SOL_ALB	Moist soil albedo
USLE_K	Erodibility factor

Source: SWAT 2012 database

- **Soil map**

In order to estimate the fraction of water run-off, SWAT model defines the HRU by combining soil map, land cover, and slope. The main soil physical and chemical characteristics required by the model are listed in the following board.

The Soil Survey Geographic database (ISIRIC) was used to provide the model with information on the soil. This later is a system for automated soil mapping based on the global compilation of soil profile data and spatial data, derived from remote sensing. Soil grids provides global predictions for standard numeric soil properties (organic carbon, bulk density, Cation Exchange Capacity (CEC), pH, soil texture fractions and coarse fragments), at seven standard depths (0, 5, 15, 30, 60, 100 and 200 cm), in addition to predictions of depth to bedrock and distribution of soil classes, based on the World Reference Base (WRB) and USDA classification systems (Hengl et al., 2017).

The soil attribute used in the presented study results from the compilation between world soil information data and the soil map taxonomy provided by the Water and Forest direction of Fez (Appendix 1).

The USLE K factor represents the inherent erodibility of a given soil and typically ranges from 0.1 (low erodibility) to 0.67 (highly erosive).It was estimated from the percent of sand, silt and organic carbon content (Wischmeier& al., 1978). The table illustrates that USLE K factors (for surface soil layers) within the Innaouene watershed range from 0.16 to 0.18, which represent low to moderately erodible soils (Wischmeier& al., 1978) (Appendix 2).

These soils are classified into Hydrologic Soil Groups (A, B, C, and D) based on soil characteristics in the ISIRIC database. Note that soils in group A have the lowest runoff potential and highest infiltration rates, while those in group D have the highest runoff potential and lowest infiltration rates.

All the correspondent's data of these maps was connected to the soil map database (shapefile), by joining their attribute tables in order to obtain one table, with the whole parametersrequire by SWAT model (Table4).

Table 4. Estimation of Innaouene watershed soil characteristics from ISRIC database and soil taxonomy map

OBJ	SNAM	SOL_ ZMX (in m)	SOL_ Z1 (in m)	SOL_ BD (in mg/ m3)	SOL_ AWC IN (mmH_2O/ mm soil)	SOL_ CBN (In %)	CLAY (In %)	SILT (In%)	SAND (In %)	ROCK (In %)	USLE_K (In %)
1	REGOSOL	192.88	192.88	1.39	0.12	0.18	24.57	40.87	34.58	17.19	0.18
2	CALCISOL	197.1	197.1	1.42	0.12	0.12	24.65	41.33	34.04	17.66	0.18
3	BRUNIZEMS	199.48	19948	1.46	0.11	0.09	26.41	37.26	36.32	14.81	0.17
4	VERTISOL	199.19	199.19	1.45	0.11	0.07	26.21	34.74	35.07	16.33	0.17
5	FERSIASOL	193.8	193.8	1.37	0.12	0.26	25.17	40.08	34.74	18.42	0.18
6	BRUNISOL	199.19	199.19	1.43	0.12	0.1	24.11	41.65	34.25	16.37	0.18

Source: *ISRIC – World Soil Information, 2013*

Figure 7. Delimitation of Innaouene watershed by SWAT model

Hydro-Climatic Data

The Hydro-climatic data were provided by the Watershed hydraulic agency of Sebou with a daily step time; the study area has four hydrometric and one hydrologic station (Figure4), with one for measuring temperature (Bab Marzouka).

In Sebou watershed there is no station for measuring the rest of SWAT parameters (Solar radiation, relative humidity, and wind speed), to overcome this problem we referred to Global Weather Data (Saha, 2010),which provides information with a daily time step from 1979 to 2014.

The Model Set Up

SWAT simulation start by watershed delineation, the basin is partitioned into a number of sub-basins that are spatially related, for example, the water from the sub-basin 1 go in the sub-basin 2. The delineation is obtained from subwatershed boundaries that are defined by surface topography. The surface area within the sub-basins is subdivided to hydrologic unit response (HRU) in order to capture watershed diversity, HRU simplifies the simulation by lumping all similar soil, slope, and land use area into a single unit. (Arnold et al., 2012).

In this study, 19 sub-basins and 1136 HRU were defined, this number could be reduced if the land use, soil, and slope diversities are limited.

The upstream of Innaouene has six types of soils; REGOSOL, CALCISOL, BRUNOZEMS, VERTISOL, BRUNISOL, FERSIASOL (direction of water and forest of Fez, 2011), six types of land use(global cover map) and 6 classes of slope defined from the digital elevation model.

Concerning water balance estimation, SCS method with calculated plant evapotranspiration was selected for the runoff assessment.Manning equation is used to estimate the flow rate and velocity through the channels, flow routing is based on variable storage method.(Neitsch et al., 2011)

SCS Method

The method was developed by the Soil Conservation Service in 1975 from infiltrometer and measured rainfall and runoff on small pots and basins,in order to estimate runoff volume and peak discharge in small to medium size ungauged basin. This method was used firstly in the United States and has been extended in other countries (David,1992). The origin and evolution of this method are described by (Rallisson, 1980) and (Miller & Cornshey, 1989), Other detailed description can also be found in (McCunen, 2005).The basic equation of volume runoff estimation was derived from the following scheme

When rainfall equals initial abstraction I_a, Runoff occurs, the residual runoff Q is the result of subtraction of infiltration or retained water F, and initial abstract I_a from the total precipitation P. The potential retention S is the sum of infiltration, with initial abstract $(F + I_a)$. The effective rainfall P_e is the total precipitation minus the initial abstraction $\left(P - I_a\right)$. This theoretical method constitutes the steppingstone of the SCS curve number method.

$$\frac{F}{S} = \frac{Q}{P_e}$$

According to figure 8, F equals ($P_e - I_a$), then

$$Q = \frac{P_e^2}{P_e + S} = \frac{\left(P - I_a\right)^2}{P - I_a + S}$$

From the analysis of data series used in the development of the SCS method, the empirical relation $P_e = P - 0.2S$ was adopted as the best proximation. Then

Figure 8. Accumulation rainfall, losses, and runoff during a uniform storm
Source: Davide R. Maidment. (1992). HANDBOOK OF HYDROLOGY, flood runoff, Chapter 9 part 21

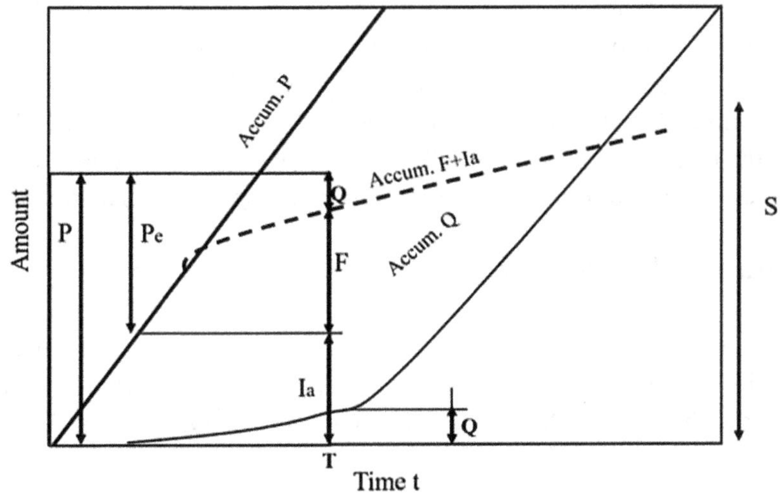

Figure 9. SWAT model set up

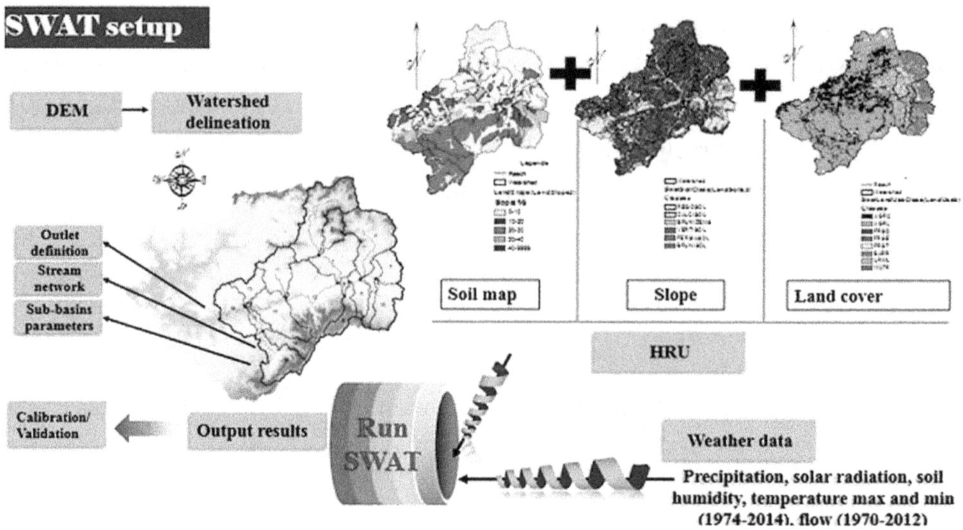

$$Q = \frac{\left(P - 0.2S\right)^2}{P + 0.8S}$$

The potential storage S is expressed by runoff Curve number CN,

$$CN = \frac{1000}{S + 10}$$

Therefore the valuation of the CN leads to the estimation of the residual runoff discharge Q. the value of the curve number depends on the soil structure, land cover, hydrologic conditions, the wetness of the watershed, and three classes of antecedent moisture condition (AMC) are defined dry average and wet (David R.1992). The value of CN for the three classes and the SCS relation between cumulative direct runoff and cumulative rain could be found in (National and Engineering Handbook, 1985).

Calibration and Validation

The calibration step aims to estimate the optimal values of non-measurable parameters, for which the observed and simulated hydrograph are well adjusted. Two types of calibration are distinguished; the manually and the automated using SWAT CUP, which is a public domain program that enables sensitivity analysis, calibration validation and uncertainty analysis of SWAT model (Abbaspour et al., 2015).The auto-calibration option provides a powerful, labor-saving tool that can be used to reduce the frustration and uncertainty, that often characterize manual calibration (Van Liew et al., 2005). The procedure is based on the optimization algorithm, that tries to minimize an objective function that expresses the deviation between a measured and a simulated streamflow series. (Abdelhamid et al. 2011) In order to underscore the influence of every single parameter on the hydrologic behavior of the watershed, the manual calibration was adopted for the period (2005-2012), by changing parameters in several attempts and seeking for the best steadiness between measured and simulated outflow.

The validation aims to assess the accuracy of the estimated parameters by running the model using other period and using the values obtained in the calibration step.

Validation of the model was carried out by using:

- Coefficient of determination R^2: measure how well the observed and the simulated hydrograph are adjusted, it ranges between 0 and 1,
- Nash-Sutcliffe efficiency (NSE):which range from $-\infty$ and 1, indicated the fitness degree between observed and simulated data,
- RSR: RMSE-observations standard deviation ratiois a statistic error index, the lower RSR the better is the simulation performance, the optimal value of this indicator is 0, (Moriasis, 2007)
- PBIAS: percent Bias measures the average tendency of the simulated data to be larger or smaller than the observed one (Gupta & al., 1999), a negative value of PBIAS indicate an accurate model when model underestimation is traduced by positive values.

A detailed description of the cited evaluation criteria is presented in Moriasis's performance rating article (Moriasis, 2007).In general, model simulation is"satisfactory"if NSE > 0.50, RSR < 0.70, and if PBIAS $\pm$ 25%for streamflow (Moriasi& al., 2014).

Results and Discussion

The entire simulation period is from 2005 to 2014, the model calibration performs at a daily time step from 2005 to 2012, the first year is devoted to the model set up. The choice of this time series period is due to data availability, on a previous study this model was performed on the same area, using one year for model set up (2003-2004), four years for calibration (2004-2008) and two years for validation (2008-2010), (Bouslihim et al., 2016), the data used in this study was provided by the Watershed agencies of Sebou watershed, the longest available data range for both hydrometric and hydrologic data, for the used stations is from 2004 to 2014.

Several attempts of the manual calibration were performed at the sub-basins number 16 (The outlet of the watershed), 14 parameters were selected for calibration, the most influential are; the Curve Number CN2, the Soil Evaporation Compensation factor ESCO, Hydraulic Conductivity K and Available Water Capacity.

With a CN2=82 the model under-estimate the outflow, a low value of ESCO decrease it to, and this is due to the dry climate of the watershed and its geological nature dominated by the clay. The estimation of hydraulic conductivity parameter scale up the model adjustment.

The model sensitivity to these parameters support the results of some study previously done (Abdelhamid et al., 2011; White &Chaubey, 2005; and Yassine Bouslihim et al., 2016).

Figure10 presents the comparison hydrographs for both periods; calibration and validation, SWAT model reproduce generally, the same edge of the reel outflow, but sometimes it under-estimates the pick flow. For example in the calibration period the high flow simulated is in 03/09/2010 with a value of 883 m^3/s, while it reaches $1000m^3/s$ in observed flow. According to a study, which tested the ability of the model to simulate water patterns, SWAT had a tendency to over-predict soil water in dry soil conditions and under-predict in wet soil conditions (Mapfumo et al., 2011). The explication of the high picks of outflow even that Innaouene climate is semi-arid can be explained by Hortonien precipitation (Horton & Robert, 1945), when the rainfall power exceeds the infiltration capacity.

The NES coefficient value of the monitoring gage ranged between 0.55 for calibration and 0.58 for validation (Figure 11 & Table 5).

The correlation between daily observed and simulated river discharge isquitewell. According to Van Liew et al., (2007), the model performance is satisfactory, but the results could be improved by reducing uncertainties.

In general, SWAT modeluncertainties are due to conceptual simplifications (e.g., SCS curve number method for flow partitioning), processes occurring in the watershed, but not included in the program (e.g., hydric erosion, wetland processes), processes that are included in the program, but their occurrences in the watershed are unknown to the modeler or unaccountable because of data limitation (e.g., dams and reservoirs, water transfers, farm management affecting water flow), and input data quality (Abbaspour et al., 2015).

Figure 10. The hydrograph comparison for calibration and validation periods

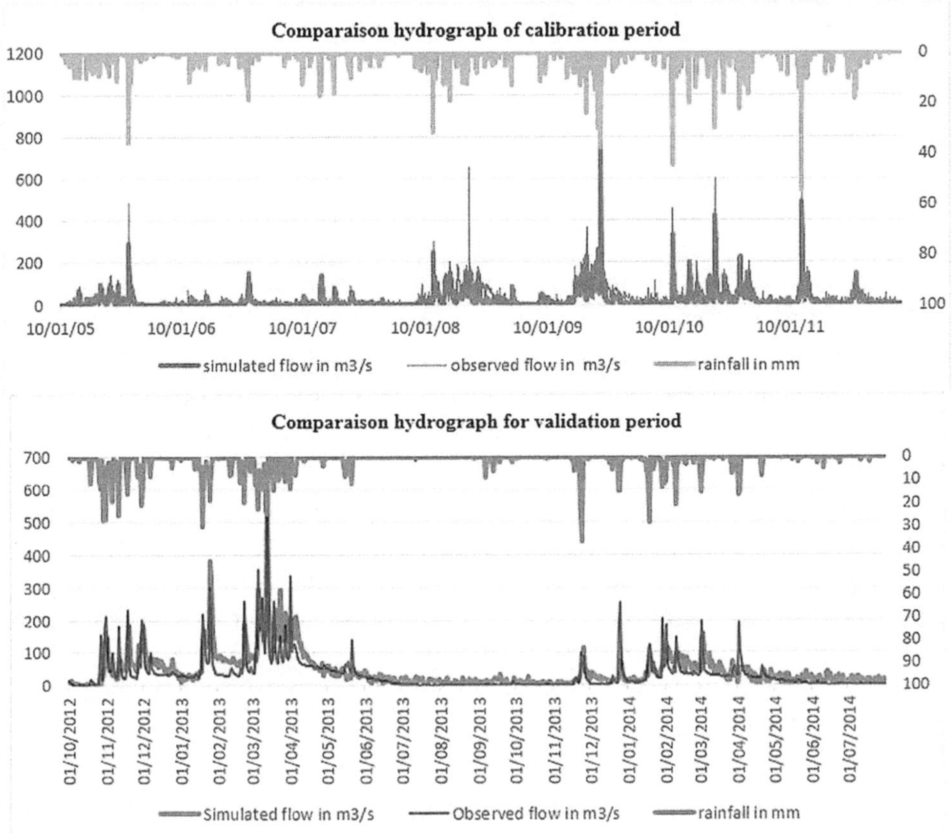

Figure 11. Correlation between observed and simulated flow for the historical period (2005–2014) at Bab Marzouka Station

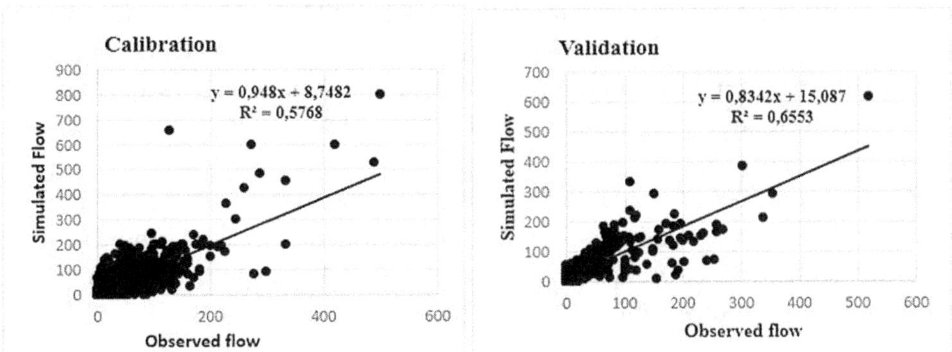

Table 5. Moriasi's performance rating

	Calibration	Validation
NSE	0.58	0.55
R^2	0.57	0.65
PBIAS	26	-31.22
RSR	0.67	0.68

Another causeis that Innaouene watershed contains karstic groundwater in its north-eastern part, after soil saturation the water cross the deep aquifer and may occur in the stream as baseflow after a rain event, the total amount of drained water will be higher, and SWAT model doesn't simulate accurately this part.

Note to mention that the graphs show the results according to the hydrological year, this latter was determinate for the first time according to the method of TONINI in 1948. Based on the characteristic time of rainfall for a certain period in which there is the best correspondence between flow and rain. It starts after the month in which the water amount is the lowest. According to the United States Geological Survey, the flow year starts in October 1st and end on September 30 thof the next year.

CONCLUSION

Hydrologic modeling plays an important role in flood risk management, the results of such models could be used as an input of hydraulic models, which aims to quantify, map and delineate floods' expansion. This worktests the capacity of SWAT model to simulate the hydrologic behavior in large scale basin in the semi-arid climate context, the model was applied on a pilot area (upstream of Innaouene watershed), GIS interface provides straightforward tools, which enable an easy configuration.

Arc SWAT translate the DEM, Land use and Soil map into data inputs.The calibration was performed on a daily time step from 2004 to 2014, the first year is excluded for the model set up, the two last years was used for the model validation, the performance examination of simulated data at subwatershed 19 denotes that SWAT generated quite good results, which go hand by hand with precedent work (Bouslihim et al., 2016).

In other words, the acceptable fitting between observed and simulated daily discharge values, at the hydrometric station of Bab Marzouka stations indicates that SWAT model with the set of optimized parameters could be applied to examine the hydrologic behavior of a semi-arid climate but not in karstic areas.

According to the quantitative statistic indicator of the simulation performance, (percent bias (PBIAS), Coefficient of determination R^2, Nash-Sutcliffe efficiency (NSE), and the RMSE-observations standard deviation ratio RSR), the fitting rate between the two hydrographs is acceptable, but the model has a trend to under-estimated the stream flow, especially after a heavy rain period of the hydrologic year.

In order to obtain better result, first the quality of the input data should be enhanced by implementing more gage station in the watershed, For Instance, in the northern part of Innaouene basin the slope is high and it decreases progressively toward the outlet, this part is drained by Oued Lahdar and it did not include any gage or dischargestation.

The biggest problem that researcher face in developing country like Morocco is data availability, furthermore if the required data are available, it contains large spatiotemporal gaps, it is recommended then to provide more resources for the study of pedology, geology and Land Use/Land cover and data actualization in order build a national database.

The result obtained using SWAT model are satisfactory but not good(Van Liew et al., 2007), considering the area is mainly karstic, the majority of water goes to deep aquifer and reapers at the outlet of watershed. SWAT misses this process. it is recommended to integrate a component for karstic system simulation into the model in order to improve its performance.

In the cradle of the presented study,a set of research could be driven from it, the flow generated in this study could be a:

- The subject of hydraulic and hydrologic coupled study,
- Study the impact of land use and climate changes on water quantities as well as quality,
- Integration of karstic component into the SWAT model.

REFERENCES

Abbaspour, K. C., Rouholahnejad, E., Vaghefi, S., Srinivasan, R., Yang, H., & Kløve, B. (2015). A continental-scale hydrology and water quality model for Europe: Calibration and uncertainty of a high-resolution large-scale SWAT model. *Journal of Hydrology (Amsterdam)*, *524*, 733–752. doi:10.1016/j.jhydrol.2015.03.027

Abbaspour, K. C., Rouholahnejad, E., Vaghefi, S., Srinivasan, R., Yang, H., & Kløve, B. (2015). A continental-scale hydrology and water quality model for Europe : Calibration and uncertainty of a high-resolution large-scale SWAT model. *Journal of Hydrology (Amsterdam)*, *524*, 733–752. doi:10.1016/j.jhydrol.2015.03.027

Abdulrazzak, M., Elfeki, A., Kamis, A., Kassab, M., Chaabani, A., & Noor, K. (2019). Flash flood risk assessment in urban arid environment : Case study of Taibah and Islamic universities' campuses, Medina, Kingdom of Saudi Arabia. *Geomatics, Natural Hazards & Risk*, *10*(1), 780–796. doi:10.1080/19475705.2018.1545705

Alaoui, M. (2013). Water Sector in Morocco: Situation and Perspectives. *Journal of Water Resources and Ocean Science.*, *2*(5), 108–114. doi:10.11648/j.wros.20130205.18

Ambroise, B. (1991). *Hydrologie des petits bassins versants ruraux en milieu tempéré, processus et modèles.* Séminaire du Conseil Scientifique du département « Science du sol » de l'Institut National de Recherches Agronomiques de Dijon.

Ambroise, B. (1998). La dynamique du cycle de l'eau dans un bassin versant. Processus, Facteur, Modèles.

Arnold, J. G., Kiniry, J. R., Srinivasan, R., Williams, J. R., Haney, E. B., & Neitsch, S. L. (2012). *Soil & Water Assessment Tool.* Academic Press.

ASCE. (2000). *Artificial Neural Networks in Hydrology. J.* Australian Government.

Attorney General's Department Flood. (2009). *Preparedness Australian emergency manual series.* Commonwealth of Australia.

Base, C. D. E., Modelisation, D. E. L. A., & Et, H. (2013). *Compétitivité économique du maroc.* Academic Press.

Berterreche, L., Champredonde, R., & Dartus, D. (2008). *Etude de la génération de crues rapides par la confrontation de deux logiciels.* Academic Press.

Beven, K. (1989). Changing ideas in hydrology, the case of physically-based models. *Journal of Hydrology (Amsterdam)*, (105): 157–172.

Bicheron, P., Defourny, P., Brockmann, C., Schouten, L., Vancutsem, C., Huc, M., … Arino, O., (2008). *GlobCover: Products Description and Validation Report.* ESA GlobCover project. Available online at: http://www.esa.int/due/ionia/globcover

DARA. (2012). DARA and the Climate Vulnerable Forum, Climate vulnerability monitor 2nd edition. In *A guide to the cold calculus of a hot planet*. DARA.

Emergency Management Australia and Australian Bureau of Meteorology. (1998). *Flood-warning, preparedness and safety*. Emergency Management Australia.

Fadil, A., Rhinane, H., Kaoukaya, A., Kharchaf, Y., & Bachir, O. A. (2011). Hydrologic Modeling of the Bouregreg Watershed (Morocco) Using GIS And SWAT Model. *Journal of Geographic Information System*, *3*(04), 279–289. doi:10.4236/jgis.2011.34024

Filahi, S., Tanarhte, M., Mouhir, L., El Morhit, M., & Tramblay, Y. (2016). Trends in indices of daily temperature and precipitations extremes in Morocco. *Theoretical and Applied Climatology*, *124*(3), 959–972. doi:10.100700704-015-1472-4

Gassman, P. W., Arnold, J. G., Srinivasan, R., & Reyes, M. (2010). The worldwide use of the SWAT Model: Technological drivers, networking impacts, and simulation trends. In *Proceedings of the Watershed Technology Conference*. American Society of Agricultural and Biological Engineers, Earth University.

Gaume, E. (2009). Hydrologie de versants et de bassins versants et modélisation pluie-débit. Master2 Sciences et Génie de l'environnement, Ecole Nationale des Ponts et Chaussées.

Gleick, P. H. (1993). *Igor Shiklomanov's chapter "World fresh water resources", Water in Crisis: A Guide to the World's Fresh Water Resources*. New York: Oxford University Press.

Grayson, R. A. (2000). *Spatial Patterns in Catchment Hydrology*. Cambridge University.

Gupta, H. V., Sorooshian, S., & Yapo, P. O. (1999). Status of automatic calibration for hydrologic models: Comparison with multilevel expert calibration. *Journal of Hydrologic Engineering*, *4*(2), 135–143. doi:10.1061/(ASCE)1084-0699(1999)4:2(135)

Hengl, T., Mendes de Jesus, J., Heuvelink, G. B. M., Ruiperez Gonzalez, M., Kilibarda, M., Blagotić, A., … Kempen, B. (2017). SoilGrids250m: Global gridded soil information based on machine learning. *PLoS One*, *12*(2). doi:10.1371/journal.pone.0169748 PMID:28207752

Horton, R. E. (1933). The role of infiltration in the hydrologic cycle. *Transactions of the American Geophysics Union, 14th Annual Meeting*, 446–460. 10.1029/TR014i001p00446

Horton, R. E. (1945). Erosional development of streams and their drainage basins; Hydrophysical approach to quantitative morphology. *Geological Society of America Bulletin*, *56*(3), 275–370. doi:10.1130/0016-7606(1945)56[275:EDOSAT]2.0.CO;2

ISRIC – World Soil Information. (2013). *Soil property maps of Africa at 1 km*. Available for download at www.isric.org

Jamal, P., Stitou, E., Messari, E., Dridri, P. A., Abdel, P., Chaouni, A., … Fès, F. S. T. (2012). *UFR: Chimie de l ' Environnement Discipline*. Géologie Présentée par Jamal Naoura N aoura.

Kannan, N., Santhi, C., White, M. J., Mehan, S., Jeffrey, G., & Gassman, P. W. (2019). *Some Challenges in Hydrologic Model Calibration for Large-Scale Studies : A Case Study of SWAT Model Application to Mississippi-Atchafalaya River Basin*. Academic Press. doi:10.3390/hydrology6010017

Khaddor, I. (2016). *Simulation numérique des écoulements à surface libre du bassin versant de Mghogha en moyen de SIG, Télédétection, HEC-HMS et HEC-RAS: Études des protections contre les inondations dans la ville de Tanger, nord de Maroc*. Universite Abdelmalek Essaadi Faculte Des Sciences Et Techniques Tanger.

Likens, G. E. (2013). *Biogeochemistry of a forested ecosystem*. Springer Science & Business Media. doi:10.1007/978-1-4614-7810-2

Luzio, M., Srinivasan, R., & Arnold, J. G. (2003). *Integration of watershed tools and SWAT model into basins*. Academic Press.

Maison, P. (2000). *Un modèle hydrologique de suivi de la pollution diffuse en bassin versant. Approche mécaniste simplifiée de la zone non saturée*. Thèse de Doctorat, L'institut National Polytechnique de Toulouse en Science de la terre et environnement, 303

Marchandise, A. (2007). *Modélisation hydrologique distribuée sur le gardon d'Anduze: étude comparative de différents modèles pluie-débit, extrapolation de la normale à l'extrême et tests d'hypothèses sur les processus hydrologiques*. Thèse de Doctorat, Univ.

Maria, A., Evangelos, A., & Vassilios, A. (2015). *Hydrology and Water Resource Systems Analysis*. Academic Press.

Meyer, V., & Schwarze, R. (2019). The Economics and Management of Flood Risk in Germany. *Volker Meyer and Reimund Schwarze*, 473–495.

Miller, N., & Cronshey, R. (1989). Runoff Curve Numbers, the next Step. In B. C. Yen (Ed.), *channel Flow and Catchment Runoff, Departement of civil Engineering, Universtity of Virginia* (pp. 910–916). Charlottesvilles, VA: Academic Press.

Moriasi, Arnold, Van Liew, Bingner, Harmel, & Veith. (2007). *Model evaluation guidelines for systematic quantification of accuracy in watershed simulation*. Academic Press.

OECD. (2017). *OECD Review of risk Management Policies Morocco*. Paris: OECD Publishing.

Perrin, C. (2000). *Vers une amélioration d'un modèle global pluie-débit au travers d'une approche comparative*. Thèse de Doctorat, INPG (Grenoble) / Cemagref (Antony).

Perrin, C., Samie, R., & Hendrickx, F. (2009). Les outils du projet: modélisation hydrologique et représentation des usages. Journée de restitution du projet.

Pierre, U., Doctorale, E., Naturelles, R., & Moussa, M. R. (2009). *Quels apports hydrologiques pour les modèles hydrauliques ? Vers un modèle intégré de simulation des crues*. Academic Press.

Rallison, R. E. (1980). Origin and Evolution of the SCS Runoff Equation. *Proc. symp. on watershed management 1980, 2*, 914-924.

Refsgaard, J. C. (1997). Parametrisation, calibration and validation of distributed hydrological models. *Journal of Hydrology (Amsterdam), 198*(1-4), 69–97. doi:10.1016/S0022-1694(96)03329-X

Saha, S., Moorthi, S., Pan, H.-L., Wu, X., Wang, J., Nadiga, S., ... Goldberg, M. (2010). The NCEP Climate Forecast System Reanalysis. *Bulletin of the American Meteorological Society, 91*(8), 1015–1057. doi:10.1175/2010BAMS3001.1

Singh, V. P. (1995). Accuracy of kinematic and diffusion wave approximations for space independent flows on infiltrating surfaces. *Hydrological Processes, 9*(1), 1–18. doi:10.1002/hyp.3360090102

United States Geological Survey. (2016). *Water science school*. Author.

U.S Soil Conservation Service. (1985). *Natioanl Engineering Handbook, Sec. 4, Hydrollogy*. Washington, DC: U.S. Department of Aggricylture.

Wang, G., Yang, H., Wang, L., Xu, Z., & Xue, B. (2014). *Using the SWAT model to assess impacts of land use changes on runoff generation in headwaters*. doi:10.1002/hyp.9645

Wischmeier, W., & Smith, D. (1978). *Predicting Rainfall Erosion Losses: A Guide to Conservation Planning*. Washington, DC: U.S. Department of Agriculture.

Yonaba, H. (2009). *Modélisation hydrologique hybride: réseau de neurones - Modèle conceptuel*. Thèse de Doctorat, Univ. Laval, Québec.

Zhou, F., Xu, Y., Chen, Y., Xu, C., Gao, Y., & Du, J. (2013). *Hydrological response to urbanization at different spatio-temporal scales simulated by coupling of CLUE-S and the SWAT model in the Yangtze River Delta region*. doi:10.1016/j.jhydrol.2012.12.040

KEY TERMS AND DEFINITIONS

Available Water Capacity: Volume of water that can be held by the soil.

Calibration: Estimation of non-measurable parameters on the field by decreasing the difference between the observed and simulated hydrograph using historical rainfall, flow, and current basin conditions.

Curve Number: A parameter that has a relation with land use and hydrologic soil properties of a watershed, and is used in the SCS method.

Hortonien Precipitation: The limiting curve when rainfall intensity surpasses the capacity of infiltration.

Hydraulic Conductivity: Indicates permeability of intercepted porous matter by the ratio of velocity to the hydraulic gradient in a given time.

Hydrologic Cycle: A series of complex processes of water movement which is continuous, with its different states on the earth's surface.

Hydrologic Response Unit: Unique combination between slope, land use and soil time which have the same hydrological behavior regarding water runoff.

Interception: The portion of rainfall that gets trapped on the vegetation surfaces and return to the atmosphere usually through evaporation.

ISRIC Database: International Soil Reference and Information Center is a dependent foundation formed by Dutch law in Wageningen University and research.

Validation: Verification of model hydrograph using other time series while maintaining the same calibration parameters.

APPENDIX 1

Figure 12. Soil characteristics required by swat model
Source: ISRIC database

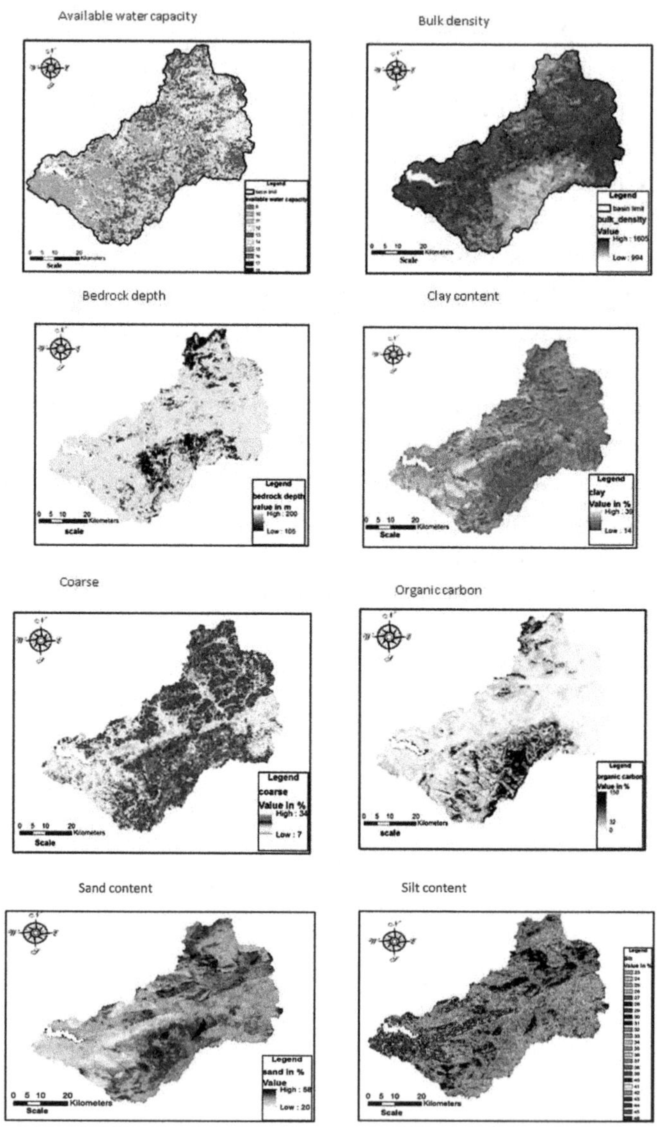

APPENDIX 2

Table 6. USLE-k factor calcul

OBJ	SNAM	SOL_ CBN1	CLAY1	SILT1	SAND1	fcsand	fsc-si	forgc	fhisand	USLE_K
1	REGOSOL	0.18	24.57	40.87	34.58	0.20159889	0.89502459	0.99981149	0.99996472	0.18039559
2	CALCISOL	0.12	24.65	41.33	34.04	0.20180604	0.90014068	0.99989437	0.99996857	0.18162893
3	BRUNIZEMS	0.09	26.41	37.26	36.32	0.20087833	0.85532323	0.99992739	0.99994885	0.17179464
4	VERTISOL	0.07	26.21	34.74	35.07	0.20085624	0.8434383	0.99994672	0.99996083	0.16939418
5	FERSIASOL	0.26	25.17	40.08	34.74	0.20145475	0.88639749	0.99965732	0.99996349	0.17850128
6	BRUNISOL	0.1	24.11	41.65	34.25	0.20179976	0.90375334	0.99991694	0.99996712	0.18235607

APPENDIX 3

Table 7. Results of calibration performed on the upstream of Innaouene watershed (2005-2012)

	Parameter	Parameter's Name	max	min	Optimal Value
Groundwater	GW delay (day)	Groundwater delay time per day	50	0	5
	Alpha BF (days)	Base flow alpha factors (1/day)	1	0	1
	GWQWM (mm)	Threshold depth of water in shallow aquifer required for return flow to occur (mm H2O)	5000	0	100
	Alpha_BF_D	Base flow alpha factor fo deep aquifer	1	0	0,01
	GW_SPYLD (m3/m3)	spesific yield of the sallow aquifer	0,4	0	0,3
	GWTHT (m)	Initial ground water high	0,25	0	1
	REVAPMN	treshold water level in shallow aquifere for revap to occure	1000	0	10
HRU	Esco	Soilevaporation compensation factor	1	0	0,95
	Epco	Plant uptake compensation factor	1	0	1
	LAT TIME	Lateral flow travel time (day)	180	0	2
	Surlag	surface runoff lag time in the HRU	24	0	2
	CANMAX	maximum canopystorage	100	0	0
	DEP_imp	depth to impervious layer in soil profile in mm	6000	0	6000
	OV_N	Mannening's value for overland flow	30	0,01	0,14
soil	Soil_K	Saturated hydraulic conductvity mm/hr	2000	0	500
	Soil_AWC	available water capacity mm H2O/mm soil	1	0	0,8
	Soil_BD (gr/Cm3)	Moist bulk density	2,5	0,2	1,42
	Soil_Z	soildepth in mm	3500	0	10
Managment	CN2	Initials runoff SCS curve number for moisture condition 2	98	35	65,8
Routing	CH_N2	manning value for the main channel	0,3	-0,01	0,014
	CH_D	average depth of main channel	30	0	1,2
	CH_K2	effectiv hydraulic conductivity in the main channel (mm/hr)	500	-0,1	150

Chapter 9
Hydraulic Modeling of Oued El Jawaher Using HEC-RAS Model for Flood Protection

Himedi Maroua
Université Sidi Mohammed Ben Abdellah, Morocco

Moumen Zineb
Université Sidi Mohamed Ben Abdellah, Morocco

Lahrach Abderahim
Université Sidi Mohamed Ben Abdellah, Morocco

ABSTRACT

Flooding has a wide range of impacts on societies and natural environments. In this sense, the city of Fez suffers from these problems reflected by the overflow of Oued El Jawaher during the rainy periods. This situation led the authors to compare between the current situation and the situation developed by the thresholds of Oued El Jawaher. HEC-RAS hydraulic model consists of 31 cross-sections, which will be used in the course of this study. The simulations will concern the current state and the developed state for flows of different frequencies. The result of the simulations confirms that the capacity of the proposed hydraulic structures is insufficient to transit and should be considered. To conclude, the development of the channel by thresholds, which serves for the creation of water plan, magnifies the risk of an overflow of the banks of the canal by the water line along with the longitudinal profile.

DOI: 10.4018/978-1-5225-9771-1.ch009

INTRODUCTION

Flooding can be defined as "a phenomenon of temporary submersion, natural or artificial, of a land space" (Helga-Jane & Richard, 2004). The latter produces very different impact on societies and natural environments. The city of Fez experienced from the middle of the 20th century (September 1950 and October 1989) excessively violent floods which caused very important damages (ABHS, 2010). The overflowing waters of El Jawaher River during rainy periods are one of the main sources of flood risk for the city of Fez, which continues to grow due to human activity.

The aim of the present research paper is to develop a hydraulic model that will simulate the flow in El Jawaher River for floods of different return periods, and make a comparison of the current state and the state of development by thresholds intended to remedy the overflow threatening the local population at the level of the city.

PRESENTATION OF THE STUDY AREA

El Jawaher River is located between the parallels 33 ° 30 and 34 ° 08 N and between the meridians 4 ° 54 and 5 ° 09 W. The city of Fez is located in the northwestern part relatively watered compared to the rest of Morocco, the climate type of this region will be defined according to the data provided by the resorts in the region. The region of the city of Fez has several weather stations enough distributed and distant from each other. Only the data of the station Fès-Saïs is selected, for the preparation of this overview, by their proximity to the study area.

The zone is a subsidence filled by deposits of Neogene age whose southern fringe comes to stumble against the Jurassic Limestones and dolomites of the middle-atlas and the northern fringe against the Pre-Rifain groundwater.

The city of Fez is characterized by a semi-arid climate; the annual rainfall regime is characterized by the existence of two distinct seasons: a wet season that runs from October to May and a dry season that extends from June to September, cold in winter and hot in summer. From the series measured by the station Fez Sais, the average annual rainfall is 494 mm, and the calculated average temperature is of the order of 17.5 ° C (ABHS, 2015).

The hydrographic network is very important, most rivers which constitute this network are supported by eternal sources. It converges towards the valley of the Wadi Fez which is thus the main collector which carries the water of the two sides which frame the city of Fez, towards the Sebou River.

Figure 1. The geographical location of the study area

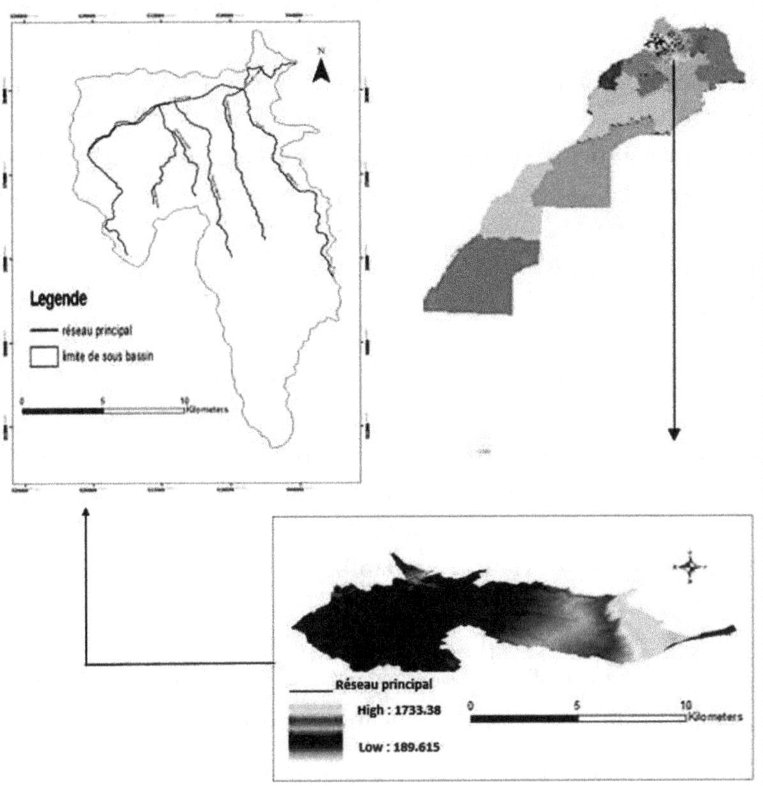

MATERIALS AND METHODS

Hydrological Study

The hydrological study is a very important phase in the study of the protection against the floods; it aims at recognizing the floods by the aspects of the peak flow. The latter is essential for the hydraulic simulation of El Jawaher River. For the present purpose, it is necessary to go through two essential steps, the estimation of the concentration times by different formulas, and finally an estimation of the peak flows.

Concentration Time Estimation

The concentration time (referred to as Tc) is the time elapsed before a water particle falling at the furthest point of the watershed reaches the outlet. It is equated with the length of time between the end of the rain and the end of the runoff.

Based on the physical characteristics of the watersheds, we calculated the concentration time by different formulas that of Ventura, the Spanish formula, Van Te Chow Formula, the Californian formula, Formula of US Corps of Engineers, Formula Kirpich, Formula Turraza-Passini and Giandotti's Formula (Musy, 1998; Musy & Higy, 2004).

Very large or very small extreme values were eliminated to leave only the central values, and the average values were retained. Table 1 summarizes the retained value of the concentration time.

Estimation of Peak Flows

The flow rates of these rivers can be calculated by several empirical formulas such as the method of Fuller 2, Hazan Lazarevick, and Mallet Gautier. As well as by the transposition from the Francou-Rodier formula commonly used in Morocco, and without forgetting the Gradex method, which relies on historical climatic observations, is to deduce the asymptotic behavior of the law of probability of volumes. Rare floods and the law of probability of cumulations of extreme rainfall (Guillot and Duband, 1967).

The El Jawaher river is the main collector of the watershed hydrographic network, the time of concentration of the rivers is very short (Table 2) which makes the forecasting and warning operations of local residents difficult. Along these rivers

Table 1. The characteristics of watersheds at the entrance to the city of Fez

Underwater Basin	Area (km^2)	Length of the Talweg	Difference in Height (m)	Concentration Time (Hours)
Boufekrane	48	28	775	3
Mehrez	122	22	700	2
El Himmer	80	36	1084	4
Ain Chkef	145	62	1450	7

Table 2. Flow rates

Watershed	T = 10 Years	T = 50 Years	T = 100 Years
Boufekrane	24	53	68
Mehrez	76	177	217
El Himmer	32	95	112
Ain Chkef	3	11	16

one encounters hydraulic structures constituting real points of constriction of the flow and can thus be at the origin of floods, resulting from the elevation of the water line, affecting the road system and a part of the Medina.

The Al Gaâda dam installed on Boufekranne Oued offers a high degree of protection, with operation conditioned by the draining of one third of its restraint (800000 m^3) (Agence du bassin hydraulique de Fès A.B.H.S., (2015)). Due to its proximity to El Jawaher River and its high waterways, it ensures the flow even during low water periods which is taken as a basis for calculation for the hydraulic study of this project (max to take into account floods, low flows for the normal operation of the thresholds).

Hydraulic Study

The hydraulic study consists in making the diagnosis of the study area for any hydraulic event the goal of realizing a hydraulic model to calculate the elevation of the water line for gradually varying flows, on the one hand, and of on the other hand to determine overflow areas.

The hydraulic analysis for the establishment of the flood zones, was based on the hydraulic modeling along the river by the use of the HEC-RAS software (Hydrologic Engineering Center, River Analysis System), which is a computer software capable of modeling free surface flows in natural and artificial channels with consideration of crossing structures. It allows to simulate the gradually varying flows in transient mode, to perform calculations of water lines in dynamic mode by simulating the various obstacles along the stream. River modeling is based on cross sections, takes into account all existing structures, and allows different roughness weightings to be defined for each section (Davis, 2008)The hydraulic simulation is built on the basis of 31 profiles made on the 1/2000 restitution plan of Fez.

Input data:

- The topography of profiles across streams;
- The distances between the profiles;
- The Manning weighting: also called roughness coefficient, the value used in our simulation was determined on the basis of the Van Te Chow table reference table (1959);
- Flow rates and boundary conditions (upstream and downstream), peak flows upstream of each section, normal slope (0.016).

Figure 2. The plane of the cross sections on the El Jawaher River

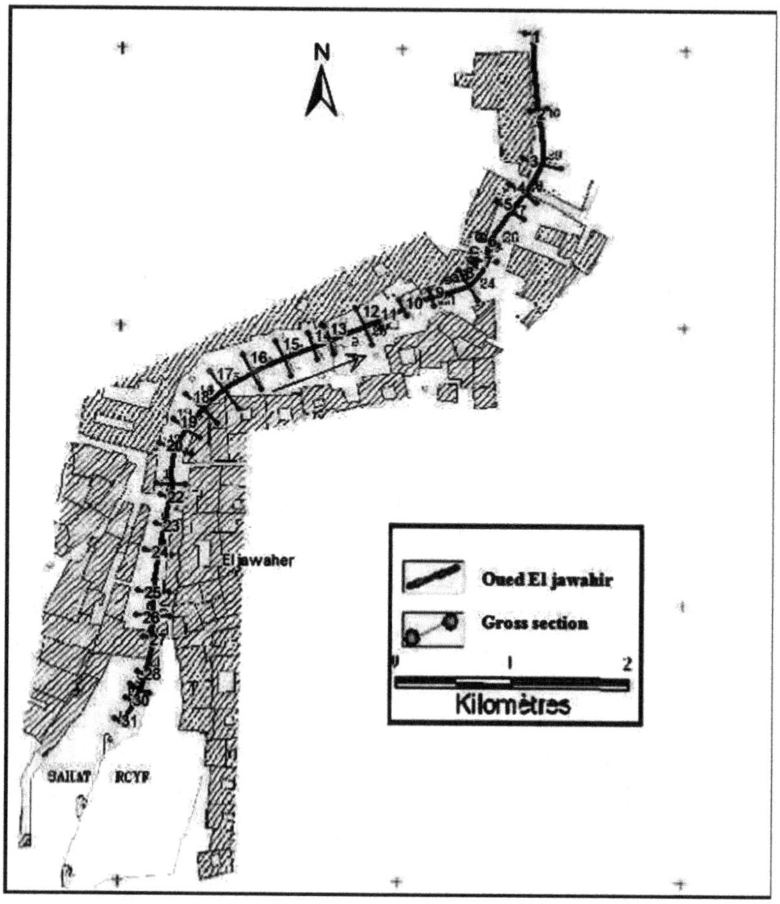

The results of calculations:

- The water and energy levels in each cross section;
- The flow velocities in each section;
- The profile along the lines of water.

RESULTS AND DISCUSSION

In order to meet the objectives of the present study, we proceeded first of all to the construction of a model HEC-RAS on a section of 400m of El Jawaher river, consisting of 31 cross sections extracted from the plan restitution of Fez to 1/2000, before moving to the execution of several hydraulic simulations in steady state, for the two current states and arranged by thresholds.

For Modeling the Current State

The modeling of the current state was done with a flow of $70m^3$ / s with a total of 31 sections across the river. From the longitudinal profile (Figure 3) which shows the level of water reached as a function of the distance between the sections across the river, it is observed that the flow rate varies according to the slope of the river. 'river, the values of these speeds show that the flow has a torrential regime, the dimension of the water line exceeds the minimum dimension of the channel with values of about 3m (The hydraulic section is insufficient compared to the flow in some places); consequently, the river overflows on both banks on certain sections of the canal studied. Flooded areas, especially downstream, are important. These overflows combined with high speeds that vary between 1.45 m / s and 11.17 m / s is a threat to the riverside homes.

State Adjusted by Thresholds

The future state modeling was performed based on different flow rate values (Table 3), the sections were modified to take into account the setting of the thresholds.

Figure 3. The profile along the water line for flow

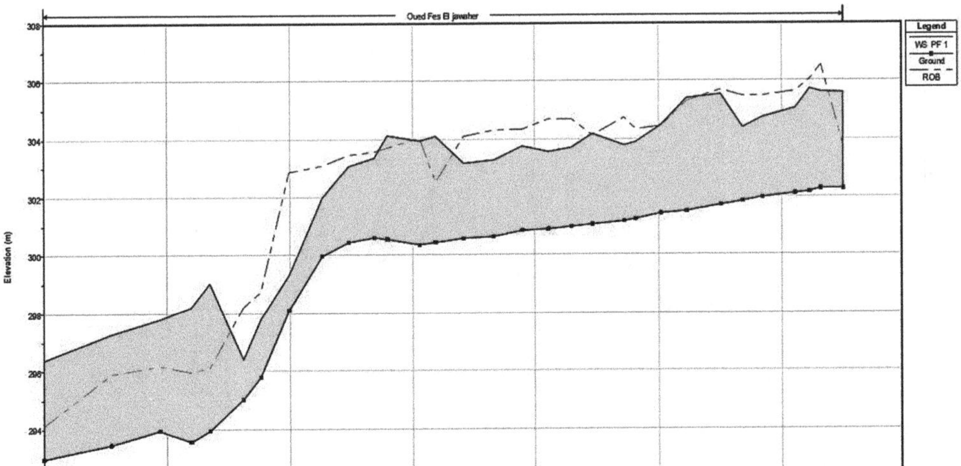

Table 3. The flow rates used in the simulation

Section	Q1	Q2	Q3	Q4
El Jawaher River	0.1 m³/S	0.5 m³/S	20 m³/S	70 m³/S

The objective of these thresholds is to create a permanent water level at the level of the canal which is integrated in the tourist and urban planning of the area of El Jawaher River. The presence of a threshold creates an elevation of the water line upstream of the threshold, a hydraulic control section at the threshold, which can lead to the creation of water bodies upstream of the structure.

The development of a threshold with HEC-RAS was carried out by the input of the geometrical data of the superficial points of section considered; it is about three thresholds with a height of 3 m (Figure 4), and variable lengths depending on the width of the bed of the river. The height of the thresholds depends on the altitude between the maximum and minimum dimension of the left and right banks of the initial sections.

These two simulations were carried out using low flow rates of the order of 0.1 and 0.5 m^3 / s. It is concluded that the flow velocities become low, often less than 1.0 m / s, from the upstream to the threshold presented in section 10, where the

Figure 4. The profile along the water line of the thresholds for a flow rate of (0.1 m^3 / s)

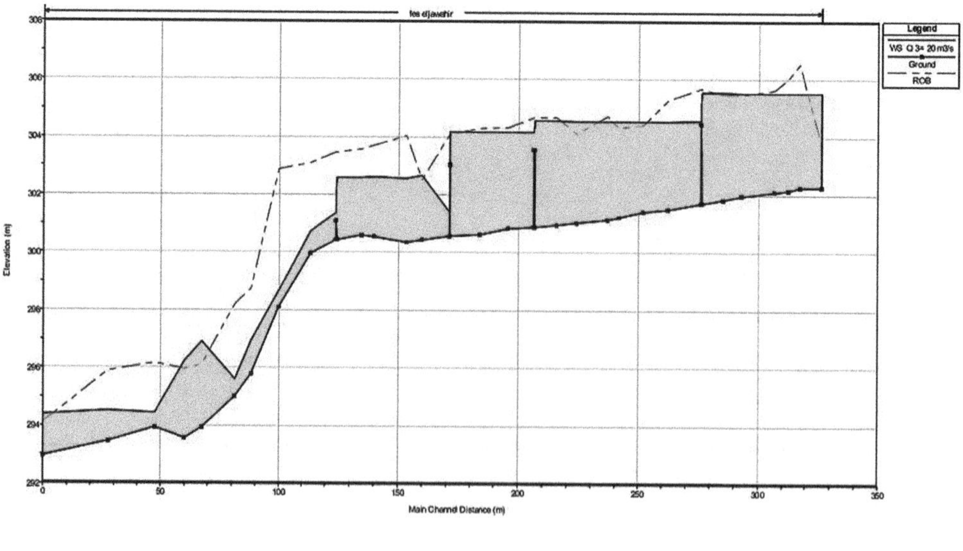

Figure 5. The profile along the water line of the thresholds for a flow (0.5 m^3 / s)

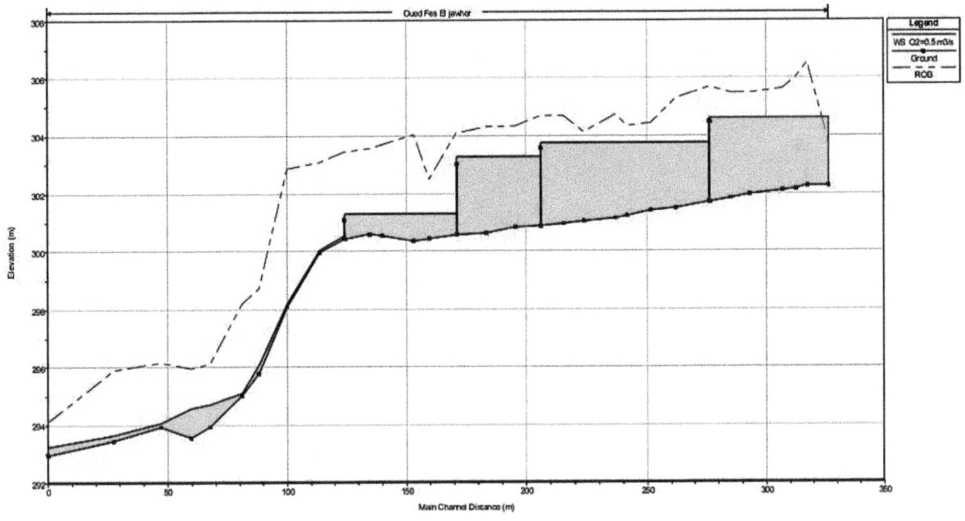

Figure 6. The longitudinal profile of the water line of the thresholds for a flow rate of (20 m^3 /s)

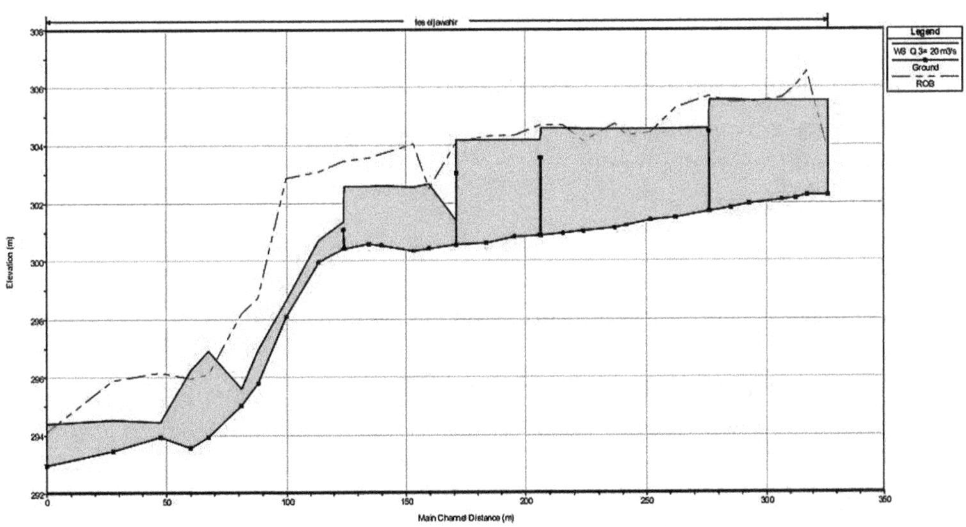

Figure 7. The profile along the water line of the thresholds for the flow rate of (70 m^3 / s)

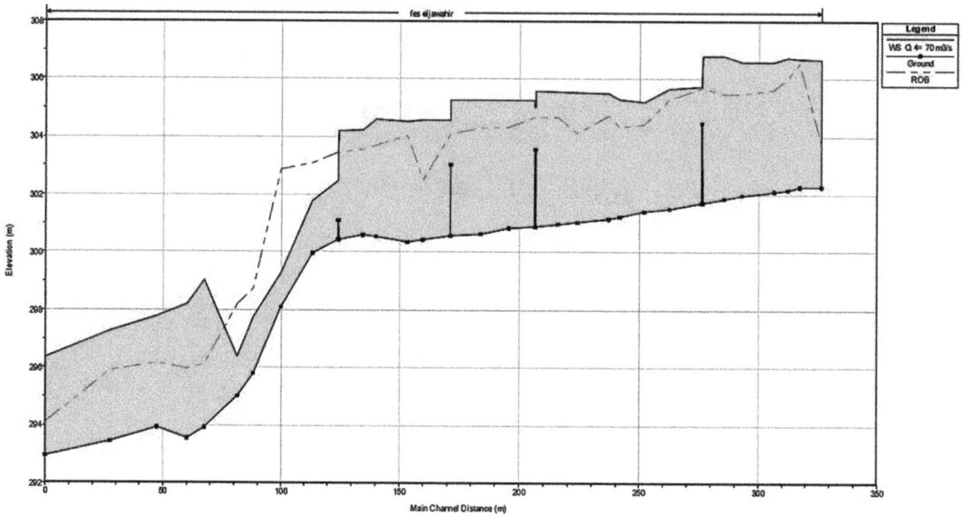

velocity becomes greater than 3.0 m / s, which can be explained by the presence of the thresholds causing the slowing down of the flow.

We can conclude that in these two cases presenting the period of low water, that the objectives of the project of development can be reached. The blades at the thresholds vary between 0.1 m and 0.2 m. This gives a visual appearance of an acceptable spill for development.

For a flow of 20 m^3 / s, there is:

- An elevation of the level of the water line at an altitude exceeding that of the thresholds of about 1 m of altitude, and consequently an overflow along the canal.
- The flow velocity values vary somewhat (not more than 3.0 m / s) going from upstream to downstream to point 15 where it increases to a value of 7.1 m / s.

A hydraulic simulation with a flow of 70 m3 / s shows, in a good way, the overflow upstream of the threshold. Comparing the results of this simulation with that of the current state, there is a slight slowdown of the speed in the last simulation, which is caused by the presence of the thresholds, and the enlargement since we are witnessing an overflow on both shores, and consequently the creation of floods.

This simulation confirms that the capacity of the proposed hydraulic structures is insufficient to transit the flood and additional developments should be considered. Hence, the need for the installation of mobile valve thresholds to allow the passage of floods.

CONCLUSION

The canal increases the flow of the centennial flood of 70 m^3 / s to the current state with a recalibration of the section just upstream of the channel's fall. The development of the canal by thresholds that were used for the creation of water body amplifies the risk of overflow on homes in the study area for high flows (two cases were simulated 20 m^3 / s and 70 m^3 / s). To overcome the problem of these overflows two solutions to consider:

- Elevate the banks of the canal by draining the water line along the longitudinal profile.
- Proceed to the installation of the mobile thresholds; these are "all or nothing" valves.

The principle of operation of these valves is based on an upstream level of the water to be detected by automatic sensors and subsequently the deletion of thresholds in a faster way. This system requires ongoing maintenance by the departments concerned to ensure its continued operation. It can be concluded that the mobile valve threshold variant is the most suitable.

REFERENCES

ABHS. (2010). *Etude de schéma directeur de protection de la ville de Fès contre les inondations*. ABHS.

ABHS. (2015). *Data*. Agence du bassin hydraulique de Fès.

Bouvard, Gaross, & Berthet. (1994). Les crues de projet des barrages méthode du gradex, editors. *Bulletin du comité français des grands barrages*, 17-23. Available on: https://hydrologie.org/BIB/Gradex/Gradex.pdf

Guillot, P., & Duband, D. (1967). La méthode du gradex pour le calcul de la probabilité des crues à partir des pluies. *Flood and Their Computation. Proceedings of the Leningrad Symposium*, *84*, 560–569. Available on: https://iahs.info/uploads/dms/084063.pdf

Helga-Jane, S., & Richard, L. (2004). *Risque d'inondation et aménagement durable des territoires*. Available at http://books.openedition.org/septentrion/15682?lang=fr

Musy, A. (1998). *Hydrologie appliquée*. Bucharest: Edition HGA.

Musy, A., & Higy, C. (2004). Hydrologie, une science de la nature. *Coll. Gérer l'environnement, Presses polytechniques et universitaires romandes*, 314.

U.S. Army Corps of Engineers. (2008). *HEC-RAS (Version 4.1) Hydraulic Reference Manuel*. Davis, CA: Hydrologic Engineering Center, U.S. Army Corps of Engineers. Available on: http://www.hec.usace.army.mil/software/hec-ras/documentation/hec-ras_4.1_reference_manual.pdf)

ADDITIONAL READING

Chow, V. T. (1959). *Open-channel hydraulics*. New York: McGraw-Hill Book Company.

Section 3

Climate Change, Natural Hazards, and Anthropogenic Impacts

Chapter 10
Flood Hazard Casting and Predictions of Climate Change Impressions

Vartika Singh
Amity University, India

ABSTRACT

Climate change is a word that we have heard hundreds of times, but what is it? Is it happening or is it something made by us? There are thousands of such questions, thoughts, doubt which come to our minds as soon as we hear the words "climate change." Even though there are hundreds of research works and many more proofs stating that the climate change is happening, there is a side which has been generally overlooked, and that is what if the climate change that we look is just something made by us. Climate change refers to long-lasting changes in temperature, clouds, humidity, and rainfall around the world. Both local and global factors cause regional climate change. This difference is significant because if a regional climate change occurs on account of local factors, then these changes can be mitigated by local actions. This chapter explores flood hazard casting prediction of climate change impressions.

DOI: 10.4018/978-1-5225-9771-1.ch010

INTRODUCTION

What is climate change? According to the wired " Climate change is the catch-all term for the shift in worldwide weather phenomena associated with an increase in global average temperatures." There are many proofs that climate change is happening and one of the significant evidence is the melting of the ice in the North Pole. Big glaciers have been melting in the North Pole making the covered area smaller day by day. But according to different studies it is seen that the North Pole has a small island like structure with open waters surrounding it, whereas the South Pole has a closed structure. Due to these water currents in the North Pole are greater in comparison to the South Pole, therefore suggesting drifting of the accumulated ice in the North Pole towards the warmer currents which then results in melting of this ice. It advised that the glaciers in the North Pole are melting but are melting due to the natural cycles of the Earth. Although due to increased global temperature because of the human activities, the rate of this natural decay have risen by 4.4% per decade in the North and 1.8% in the South.

Another proof for climate change is the warming of the global temperature. Even though there is a cycle of the Earth, where it cools and then again warms up, but the global temperatures have risen dramatically. Not only does the Earth have cooling and warming cycles but our sun have these cycles as well, and they are known as the solar maxima and the solar minima. These cycles for the sun lasts for about 11 years at a stretch whereas the cooling and warming cycles of the Earth lasts about 80-90000 years for the cold period and 10-20000 for the warming periods. Even though we are in the warming period, but the temperatures of the Earth have risen dramatically, therefore causing concern for the scientists and people worldwide. Human activities have worked as the catalyst in the process of the warming cycle of the Earth. The current environmental situation of the world proves that the climate is changing but if we think for a moment this catalytic reaction in the fast pacing of the warming of the Earths warming period means one more thing and that is, the cooling period of the Earth will also begin earlier. But this being a far-fetched theory with no proves have been discarded by the scientists.

The second problem which has widely shown as a proof for the climate change would be looked in this extract, and that is the flooding or uprising of the water due to change in the climate and global warming, refers to the observed century-scale rise in average temperatures of the Earth's climate system. Climate change was caused happened by factors like biotic processes, variations in solar radiations received by Earth, plate tectonics, the concentration of carbon dioxide, global temperatures, and volcanic eruptions.

We can also say that a strong correlation between higher temperatures and more carbon dioxide. The increase in surface temperature and changes in rainfall patterns may have an impact on vector-borne diseases. Forcing mechanisms is a group of factors which can shape climate, sometimes we also called it climate forcing, it also can be internal or some time external. Inside the climate system, internal forcing mechanisms are natural processes; for example, thermohaline circulation. External forcing mechanisms can be either anthropogenic – caused by humans, for example, increased emissions of greenhouse gases or natural for example, changes in volcanic eruptions, Earth's orbit, etc. The most significant human influence has been the emission of greenhouse gases such as carbon dioxide, methane, water vapor, and nitrous oxide. Climate change is approached primarily as a developmental challenge and marked by determinations to explore how various objectives of development, equity, and climate mitigation can simultaneously be met. Therefore climate change will add the new dimension to the problems produced by uncontrolled population growth and improper development policies in the country. It not only occurs in the riverine villages but also occurs in the seaside area and the island nations as well. One of the leading cause for the flooding has been seen as the melting of snow and glaciers, then be it the polar regions or the hilly regions or the regions of the world with accumulated snow such as Greenland. Flooding is a problem which is harming us no matter what side we are on, as it is affecting us directly and no one can deny it. Due to these sudden rises in temperatures, higher evaporation rates, and sporadic rainfall have been seen. Not only have that rate of melting increased throughout the world, therefore, increasing the water in almost all of the water bodies. Another cause for the flooding is the increase in the population due to which settlement in flood lands have been more prevalent nowadays, leading to more floods these days.

However, observed changes in regional climate cannot be attributed confidently to one cause. This is because regional climate change is caused by both local and global factors. There have been large changes observed in pollution profile and land –use pattern in the affected area. This aspect is usually discarded by the science community as the majority of the people stand with the idea that climate change is occurring. Even I lie in the same sect as the other scientists, researchers, stockholders, decision makers and go with the majority, but some of the aspects which we cannot overlook and those things are the ones which I'll be mentioning in this particular extract.

THE EVOLUTION OF FLOOD RISK

Climate change is happening, and that is a fact which cannot be denied. There are some aspects of the climate change which we are overlooking like natural cooling and warming periods of the Earth and the melting of snow in the North pole is higher than the South pole etc. These activities of nature are showing that the topic of climate change we have half of the knowledge, or we can say the government only provided half information to the general public, and the other half is unseen for some random agenda. Flooding is demarcated as the dry land near the alluvial area or water bodies around which societies and human are developed. The presence of humanoid increasingly interconnected with this phenomenon. Currently, the most densely settled and growing areas are situated neighboring water body, river or at the coast area. With the increasing level of sea water and change of rainfall concentration due to climate change, frequent flooding happens all around the world. Flooding is known as the most dangerous and repetitive natural hazard that people face day to day. Flood risk is driven by the three components of risk: hazard, exposure, and vulnerability (IPCC, 2012). We can measure the intensity of the flood with the velocity of flow and undulation of depth. If the flood is in the barren area or non-populated area it is not harmful or not come under the hazard. If the flood affected the human, animal, agriculture land and make the big economic loss, then flood becomes the hazard. Flood risk (fig.1) drivers are affected by economic growth, human activities, climate change, and disaster risk management. Due to global warming or temperature rising, sea level is gradually expanding, after effect of that frequency of coastal and river flood is increasing. With the growing number of population and growing city outlet, exposure to the flood frequency rises around the worldwith economic development, changing building quality, and planned risk management. The vulnerability of the exposed elements has changed due to the economic development of the area, changes in quality of building material, and planned interventions of risk management.

HAZARD: CLIMATE AND FLOODING

For the batter assessment of flood hazard, we have a clear understanding of the whole process of flood and the main drivers which lead the flood process. We also know the actions of flood drivers concerning space and time. Usually, we can divide flood processes into three subdivisions: pluvial flood, river flood, and coastal flood.

Figure 1. Process of evaluation of flood risk

Now we have discussed the process of following three floods and significant climate conditions which are playing a leading role in flooding. In the delta region effect of these factors (fig.2) easily observed. For example, Wahl, Jain, Bender, Meyers, and Luther (2015) showed that this interaction exists across a major part of the U.S. coastline and that these interactions have been increasing over the past decades.

Flash and River Flooding

In the present world's dominant flood processes are flash and river flooding. Trigger point of river flood is heavy rainfall in the river catchment areas. Flash floods generally occur in lesser headwater rivers with comparatively small upstream catchment areas. This type of topography we find in higher altitude area within a short period. The main trigger factor for floods is heavy rainfall in a limited time which is governed by thunderstorms which carry a large amount of precipitation within a river basin. Intensities of the rainfall complemented with the amount of moisture and intense rainstorms, it is also known atmospheric moisture content. As this is nonlinearly dependent on temperature, the heavy storms arise with the

Figure 2. Climates and flooding as hazard

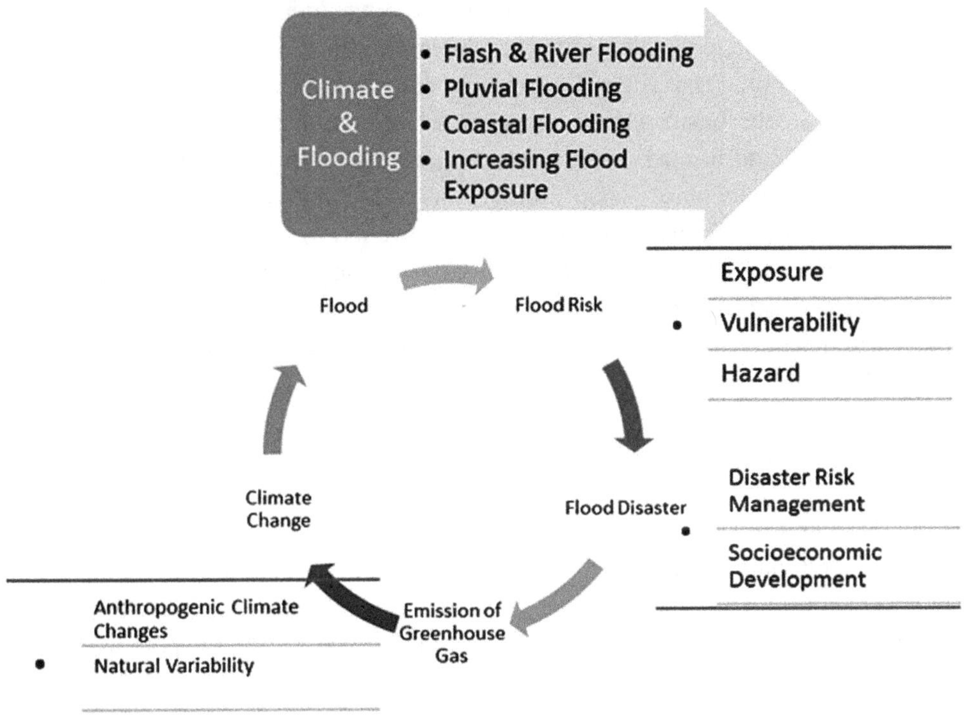

high temperature of the air. Nowadays we observe with rising temperature, days are extremely hot, and the frequency of thunderstorm increase at the global level, the intensity of that thunderstorms are also increasing. This is shown, for instance, by Trapp et al. (2007) over the United States. At the local level, intense rainfall events even scale more nonlinearly potential towards atmospheric moisture content. This is mainly related to local pluvial flood hazard, described in the section "PLUVIAL FLOODING" (Lenderink, Mok, Lee, & van Oldenborgh, 2011; Lenderink & van Meijgaard, 2008).

River flooding signifies a larger scale of flooding that happens due to heavy and intense rainfall and falling over a larger upstream basin area. Therefore, the main cause of these type of floods is advective storms or seasonal rain. For instance, to assess flood hazard in the Netherlands, 10-day rainfall accumulation is often used as a proxy (see, e.g., Kew, Selten, Lenderink, & Hazeleger, 2013). With the after effect of climate change, river flood hazard also changes, therefore, are ambiguous and difficult to model. Although in many areas, flood models forced by climate model outputs show increases in flood hazard, other areas may see decreases in flooding under climate change (Hirabayashi et al., 2013; Winsemius et al., 2015).

It should be noted that river flood hazard also may strongly depend on human interventions in the basin. For instance, reservoir operations may either lead to a reduction in flood hazard, when a reservoir is operated with the target to store floodwater, or an increase when storage has a use for other purposes such as hydropower. The preconditions of heavy rainfall events are essential in these cases, as they may cause a higher or lower probability that a reservoir is capable of capturing the additional floodwaters. For instance, reservoir conditions and operations are believed to have played a significant role in the Brisbane floods in 2011 (see, e.g., van den Honert & McAneney, 2011).

Pluvial Flooding

Pluvial flooding is flooding that takes place because of very local rainfall. Most instances of pluvial flooding can be located in low-lying, flat polder areas and urban facilities. Within cities, the incidence of pluvial flooding firmly relies on two principal matters. The first detail includes the amount of urbanization and the potential of the soil to take in rainfall (the degree of accommodation to address nearby rain with the aid of either conveying the floodwater downstream with drainage and sewerage infrastructure or quickly storing it in storage ponds or underground sewerage basins). As inside the case of flash flooding, heavy rainfall occasions that are local each in space and time are the most applicable phenomena causing pluvial flooding. These, again, commonly arise at some point of nearby thunderstorms, and due to the higher convective capability, the maximum extreme rainfall activities correlate with days with very excessive temperatures and capacity atmospheric water vapor content. Lenderink et al. (2011) and Lenderink and van Meijgaard (2008) show that the increase in potential rainfall intensity with temperature can be as high as between 7%–14% per degree Celsius. Hence, as climate change progresses, more intense rainfall intensities and more frequently occurring pluvial floods may be expected. This has implications for the climate-robust design of urban and polder infrastructure.

Coastal Flooding

Extreme water levels and flooding along the coast occur due to high tides, storm surges, high waves, or a combination of these. A storm surge is a rise in the sea surface caused by storms with low-pressure and strong winds, like extratropical cyclones or tropical cyclones, such as hurricanes and typhoons. Tropical cyclones have lower interior pressures and higher wind speeds than extratropical cyclones, and as such, they typically produce much higher surges than extratropical storms. In addition to the storm characteristics (wind speed, pressure, angle, and size), the height of the storm surge is strongly influenced by the local characteristics of coastal areas, like the nearshore bathymetry and the geometry of the coastline (Resio & Westerink, 2008). Essentially, shallow areas with a wide continental shelf, such as the east coast of the United States or Germany, will experience much higher storm surges than areas with steep offshore slopes, such as Caribbean islands.

Increasing Flood Exposure

Increasing exposure due to rising population and urbanization in flood-prone areas is considered the main driver of increasing flood losses over the past decades (Bouwer, Crompton, Faust, Hoppe, & Pielke, 2007). The world's population rose from 3 billion in 1960 to over 7.5 billion in 2017. By the year 2100, the global population is expected to reach between 9 and 13 billion (Gerland et al., 2014). A large part of the population growth takes place in flood-prone areas. As a result, the global population exposed to a river and coastal flooding is estimated to have increased from 520 million in 1970 to 1billion in 2010 (Jongman et al., 2012), and this trend is likely to continue. Due to the attraction of water bodies, the growth of population and cities within flood-prone areas is estimated to be higher than outside of flood-prone areas in many world regions. In most regions of the world, and most profoundly in developing countries in Africa and Asia, population growth is combined with high levels of urbanization as people move to cities for jobs and markets. Cities are dense, highly concentrated locations of exposure and are traditionally often located near rivers or the coast. Hallegatte, Green, Nicholls, and Corfee -Morlot (2013) estimates that the growth of exposure in coastal cities, even without climate change, may lead to a tenfold increase in economic losses by 2050 compared to 2005.

The rapid development of socio-economic activities in flood-prone areas, including industrial, service, and trading, drive substantial increases in economic exposure. Another factor that contributes further to increasing exposure is the so-called levee effect (Di Baldassarre, Kooy, Kemerink, & Brandimarte, 2013). As flood defenses are protecting developed areas, the perceived increased level of security may cause even more people to move to that specific area. Population, urbanization, and flood protection, therefore, are observed to reinforce one another and lead to continuously increasing exposure in flood-prone areas.

AUTONOMOUS ADAPTATION TO CLIMATE CHANGE

In the face of climate change the global community, nations and local communities are undertaking action along two primary tracks: mitigation – the process of reducing greenhouse gas emissions and, thereby, associated climate change; and adaptation – the process of adjusting in response to, or in anticipation of, climate change.

Adaptation is not a new concept. Traditionally employed by ecologists, it has referred to the evolutionary process by which living organisms mold into a new environment. By broadening the scale of reference, the concept of adaptation can be used to describe how systems, both natural and human, evolve when faced with environmental changes. Most spontaneous or autonomous adaptation has taken place as part of the evolutionary process through which biotic communities have migrated or modified their structure and function to accommodate shifts in temperatures, rainfall, available nutrients, and habitat.

Adaptation is the process of adjusting in response to, or in anticipation of, changed conditions. In the climate change context, more specifically, it is an on

In climate change, Autonomous adaptation is an insentient method of system-wide coping and easily understood or famous as adjustments of the ecosystem. As very clear with the name reactive adaptation involves a thoughtful response to the climatic

Impact or shock and also prevent the same effect in the forthcoming year. In the advance climate change study's we introduce the preemptive adaptation action plane for advance preparedness and reduce its potential impacts. The main aim of this type of action is enhancing the protection capacities of natural systems and also affected in extreme climatic conditions. For the assessing of economic proficiency of adaptation strategies, analysts differentiate midst 'co-benefits' and 'no-regrets' measures. We must have a clear view between these two strategies."No-regret" strategies reducing the vulnerability of non-climate change and related benefits of the strategies exceed the costs of execution. These strategies are required whether or not climate change has taken place. To reduce the vulnerability of current climate extremes and their related variability flood control structures are a good example of no-regret strategies. These types of structures reduce the weakness of shifts in risk due to climate change. On the other hand,"Co-benefit" strategies, are designed exactly to reduce vulnerability related to climate change and also to create corollary profits that are not directly related to climate change.

In spite of national and international aid to cope with natural disasters nearly all such programmes have been reactive and cannot be an appropriate solution to tackle the 'dangerous climate change.' Community adaptation measures like changing sowing time for cultivation crops etc. needs is to be upgraded with modern scientific base and technological support. This study aims to develop an adaptation framework of local communities of the Himalayas which could be a pragmatic way to build upon the existing momentum of mitigation of climate change effects on livelihoods.

THE VULNERABILITY OF THE ECOSYSTEM TO CLIMATE CHANGE AND EFFECTS ON LOCAL COMMUNITIES

Intergovernmental Panel report of Fifth Assessment on Climate Change (IPCC) shows that the warming in the climate at the global level is unequivocal and since 1950s many observed changes are unprecedented. The ocean and surrounding atmosphere have warmed, the amount of ice and snow have reduced, risen of sea level, and increased concentrations level of greenhouse gases. It observed that the concentration level of Carbon dioxide is increased by 40% since pre-industrial times primarily from fossil fuel emissions (IPCC WG I 2013).

In many parts of the world, Mountains are responsible for the rapid change of climate. Some of the changes in the environment we can observe very clearly like change in the hydrological cycle may affect river runoff, accelerate water-related hazards, and affect agriculture, vegetation, forests, biodiversity, and health (Beniston 2003). Climate change is a significant concern in the higher Himalayas because of its Impending impacts on the ecology of the area, economy, biodiversity, and environment of the Himalayan region and surrounding areas.

Local communities dwelling in the Himalayan region are accustomed to harsh climatic conditions. These communities wholly depend on agriculture and NTFP collection for the livelihoods. With limited land holding and high rate of population growth these communities are already struggling to sustain their livelihood. Now with climate change, these communities would have to devise long term adaptation strategies to cope with the changes that are more drastic and are supposed to occur in a shorter period. There is a large pool of scientific literature that shows a decline in the yield of agriculture and horticulture produce due to climate change. Therefore efficient adaptation strategies have to be developed to boost up the resilience of the communities to climatic changes. Sustainable livelihood and economic upliftment of the communities should be on the top of the priority list of an adaptation framework.

REDUCING VULNERABILITY THROUGH SUSTAINABLE LIVELIHOODS

The starting point of the adaption strategy is the dropping situation of current vulnerability; reduction in poverty is crucial to the process. By research, we can say poverty is an important element and condition of vulnerability. Reduction of poverty requires a clear understanding of how native livelihood conducted there day to day activities and sustained properly. Resources and proficiencies are two parameters on which compare the livelihood of people and parameter can give the shape of scarcity and also help to reduce the scarcity from there life. To study the overall livelihood of people, we can see the effect of climate change on their life. Some of the factors which we cannot ignore during the study are how the people response to climate change with their present available resources. These present conditions are responsible for the success of adaption strategies.

Poor people depend on the environmental services for their survival and central element in this is the approach towards adaption comes under the management of an ecosystem. Under the management activities, some important practices are the restoration of water, watershed management, landscape restoration, and agroforestry. These types of activities are good support for livelihoods, and poor people maintain their life and safety. To adopted all these natural activities we safe our natural resources, reduce the level of natural hazard, developed a diversified livelihood and ready for the upcoming threat of climate change. If we fallow all these steps in a particular area, we see instant development in the area, and it is also contributing to long term capacity building. It also reduces the upcoming vulnerabilities in the area.

Sustainable Development Boon Or Problem. At recent times the biggest problems faced by humans is the depletion of the natural resources. No matter what country no matter who it is, everyone is facing the same problem. Natural resources ranging from fossil fuels to water everything are becoming scarcer day by day. In certain places, the situation is so adverse that water is almost over and the countries have to buy water from other countries. Not only water but food has been one of the issues as well, lots of countries buy vegetables and other raw materials from other countries just to meet the demands of the people. The rate at which the exploitation is going at present is so high that to meet the demand nothing is viable. Now the question is – what is the way out of the problem? Most of the countries are turning to sustainable development and not only that sustainable, eco-friendly villages and cities have been developed. More and more present-day cities and countries are developed in a way so that the risk to the environment is minimized and maximum output is attained from it, although there is a long way to go the work has already begun, and the risk to the environment has reduced.

Change in the way we harness energy has already seen a big leap especially in the sector of vehicles. Vehicles have changed from using fossil fuels to a more environment-friendly way of electricity. Harnessing the electricity from burning coals have changed to wind, sun, and water. But are these efficient? Are they a hundred percent fail-proof with no harm to the environment? Or are they creating more adverse problems for humans? There is not one but hundred different problems which have been created due to the use of the "so-called" environment-friendly methods, by solving one problem we have created hundred more and are looking to develop even more.

Although, I am not denying that these methods are better than the ones which we have been using previously these methods have their problems and in some situations are way worse than the prevailing practices. For example, the use of dams to generate electricity creates a lot of problems. Even though by using barriers we are reducing the pollution caused by the burning of fossil fuels and even the extraction of the fossil fuels is reduced but there other problems associated with the use of dams. Some of the issues include microclimate changes, flooding of forest areas, therefore, creating a risk for biodiversity and also reducing the lungs of the Earth, not only that biodiversity in the river is also disturbed and the water quality is harmed. Due to the dams sediments which are deposited in the riverside are stopped which in return causes soil erosion and degraded soil quality, resulting in poor crop yield, therefore effecting everyone at large. So just to solve one problem hundred more are created, not only that the major underlying problem of air pollution is countered by the fact that the forest is which is responsible for the purification of the air is effected gravely and thus is not able to purify the air.

But all the sustainable methods are not as bad as most of them help us in a lot of ways rather than hurting us. For example: by harnessing the solar energy a lot has been achieved, and it is one of the energies which is available to us in ample of amounts with almost no to very few side effects. Using such methods such as solar energy is good for us. To develop sustainably, we have to think and plan properly as everything is not good and doing certain things too much is harmful as well. The risk to humanity at large is reduced by developing sustainably but overdoing just creates more problems for us. Instead, we have to integrate the small aspects of sustainable development for the benefit and welfare of humanity. It is using creepers on the walls of the building to reduce the heat. It also to improve the aesthetics of the buildings. For this, we use self-watering plants in major cities on the side of the roads. We also encourage the public to use public transport, and electric vehicles are some of the examples which are helping the environment and is not creating more problems for us. In metropolitan cities and villages have been able to implement such methods perfectly and have already begun to carve the way to sustainable development for us. Although there are hundreds of factors which one have to look upon to see if the method proposed or which is being used to reduce the impact on the environment is viable or not. But one step toward environment protection is the right step taken as it shows the vulnerability due to the bad environment is reduced therefore giving us more chances to save the Earth from the extinction/ unfit for us to survive, for a matter fact any of the species to survive.

Sustainable livelihood is one of the ways to go as it reduces our chances to pollute Earth and therefore change the cycles happening on Earth. It also helps us to reduce overexploitation of the resources and therefore is one the best ways to help the environment. Although there might be issues with some of the methods if proper planning and implementations of things are done then the problems associated with these methods would be almost some to none. So sustainable livelihoods reduce the vulnerability caused by inefficient nonenvironmental methods.

VULNERABILITY AND ADAPTATION STRATEGIES

Climate change societal response involves lots of new concept and science. Along with all these activities, new terminology has also come in a frame. Here we discussed some selected concept and terminology. Intergovernmental Panel on Climate Change briefly explains the concept of vulnerability to climate change (IPCC 2001, 2007) as Climate Change Vulnerability - The degree to which a system is prone to - and not able to cope with - adverse effects of climate change; including climate variability and extremes. The vulnerability is explained as a function of a character, rate of impact and quantity of climate change. Vulnerability also shows the variation, adaptive capacity, and sensitivity towards surrounding, which help to expose the system (IPCC 2007). All thesis explanations describe the major contributing components of climate change vulnerability which is frequently used in modern science. These contain climate-change concepts, exposure of climate, sensitivity towards nature, and adaptive capacity of climate.

Following the terms has defined as:

- **Exposure**: Grade of climate stress on a particular unit analysis and it may be characterized as long-term climate change conditions or climate variability changes, magnitude, and rate of extreme events.
- **Sensitivity**: Rate at which a system will be affected by, or responsive towards climate incentives.

- **Adaptive Capacity**: Capability and potential of present nature to adjust with climate change activities and also include the climate variability and extremes situations. The capability of enough potential show towards damages, cope with consequences and take benefit from opportunities.

To evaluate the climate change exposure we have to assess the climate-related information. In this, we have included the past, current, and forecasted future conditions, for this we only use the reliable resource of information. We can apply these analyses at the local level, continental or spatial level; it is based on the resources we have. To judge the sensitivity of climate change, we must have background knowledge about community ecology, species biology. All these are helpful to calculate the probable effects of climate change revelation. We can assess the adaptive capacity of ecology knowledge of different communities- that they may or may not be sensitive towards climate change and have been identified. For the better understanding of vulnerability climate change mechanism, the concern of resource, manager of source and decision-makers are better positioned to evaluate alternative actions to respond to climate change, even in the face of considerable uncertainty (Nichols et al. 2011). These opportunity actions are referred to as climate trade model strategies.

Climate Change Adaptation includes actions that enable species, systems, and human communities to better cope with or adjust to changing conditions. These strategies may take some forms. Some have categorized strategies into three areas, including resistance, resilience, and facilitated transformation (Biringer et al. 2003, Millar et al. 2007, McLachlin et al. 2007). Resistance strategies for adaptation aim to prevent the direct effects of climate change. Frequently cited examples include building sea walls and coastal hardening to Avert effects of coastal sea-level rise (Klein and Nicholls 1999). Preventive measures to go off results of invasive species, or uncharacteristic panorama-scale fires, can also fall into this class. Resilience techniques goal to comfortable the capability to deal with the impact of climate alternate via ensuring that critical ecological technique – as presently understood – are restored to an excessive stage of feature or integrity.

For example, by way of securing huge and interconnected natural landscapes, styles of species dispersal and migration are secured to defend food-internet dynamics. Facilitated Transformation strategies anticipate the nature of climate-change-induced transitions and, Operating with these anticipated developments, encompass actions that facilitate transitions which can be congruent with future weather situations even as minimizing ecological disruption. Somewhat radical expressions of these techniques may include assisted migration of touchy species from modern habitats to locations wherein changing climates would possibly provide new habitat into the destiny (McLachlin et al., 2007, Milly et al. 2008). Some have characterized these resistance and resilience techniques as 'retrospective' due to the fact they emphasize the use of knowledge about ancient or modern ecological sample and procedure; i.E., protection, and recovery of herbal situations as they may be currently understood. Facilitated Transformation is, therefore, a 'potential' set of strategies in that they're primarily based on the hypothesis of destiny situations(Magnuss et al. 2011). Finally, there may be an essential temporal measurement to version strategies. Conservation choices are made by way of people, often inside the policy constraints of cutting-edge regulation and institutions. While conventional herbal resource management has been 'retrospective' – utilizing knowledge of past and modern-day conditions to inform contemporary control movements – planners are an increasing number of required to forecast destiny conditions scrupulously (see, e.g., Comer et al. 2012). This forecasting must try to decide the nature and importance of change likely to occur and translate that understanding to modern selection-making. It is no longer sufficient to assess "how are we doing?" and then decide what moves ought to be prioritized for the upcoming five-15 year control plan. One should now ask "in which are we going, and using while?" after which translate that know-how again into movements to take inside the close to-term, or medium-term, or the ones to display and count on taking over multiple making plans horizons. Considerable new science and policy could be required to help this new kind of natural resource decision making.

REFERENCES

Beniston, M. (2003). Climatic change in mountain regions: A review of possible impacts. *Climatic Change*, *59*(1/2), 5–31. doi:10.1023/A:1024458411589

Biringer, J. L., Hansen, L., & Hoffman, J. (2003). *Buying Time: A User's Manual to Building Resistance and Resilience to Climate Change in Natural Systems*. World Wildlife Fund.

Bouwer, L. M., Crompton, R. P., Faust, E., Hoppe, P., & Pielke, R. A. Jr. (2007). Confronting disaster losses. *Science*, *318*(5851), 753. doi:10.1126cience.1149628 PMID:17975052

Comer, Crist, Reid, Hak, Hamilton, Unnasch, ... Kutner. (2012). A *Rapid Ecoregional Assessment of the Mojave Basin and Range Ecoregion*. Bureau of Land Management.

Di Baldassarre, G., Kooy, M., Kemerink, J. S., & Brandimarte, L. (2013). Towards understanding the dynamic behavior of floodplains as human-water systems. *Hydrology and Earth System Sciences Discussions*, *10*(3), 3869–3895. doi:10.5194/hessd-10-3869-2013

Gerland, P., Raftery, A. E., Sevčíková, H., & Wilmoth, J. (2014). World population stabilization unlikely this century. *Science*, *346*(6206), 234–237. doi:10.1126cience.1257469 PMID:25301627

Hallegatte, S., Green, C. H., Nicholls, R. J., & Corfee-Morlot, J. (2013). Future flood losses in major coastal cities. *Nature Climate Change*, *3*(9), 802–806. doi:10.1038/nclimate1979

Hirabayashi, Y., Mahendran, R., Koirala, S., & Kanae, S. (2013). Global flood risk under climate change. *Nature Climate Change*, *3*(9), 1–6. doi:10.1038/nclimate1911

Intergovernmental Panel on Climate Change (IPCC). (2012). *Managing the risks of extreme events and disasters to advance climate change adaptation. A special report of Working Groups I and II of the Intergovernmental Panel on Climate Change*. New York: Cambridge University Press.

IPCC. (2007). *Synthesis Report*. IPCC.

IPCC. (2007). *Impacts, adaptation, and vulnerability. Contribution of working group*. IPCC.

IPCC. (2013). Climate Change 2013: The Physical Science Basis. Cambridge University Press.

Jongman, B., Ward, P. J., & Aerts, J. C. J. H. (2012). Global exposure to river and coastal flooding: Long-term trends and changes. *Global Environmental Change*, *22*(4), 823–835. doi:10.1016/j.gloenvcha.2012.07.004

Kew, S. F., Selten, F. M., Lenderink, G., & Hazeleger, W. (2013). The simultaneous occurrence of surge and discharge extremes for the Rhine delta. *Natural Hazards and Earth System Sciences*, *13*(8), 2017–2029. doi:10.5194/nhess-13-2017-2013

Klein, R. J. T., & Nicholls, R. J. (1999). Assessment of Coastal Vulnerability to Climate Change. *Ambio*, *28*, 182–187.

Lenderink, G., Mok, H. Y., Lee, T. C., & van Oldenborgh, G. J. (2011). Scaling and trends of hourly precipitation extremes in two different climate zones—Hong Kong and the Netherlands. *Hydrology and Earth System Sciences, 15*(9), 3033–3041. doi:10.5194/hess-15-3033-2011

Lenderink, G., & van Meijgaard, E. (2008). Increase in hourly precipitation extremes beyond expectations from temperature changes. *Nature Geoscience, 1*(8), 511–514. doi:10.1038/ngeo262

Magnuss, D. R., Morton, J. M., Huettmann, F., Chapin, F. S. III, & McGuire, D. (2011). A climate-change adaptation framework to reduce continental-scale vulnerability across conservation reserves. *Ecosphere, 2*(10), 112. doi:10.1890/ES11-00200.1

McLachlan, J. S., Hellmann, J. J., & Schwartz, M. W. (2007). A Framework for Debate of Assisted Migration in an Era of Climate Change. *Conservation Biology, 21*(2), 297–302. doi:10.1111/j.1523-1739.2007.00676.x PMID:17391179

Millar, C. I., Stephenson, N. L., & Stephens, S. L. (2007). Climate change and forests of the future: Managing in the face of uncertainty. *Ecological Applications, 17*(8), 2145–2151. doi:10.1890/06-1715.1 PMID:18213958

Milly, P. C. D., Betencourt, J., Falkenmark, M., Hirsch, R. M., Kundzewicz, Z. W., Lettenmair, D. P., & Stouffer, R. J. (2008). Stationarity is Dead: Whither Water Management? *Science, 319*(5863), 573–574. doi:10.1126cience.1151915 PMID:18239110

Nichols, J. D., Koneff, M. D., Heglund, P. J., Knutson, M. G., Seamans, M. E., Lyons, J. E., ... Williams, B. K. (2011). Climate change, uncertainty, and natural resource management. *The Journal of Wildlife Management, 75*(1), 6–18. doi:10.1002/jwmg.33

Resio, D. T., & Westerink, J. J. (2008). Modeling the physics of storm surges. *Physics Today, 61*(9), 33–38. doi:10.1063/1.2982120

Trapp, R. J., Diffenbaugh, N. S., Brooks, H. E., Baldwin, M. E., Robinson, E. D., & Pal, J. S. (2007). Changes in severe thunderstorm environment frequency during the 21st century caused by anthropogenically enhanced global radiative forcing. *Proceedings of the National Academy of Sciences of the United States of America, 104*(50), 19719–19723. doi:10.1073/pnas.0705494104

Van den Honert, R. C., & McAneney, J. (2011). The 2011 Brisbane floods: Causes, impacts, and implications. *Water (Basel), 3*(4), 1149–1173. doi:10.3390/w3041149

Wahl, T., Jain, S., Bender, J., Meyers, S. D., & Luther, M. E. (2015). Increasing risk of compound flooding from storm surge and rainfall for major US cities. *Nature Climate Change*, *5*(12), 1093–1097. doi:10.1038/nclimate2736

Winsemius, H. C., Aerts, J. C. J. H., van Beek, L. P. H., & Ward, P. J. (2015). Global drivers of future river flood risk. *Nature Climate Change*, *6*(4), 381–385. doi:10.1038/nclimate2893

KEY TERMS AND DEFINITIONS

Climate Change: When we observe significant change over long time period or decade, in heat, rainfall, wind pattern, and other climate conditions, we can say climate change is happening. It could be an amendment in Earth's normal temperature.

Flood Valuation: Flood valuation is an estimation of monetary value which we use to control the natural flood, it's helpful in management of catchment area flood risk planning of the catchment and also helpful in predation of flood risk guideline.

Hazard: The risk or threat to human life is hazard. It refers to any vulnerability present in an area or region that may have the potential to cause harm. Hazard happens due to any kind of natural or man-made activities, which may lead loss and dangers for environmental properties and human life.

Livelihood: Way to earn money for food, cloth, place and basic needs is known as livelihood. It depends upon the capability of an indusial person, efforts, and hard work. Livelihood should be sustainable if people enhance their well-being and proper use of natural resource.

Risk Assessment: Risk assessment is a process which helps to identify the foremost factors of risk, major the intensity of hazard, analysis of risk which is linked with hazard, control the risk factors, and reduce the effect of hazards.

Sustainable Development: Sustainable development is an expansion which meets the needs of the present world with the ability of future generations to meet their requirement.

Vulnerability Assessment: Vulnerability assessment is a well-defined process which helps to ranking and classifying the problem or target. It provides the basic knowledge, background of risk, infrastructure information, awareness, environmental threats, and proper action.

Chapter 11
Climate Change-Induced Flood Disaster Policy Communication Issues for Local Community Adaptation Resilience Management in Uganda:
Climate Information Services for Effective National Flood Risk Assessment Decision Communication

Wilson Truman Okaka
Kyambogo University, Uganda

ABSTRACT

Effective climate change and disaster policy communication services are vital for enhancing the adaptive resilience capacity of the vulnerable local communities in poor countries like Uganda. This chapter focuses on the effectiveness of the Ugandan national climate change and disaster policy information communication strategies in addressing national flooding disaster risks, highlights the recent trends of knowledge based responses to climate change induced floods, assesses the impact of the flood on the socio-economic well-being of local households and communities, and determines the vulnerability issues with corresponding adaptation strategies to floods in the flood prone country. Climate change flood risks have continued to

DOI: 10.4018/978-1-5225-9771-1.ch011

exact huge socio-economic loss and damage effects due to the vulnerability and weak adaptation strategies to floods. The national meteorological services tend to forecast seasonal flood events; some flood forcing factors; and the impact of floods on social, economic, ecological, and physical infrastructure are on the rise in some parts of the country.

INTRODUCTION

Uganda has experienced a flurry of serial climate change and variability induced flood disaster episodes that have caused wanton loss and damage of unquantifiable socio-economic and environmental impacts at all levels of local communities (2010).The country has often suffered prevalent: multiple civil strife; famines, drought; transport accidents, earthquakes; epidemics of disease; flooding, landslides, environmental degradation, technological accidents, crop pest infestation, livestock and wildlife disease epidemics from time to time, as a result of extreme climate events in all regions of the country(GU, 2010).

As a result, Uganda has often suffered from negative effects like; declining crops yields and increasing food insecurity; melting of snow caps and glaciers on Rwenzori Mountain; increased frequency and intensity of droughts and floods; reduced water supply; increase in pests and diseases for livestock, wildlife and crops; increase of vector-borne diseases, including malaria and rift valley fever, water-borne diseases like dysentery, bilharzias, cholera, and typhoid; increase in invasive species; declining levels of fresh water resources; rising sea levels, leading to displacement of people, and disruption of both; terrestrial and marine ecosystems and other important natural habitats; and natural resource based conflicts due to water floods that are caused by sudden heavy downpours (ICSU, 2008).

According to the Ugandan environment state agency, climate is a vital natural resource necessary for socio-economic development because the influence of climate variability on agricultural production shows that we depend on rain-fed agriculture for livelihoods and food security and climate change flood assessment strategies consider that (NEMA, 2014):

- Climate is a vital natural resource which should be well harnessed for socio-economic development;
- The utilization of the climate and atmospheric information is critical in aviation safety, agriculture and the efficient management of the environment;
- Resource users like farmers should participate in the monitoring and dissemination of climatic information;

- The promotion of international cooperation for smooth exchange of climatic information and control of transboundary atmospheric air pollution is important in the management of the resource; and
- Access to climatic data/information should be guaranteed on terms determined by the relevant authority.
- Improve coordination and exchange of meteorological information among various stakeholders;
- Strengthen the national meteorological monitoring networks with capacity to process data;
- Improve awareness among potential users and decision-makers of climatic and atmospheric information;
- Strengthen the infrastructure and man power for climate, meteorology and meteorological studies;
- Strengthen the Early Warning Information System for effective disaster preparedness and response to extreme climatic events or accidental hazardous emissions into the atmosphere;
- Strengthen national, regional, and global cooperation for climate and weather management facilities.

Furthermore, the increasing frequency and intensity of natural disasters like floods, droughts, and landslides are among the top climate change risks in the East African sub-region. Adverse impacts of climate change in the area include: sea level rise which also leads to infrastructure destruction along the coasts, submerging Indian Ocean small islands, salt water intrusion, contamination of fresh water wells along the coasts in Tanzania, beach erosions in Mombasa, Kenya, rampant floods, and droughts (GoT, 2006).In the past decades, Uganda has experienced an increase in the frequency and intensity of extreme weather events like heavy floods with huge socio-economic consequences (GoU, 2010). With rampant poverty, weak institutional capacity, lack of skills on climate change adaptation and mitigation, inadequate skills in disaster management, lack of technology, inadequate funds, and an economic dependence on natural resources.

Most local community Ugandans are too vulnerable to disastrous impacts of climate change relate. Poor climate conditions will continue to wipe the agricultural outputs, leading to higher food prices, dwindling national come, and worsening export trade. Low or lack of awareness of climate change global issues requires a communication strategy on the global environmental conventions of climate change and funds need to be allocated (GoU, 2010).At least 12,000 people were affected by one of a series of incidents of heavy flooding and landslides in Bukalasi and Buwali sub-counties in Bududa district in a natural flood disaster after a few days of heavy rainfall in the areas around Mount Elgon National Park with over 858 displaced persons and several cases of flood inflicted death (ACAPS, 2018).

DISASTER RISKS POLICY FRAMEWORK

The national flood disaster preparedness and management policy for Uganda seeks to: establish a framework for a prudent sectoral and cross-sectoral objectives, principles, and strategies for an integrated disaster management (2010). In addition, it promotes positive behavioural or attitudinal change towards disaster management; legal framework; institutional flood assessment, monitoring, and evaluation system; and effective information services for flood data collection, storage, analysis and dissemination (2010).

The key issues related to climate change induced flooding include: inadequate disaster risk management as a result of impacts made worse by climate change; Uganda's position in international climate change negotiations is not strong enough to represent and effectively articulate and influence the global negotiations the interests of Uganda; water supply endangered in quality and quantity because of climate change; and inadequate mainstreaming of climate in other important sectors such as communication, energy, food security, and agriculture. The major national constraints in especially the enabling flood disaster assessment management environment in Uganda include: conflicting inter-sectoral policies and legal instruments, conflicting interests of involved entities, media less interested in covering climate change policy issues, climate change is given low priority by policy and insufficient allocation of resources, poor public information and transparency, awareness of climate change induced flooding challenges are often quite low or biased cooperative sharing of responsibilities and institutionalized gender equality mainstreaming for inclusion

Uganda has accepted that climate change is one of the greatest challenges facing humanity this century. Uganda's sustainable development largely depends on sustainable use environment a land natural resources in order to avert the increasing climate frequency and intensity of variability and climate change impact on Uganda's socioeconomic development and the livelihoods of its people (GU,2010; GU, 2015) with the objective to ensure harmoniously coordination of climate change causes and effective adaptation and mitigation principles and strategies (NDP I,2010; NDP II, 2015).

- Promote an integrated approach to address the effects of climate change;
- Re-define climate change as a development issue;
- Mainstream climate change in all development policies; programs and projects;
- Provide and promote incentives for clean development mechanisms(CDMs);
- Promote effective response to climate change, induced disasters; and
- Promote implementation of climate change conventions;
- Develop institutional capacities for climate change management in Uganda;
- Strengthen resilience and adaptive capacity to climate induced hazards and natural disasters;
- Integrate climate change adaptation and mitigation into national strategies and plans;
- Strengthen education, awareness raising on climate change mitigation, impact reduction, and early warning;
- Strengthen advocacy and mobilization of human and financial resources to address climate change;
- Develop strategies for the transfer, acquisition and adaptation of relevant technologies to alleviate the pressure on fragile ecosystems and natural resources and contribute to mitigation of climate change;
- Develop and implement a capacity building program for climate change induced disaster prevention and response at national, local and community levels;
- Support Development and implementation of catchment based management and restoration plans;
- Support mapping out climate disaster prone areas to guide adaptation and mitigation efforts;
- Support scaling up of ecosystem based adaptation (EBA) to climate change; and
- Develop and implement mechanisms for harnessing opportunities for carbon financing.

Another Uganda district, Ibanda district (Western Uganda) is among the few Ugandan local government districts adopt climate action into its local community focused district develop plan II (2016-2020), community awareness of climate induced flood management adaptation strategy for better climate smart agriculture outputs for food security is still a work –in-progress due to implementation gaps. Ibanda district local households and communities are quite vulnerable to the hazards of climate change risks that exacerbate the poverty ridden communities across several households and community social strata (ILGDDP, 2016).

ISSUES AND OPTIONS IN POLICY COMMUNICATION STRATEGY

Flood risk assessment in Uganda is premised on the idea that, effective disaster preparedness and management depend on accurate information, projections or forecasts based on scientific precision will enhance monitoring for responses (2010). Effective communication is a key pillar in successful disaster preparedness and management operations. The media have a vital role climate information services (CIS) to link weather forecast centres to the public to provide early warning information for state and non-state actors to facilitate both private and public decisions (GU, 2010). Uganda ratified the UNFCCC in 1993, and acceded to the Kyoto Protocol in 2002. Uganda is also a partner state in implementing the nascent East African Climate Change Policy 2011. In a move to guide Uganda's immediate climate change adaptation actions, a national adaptation program of action (NAPA) was developed as early as in 2007.There is a national climate change policy for Uganda (GU, 2015) to tackle adverse climate change impacts like: droughts, high temperatures, heavy rains, hail storms, floods, and landslides (MWE, 2013).

FLOOD DISASTER KNOWLEDGE MANAGEMENT

Uganda has a flood disaster assessment policy framework at central or national, regional, and decentralized local government district disaster risk information management systems, that need to be strengthened for institutional effectiveness (2007). The information services cover climate change induced incidents of severe floods, mudslides or landslides, droughts, environment degradation, wetlands degradation and epidemic diseases (2007). The current national flood disaster information is collected by local community workers in the field and transmitted to the center for analysis and disseminated by the media, community leaders, and politicians (2010). Following on the National Development Plan I, the NDPII (2015/16-2019/20) (Uganda, 2015b) aims to achieve the objectives of Uganda's Vision 2040.ThePlanaimsto strengthen Uganda's competitiveness for sustainable development creation, employment for inclusive economic growth under these five priorities(GU, 2015):

1. Agriculture
2. Tourism
3. Minerals, oil, and gas
4. Infrastructure development
5. Human capital development

The Ugandan government has declared that the public sector must aim to (GU, 2015):

1. Develop disaster risk profile and vulnerability map of the country
2. Coordinate the development and implementation of disaster mitigation for mitigation, preparedness and response to natural and human induced disasters
3. Coordinate regular disaster vulnerability assessment at community level, hazard forecasting, and dissemination of early warning messages
4. Resettle landless communities and victims of disasters
5. Coordinate timely responses to disasters and emergencies
6. Provide food and non-food relief to disaster victims
7. Coordinate state and non-state actors in the IR mandates towards disaster issues
8. Develop and implement humanitarian interventions and support livelihoods of disaster

In the process, the following national project outputs are expected to emerge from the planned project inputs: sustainable management of environment, natural resources, and land use; enabling policy for effective environmental management/ecosystem established; economic productivity enhanced with environmental and natural resources; capacity at national, district, and community levels restored and protected ecosystems of national and global importance against degradation; climate resilient policies and measures; and financial options for national adaptation costs expanded at local, national, sub-regional, and regional levels. It is assumed that the consequences of climate change disaster risks, from the sub-regional social, economic, political, and environmental (natural resources), are quite dire.

In the recent past, frequent heavy rains had often occurred in Kasese district [Western Uganda region] and caused the many local rivers like Nyamwamba and Mobukuto burst their banks which and caused heavy floods in almost all nine sub counties of the district. Many people die and several thousands were displaced as a consequence of one of the floods (URCS, 2014). In addition, the livelihoods and infrastructure were destroyed, especially around the Kilembe copper mines. Furthermore, several bridges and houses were destroyed by the fast running water and boulders from the rivers Nyamwamba and Mobuku (URCS, 2014).

On the other hand, the Northern Uganda region's district of Adjumani has experienced adverse effects of climate change impacts: landslides, hail storms, heavy lightning incidences, increased accumulation of nitro oxide in the atmosphere, prolonged droughts, proliferation of new animal and plant diseases, insects attacks on plants like shrubs or trees, and sudden downpours of unpredictable heavy rains. A community communication strategy was been adopted to promote climate information services in the district (ILGDDP, 2016). Heavy flooding usually paralyses most socio-economic development activities across the entire local government district administration services.

PUBLIC INFORMATION COMMUNICATION SERVICES

All the regions (western, northern, eastern, and central) of Uganda are intermittently flood-prone due to extreme climatic and weather events which are associated with the impacts of climate change in the country and the government of Uganda has developed and adopted the flood disaster mechanisms and strategies across different strata of the vulnerable local communities at flooded ruaral, peri-urban, and urban areas (GU, 2010).Hence, media personnel need to be trained on the techniques of climate science reporting for disaster preparedness and resilience (GU, 2010). In addition, the national interagency technical committee does liaise with the ministry responsible for information and the private media to ensure accurate, consistent, and coordinated information and education flow (2010).

The regional community and mass media are not yet fully engaged in covering climate change science and issues. Training programs to assist both journalists and editors are essential, but civil society organizations must also improve the way they engage with the media, packaging information in a clear and simple way and actively attracting media attention. Local languages lack terms for many key concepts involved in climate change –including 'climate change' itself. Communicators should attempt to explain climate change using terms that already exist, using graphic examples of local environmental problems and innovative communication methods to get the message across.

At the same time, public figures are not being held to account for taking action on climate change due to low or no awareness of the causes and effects of climate change. Raising awareness of climate change is critical. Local and national politicians are ill informed about climate change although environmental services are decentralized under local governments. Needless to say, awareness campaigns should focus on local politicians to act on climate change. In Uganda, the members of parliament (MPs) have recently formed a special parliamentary committee on climate change.

Over 20 million Ugandans (68.5%) are classified as food insecure with the major cause of food insecurity in Uganda attributed to climate change effects that are manifested in forms of extreme weather conditions like: drought; shortage of water and pasture, crop failure, famine, increased food prices, food/Emergencies, inter district migrations, economic loss/loss of income, high temperatures; lead to escalating vectors (pests and diseases), crop wilting, poor yields, heavy rainfall; crop destruction, soil erosion and leaching, contamination of water sources, livestock and crop diseases, flooding; leads to increased crop, livestock, and human diseases; loss of lives and livestock; destruction of crops and infrastructure, post harvest losses, water pollution (GoU, 2010).

INCIDENTS OF NATIONAL FLOOD DISASTERS

Climate change induced floods have ravaged parts of Uganda, several areas experienced severe effects of floods that displaced up to 300,000 people (UMD, 2007) and the Uganda Meteorological Department, above normal heavy rains that affected the whole country for many weeks. The worst affected areas were in the Northern and Eastern parts of Uganda included the districts of Mbale, Manafwa, Bukeda, Budadu, Kumi, Soroti, Katakwi, Amuria, (Western Uganda region). Lira, Pader, Kitgum, Nebbi, Gulu(Northern Uganda). and scattered areas of the Central region districts of Uganda (UMD, 2007). Besides, up to 40 people died of mudslides and floods when heavy rains affected homes in six Bududa district villages, eastern Uganda (Adibayo & Ntale, 2018).

The adverse impacts of climate change are a major challenge to socio-economic development globally (EAC, 2011). The EAC sub-region is vulnerable to impacts of climate change, affecting key economic drivers like water, agriculture, energy, transport, health, forestry, wildlife, land use, infrastructure, and disaster risk management among others. The impacts include water stress and scarcity food insecurity diminished hydropower generation potential; loss of biodiversity and ecosystem degradation; increased incidence of disease burden; destruction of infrastructure; high costs of disaster management as result of increased frequency and intensity of droughts, floods, and landslides associated with the El Niño. The process of developing the EAC climate change policy was initiated. The summit directed the development of a regional climate change policy and strategies to urgently respond to the adverse impact of climate change, including addressing the challenge of food insecurity as a result of climate change.

El Nino climate change induced rains caused two local lakes, Kayanja and Akageti, to be dramatically formed overnight after heavy rains due to incidents of extreme climate events that affected Kiruhura district in South Western Uganda sub-region.

The flood severe socio-economic adverse effects to individuals, households, and the local communities when it displaced at least 170 families form the vulnerable local communities (The Viision Group, 2009).All the local government decentralized districts have been facilitated by the National Planning Authority (NPA) to develop a community strategy in which climate action planning took community participation. Isingiro district shares a common border and climate change regime with Kiruhura district (Western Uganda region) where a deluge of flood occurred in May 2009 and two dramatic fresh water lakes emerged overnight(The Nation Group, 2009). The lakes were later on used by the local pastoralists who used the El Nino formed lakes for watering over 3000 livestock. The floods affected about 170 vulnerable local community households which suffered devastating socio-economic loss and damage to livelihoods (The Nation Group, 2009).

Uganda's capital city, Kampala, is home to the majority of the country's built infrastructure and the scientific projections of future climate in the city suggest there will be a higher incidence of rainfall, putting Kampala at risk of flooding (GU, 2015).The cost of inaction is high, with estimates for the cost of flooding alone suggesting annual damages rising fromUS$1–7min2013toUS$33–102mby2050. And the adaptation measures would mitigate some of these costs considerably and the current building codes be revised for effective flood disaster resilience (GU, 2015).The study concluded that immediate steps must be taken to increase the climate resilience of Uganda's existing and new infrastructure. Theses hold include:

- Climate-proofing public buildings
- Developing standards for transport and infrastructure planning
- Integrating climate resilience standards into infrastructure risk assessment guidelines.

The national climate change policy costed implementation strategy outlines a range of measures for climate adaptation in new infrastructure. (GU, 2015). The economic assessment places these in order of priority as follows: Integrate climate change into the existing infrastructure risk assessment guidelines (GU, 2015).

The policy elements are grounded on three key pillars: adaptation, mitigation and climate change research. The pillars need capacity building; technology development and transfer; finance; education, training, and public awareness based on information and knowledge management. The East African climate change policy aims to create, develop, and sustain adaptation and mitigation capacity at all levels. Adaptive capacity refers to the potential or capability of a system to adjust to climate change, including climate variability and extremes, so as to moderate potential damages, to take advantage of opportunities, or to cope with consequences (Smit and Pilifosova, 2001). As the name suggests, adaptive capacity is the capability of a system to adjust to impacts of climate change. The following seven factors were identified in determining climate change impacts adaptive capacity: wealth, technology, education, institutions, information, infrastructure, and social capital.

The Uganda National Metrological Authority (UNMA, 2018) reported floods and landslides in low lying and mountainous areas, with more rains in most parts of the country. With lightening, floods, or landslides affected parts of Bududa, Bulambuli, Teso (Eastern Uganda), Bundibugyo, Kasese, Kabale, Rubanda, Kisoro (Western Uganda), Kapelebyong (North-Eastern)district (UNMA. 2010). Eastern Uganda has always been prone to serial landslides during heavy rains and some oof the worst flood disasters occured in March 2010 when100 people died in the villages of Nameti, Kubewo, and Nankobe in Bududa district (Eastern Uganda) and recently, heavy rains flooding and landslides caused harvocs in Bududa, Bulambuli (Eastern region), Bundibugyo, Kasese, Kabale, Rubanda, and Kisoro (Western Uganda)districts (UNMA, 2018).

FLOOD DISASTER INSTITUTIONAL COORDINATION

Uganda has national flood awareness campaigns programs on disaster risk reduction held every year in a national seminar which school children and the mass the media are involved using school music festivals, debates, discussions in the mass-media, posters, road safety week, Uganda Red Cross week, Environment week, among others (GU, 2005). The country has a disaster risk information management system on drought, floods, landslides, environment degradation, wetlands degradation and epidemic diseases (GU, 2005). Information is collected by community level workers in the respective field, analysed centrally and disseminated by the media, community leaders and politicians (GU, 2005).

The climate induced flood disaster preparedness management assessment responsible institutions are (2010).

- Ministry of Water and Environment (Lead Institution)
- Ministry of Agriculture, Animal Industry and Fisheries
- Ministry of Lands and Housing and Urban Development
- Ministry of Local Government

- Office of the Prime Minister
- Local Governments
- Community and Private Sector
- UN Agencies and NGOs
- Ministry of Health

A set of locally-driven criteria determined the selection of priority adaptation activities. They include (UNFCCC,2002): level of adverse effects of climate change; poverty reduction to enhance adaptive capacity; synergy with other multilateral environmental agreements; and cost-effectiveness. Ugandan marine police recently rescued many people and recovered dead bodies after the flash flooding which burst the banks of river Mayanja in Wakiso district (URN, 2018). River Mayanja burst its banks around Kawanda-Matugga area along Bombo road soon after a heavy downpour of torrential rains that caused heavy damage in Gobero, Gombe, Masulita, and Kakiri town council (URN, 2018). Several people were reported missing following the floods, raising fears that they could have been swept by violent floods.

The media disseminate government disaster preparedness and management plans in the event of an imminent disaster (2010). Likewise, other key flood disaster response actors are the national telecommunication companies. Given that information and modes of communication are critical in disaster preparedness and management, government employs the principle of public-private partnerships [PPPs] to reach out to telephone companies, internet service providers, and other communication channels to ensure effective delivery of accessible information to the vulnerable communities (2010).

Uganda Red Cross Society (URCS) is collaborating with MoWE to establish monitor flooding system in Manafwa River Basin. Sofar, Makerere University in partnership with Massachusetts Institute of Technology has completed modeling the flood characteristics. MoWE is currently installing and calibrating river gauges with plans to provide six community radios to assist communities to broadcast flood warnings (Atyang, 2014). Since 2011,URCS has collaborated with UNMA to broadcast seasonal weather forecasts to communities in selected areas using community radio network of URCS volunteers (Atyang, 2014). The weather information is obtained from regional meteorological offices in Lira and Soroti for communities in Apac and Katakwi, respectively. The community radios were setup by URCS but are managed by the communities (Atyang, 2014).

NATIONAL FLOOD DISASTER INCIDENTS

According to the Ugandan government flood assessment policy, a flood occurs when large amounts of water cover a place that is meant to be dry on temporarily or permanently (2010).Floods build up slowly. They are seasonal and usually occur in periods of intense rainfall and el-Niño phenomena. Besides causing death due to drowning, floods destroy public health facilities such as water sources and sanitation facilities(2010).The ultimate objective of the UNFCCC is to achieve stabilization of greenhouse gas concentrations in the atmosphere at a level that would prevent dangerous anthropogenic interference with the climate system within a timeframe sufficient to allow ecosystems to adapt naturally to climate change, to ensure food security and sustainable economic development (UNFCCC, 2005).

Uganda was among the flood pruned eastern Africa countries which bore the brutality of severe flooding and the emergency appeal sought support of USD 1,061,020or EUR 749,788) in cash, kind or services to support Kenya, Uganda, Tanzania, Rwanda, and Burundi National Societies for25,000beneficiariesof floods for two months(IFRC/RCS, 2009). While the El Nino effect is not expected to producehighlevelsofimpactthatwereseenin1997,exceptionalfloodingafter heavy rainfalls in the in the Eastern Ugandan districts of Soroti, Amuria, Katakwi, and Mbale, and Lake Victoria Basin (IFRC/RCS, 2009).

1. Damage to housing (landslides and destruction) resulting in widespread displacements.
2. Surface and ground water pollution that caused waterborne disease outbreaks.
3. Damage to crops (also increased production in some areas).
4. Livestock deaths and disease.
5. Outbreaks of disease, especially cholera, malaria, and Rift Valley Fever.

Basedonthepredictedimpactsofthefloodsandthepreviousexperienceof National Red Cross Societies in the El Nino affected East African region, determined that the emergency needs of flood victims were focused in the following sectors (NRCS, 2009).:

1. **Shelter**: Temporary for families displaced from damaged houses and transition and/or permanent housing for families whose houses are completely destroyed.
2. **Relief**: Non-food items distribution to most affected families as well as small quantities of food to replace food stocks.

3. **Water and Sanitation**: Emergency action to restore damaged water supplies and sanitation facilities and provide safe drinking water as well as hygiene promotion to reduce risk of disease outbreaks.
4. **Health**: replacement of mosquito nets at house-hold level and essential medical supplies at health centers and clinics.

Likewise, another Ugandan decentralized district, Adjumani local government administrative district (Northern Uganda region) has developed and adopted a decentralized climate action plan for adaptation resilience to the perceived climate change impact on the vulnerable local communities: severe disruption of transport and health services; dry spell of meteorological, agricultural, hydrological droughts; increased outbreaks of pests or diseases vectors; lack of pasturelands; agriculture production or productivity decline, raging famines due to food insecurity; extreme climatic events; water-borne diseases; diseases related to toxic algae, food security decline, heat waves, stress, air pollution; and severe outbreaks of communicable and non-communicable diseases; malaria cases; injuries, psychological impacts, death attributed to extreme disasters and lack of basic needs (ALGDDP, 2016).

FLOOD DISASTER POLICY GOAL AND OPTIONS

The overall Ugandan national flood disaster preparedness policy goal is to promote national flood vulnerability assessment, risk mitigation, disaster prevention, preparedness, effective response and recovery in a manner that integrates disaster risk management with development planning and programming(GU, 2010).This approach promotes local capacity building that enables communities to minimize the serious social and economic disruptions as a result of climate relate devastating nationwide flood disaster events (GU, 2010).

In addition, the targets are to: promote public awareness and socio-economic importance of climate change, including vulnerability, impacts, risks, and response measures to promote capacity building efforts through, inter alia education, training, research, technology development and transfer, information and knowledge management; promote climate change research and observations(2015).These involve effective monitoring, detection, attribution and model prediction for flood disaster preparedness. It is vital need to integrate climate action into national development planning for flood disaster risk management, gender equality mainstreaming; resource mobilization to implement regional climate change policy framework for flood disaster assessment planning or management (GU, 2016).

Kampala city [Central Uganda district] suburbs like Bwaise parish, have always been an area prone to flooding with ramshackled housing distressed by natural calamity whenever there is a heavy downpour (Ngwomoya, 2018). The floods get worse when water spills in the house, leaving us stranded the whole night. Victims flood prone city suburbs like: Kasubi, Katanga, Kisenyi, Nalukolongo, Kabuusu, Katwe, Namungoona, Kinawataka, Queen's Way, SsebaanaKizito Road (old Nakivubo Mews), Kabuusu junction, Jinja Road roundabout, and Kyambogo-Banda T-junction. City floods often leave dozens dead. Since its establishment in 2011, Kampala Capital City Authority (KCCA) upgraded many drainages to control floods.

In this connection, the following policy issues and actions should be performed at national levels: enhancing capacity of regional institutions in to carry out climate change related research including climate change monitoring, detection, forecasting interventions; promoting climate change issues in education and learning curricula; promoting development of climate change tools, methods; technologies; development, deployment, adoption, diffusion; developing climate change knowledge sharing and management tools. These tools include: databanks, regional network for sharing lessons, experiences and best practices amongst states and other countries; harnessing and integration of indigenous technical knowledge in modern knowledge; promoting climate change national and regional institutions to strengthen capacities; developing human and technical skills in adaptation and mitigation. The followings are some of the approaches that should be employed to implement the policy: training, education, and learning; forming a regional climate change negotiation platform; capacity building on carbon funding mechanisms through the current global funding mechanisms; strengthening participatory planning, and decision making.

Ugandan government's national climate change policy (GU, 2015) advocates effective decentralized climate action planning for climate change and variability disaster risks' resilience adaptation at all local government district development planning levels. For example, Isingiro district(Western Uganda region), is one of the first Ugandan local government districts to integrate climate change into its key priority development sectors: agriculture, education, administration, community services, water sources, natural resources, health, internal audits, finance, roads and buildings, legal, planning sectors. The decentralized local community flood disaster preparedness climate action plan is to be implemented for the first five years (ILGDDP, 2016).

CLIMATE CHANGE INDUCED POLICY ISSUES

Flood disasters are also trigger outbreaks of water-borne diseases and malaria, hence compounding community vulnerability to health hazards (GU, 2010).They also cause physical damage by washing away structures, crops, animals and submerging human settlements. The risks of floods can be minimized forecasting, studying seasonal patterns as well as the construction and maintenance of sufficient drain age systems (GU, 2010).Floods could be properly managed through flood plan mapping and surveys by air and land be casue Uganda experiences both the flash and slow onset floods which are common in some urban areas, low lying areas and areas a long river banks and close to swamps. The areas prone to them are Kampala, Northern, and Eastern Uganda(GU, 2010).

The national and local community based flood actions in Uganda were designed to (2010):

1. Create awareness in the communities on flood risk reduction measures.
2. Enforce riverbank management regulations
3. Protect and restore wetlands
4. Ensure proper physical planning of rural and urban settlements.
5. Gazette flood basins

Similarly, the climate of Africa is diverse, and controlled by complex interactions between the oceans, land, and atmosphere at local, regional, and global scales (ICSU, 2008). On average, Africa is hotter and drier than most other regions of the world, and has a less dependable rainfall. As a consequence, and considering the fact that livelihoods at all levels – from the individual household to the regional economy – depend heavily on climate, several studies have concluded that Africa is among the most vulnerable continents to the climate changes that threaten even higher temperatures and greater variability in future (ICSU, 2007). Capacity building faces key challenges in the sub-region. Knowledge, technology, and capacity gaps with a few exceptions, countries in sub-Saharan Africa lack the capacity to conduct research on natural and human-induced hazards and disasters, or to apply the knowledge and deploy technologies to mitigate disasters (ICSU, 2007).

The mass media is more influential in spreading awareness about adopting new possibilities and practices of innovations to justify the importance of information dissemination as a precondition for awareness, attitudinal, and behaviour change for adoption of climate change innovations or technologies by the audiences ((Rogers,

1995; Okaka, 2010).At the same time, climate change (Panos, 2012) poses a significant threat to lives and livelihoods in Uganda. Government policies, low-carbon technologies and financial support from international donors will all play a role in Uganda's response to climate change. Central to the fight against climate change in the sub-region is effective communication and public engagement. At every level of society – from ordinary citizens and farmers, to the media, civil society organizations and local and national governments on the need for accurate and reliable information about climate change is very high since little is known about how to communicate climate change (Panos, 2012).There is a major lack of national co-ordination in the communication of climate change policy information with respect to flood events.

Climate change induced flood catastrophes are significant challenge to indigenous understanding the weather and farming. Local communities must be supported with climate change knowledge and information. Public awareness must be raised about the emerging carbon trading since carbon trading delivers incomes to individuals, families, and companies. Climate action dialogues on the costs- benefits of carbon trading for Uganda will promote understanding of the green fund access procedures and benefits for local community carbon trading (Panos, 2012).

There is a local need for the policy information gaps to be plugged and dissemination of information to be refined for climate change policy to have impact in Africa (Okaka, 2011). Most of the severe problems of the increasing vulnerabilities to the impacts of climate change among the indigenous communities in Uganda have come about because there are still information gaps regarding the functions, values and importance of the wise use of natural and environmental resources by communities, institutions, and industries. The governments, researchers and research institutions, research networks, civil society organization (NGOs), communities, and external development partners in the EAC are aware of this fact. Many policies, initiatives, programs, projects have been unsuccessful in developing behaviour communication strategy for climate mitigation and adaptation.

There is a need for flood information gaps to be plugged and dissemination of information to be refined for climate change adaptation policy to triumph in Africa. Uganda has climate change adaptation policy advocacy strategy on the hazards of climate change induced flood disasters in tandem with climate research, international cooperation, and gender equality (UNFCCC, 2002, Okaka, 2010). Studies of communication strategies for policy leaders found high demand for radio, TV, libraries, radio, books, reports, NGOs, newspapers, magazines, professional journals, internet, colleagues, telephones, and report reading on climate change and global warming (Okaka, 2010; 2013).

CONCLUSION

All Ugandan (Eastern, Western, Central, and Northern) have experienced the vagaries of climate change induced flood incidents which have prompted the national government to develop and approve a national disaster preparedness policy with a well-coordinated institutional communication and assessment framework. Climate change or variability floods have caused serial disastrous social, economic, political, and environmental consequences due to climate change, with significant threats to lives and livelihoods. Ugandan government access to the green climate fund for early climate action for low-carbon technologies and multilateral financial support from international partners will enhance Uganda's response to climate change induced flooding. Climate change policy management agencies for green fund and climate information services in Africa have developed strategies for flood assessment, research, and adaptation capacity building measures for local communities, There are effective collaborations among the African Union Commission (AUC), African Development Bank (AfDB), United Nations' Economic Commission for Africa (UNECA), African Climate Policy Centre (ACPC),and the East African Community (EAC), that have enhanced the role of effective communication information services for climate change policy sustainable flood disaster control management in Africa. Their mandates on climate change mitigations, adaptation, and financing policy issues are vital urgent climate action in Africa.

REFERENCES

ACAPS Briefing Note, . (2018). *Uganda - Flooding and Landslides in Bududa District*. Kampala, Uganda: ACAPS.

Adebayo, B., & Ntale, S. (2018). *Uganda mudslides, floods spur deaths, destruction*. Mbale, Uganda: CNN.

Atyang, A. (2014). *Study on Early Warning Systems in Uganda. A consultancy report submitted on 3October 2014 tor the Office of the Prime Minister of Uganda, Department for Disaster Preparedness and Management with UNDP funding*. Kampala, Uganda: Government of Uganda.

East African Community Secretariat. (1999). *The Treaty for the Establishment of the East African Community*. Arusha, Tanzania: East African Governments.

Eyotaru, O. (2013, July 23). East African integration hindered by information gaps – minister. Daily Monitor Newspaper, p. 1.

Government of Rwanda. (2012). *Rwanda and UNDP climate change project*. Government of the Republic of Rwanda and UNDP Rwanda.

Government of Tanzania. (2006). *Tanzania NAPA guiding principles*. Government of the Republic of Tanzania.

Government of Uganda. (2005). *Uganda National Report And Information On Disaster Risk Reduction Efforts For The World Conference On Disaster Reduction*. Government of Uganda.

Government of Uganda. (2005). *Uganda National Report And Information On Disaster Risk Reduction Efforts For The World Conference On Disaster Reduction*. Kampala, Uganda: Government of Uganda.

Government of Uganda. (2010a). *Climate change policy draft 2012*. Kampala, Uganda: Ministry of Water and Environment.

Government of Uganda. (2010b). *The National Policy for Disaster Preparedness and Management*. Kampala, Uganda: Government of Uganda.

Government of Uganda. (2015a). *Economic assessment of the impacts of climate change in Uganda: Key results*. Government of Uganda.

Government of Uganda. (2015b). *The National Development Plan II (2016-2020)*. Kampala, Uganda: Government of Uganda.

IFRC/RCS. (2009). *East Africa: Preparedness for El Floods Emergence Appeal of 13 October 2009*. IFRC/RCS.

International Council for Science Regional Office for Africa (ICSU-ROA). (2007). *Science plan: Sustainable energy in sub-Saharan Africa*. Pretoria, South Africa: ICSU-ROA.

International Council for Science Regional Office for Africa (ICSU-ROA). (2008). *Science plan: Global environmental change (including climate change and adaptation) in sub-Saharan Africa*. Pretoria, South Africa: ICSU-ROA.

Kiva, K. (2018). *September-December Rains to Trigger Floods, Landslides - Metrological Authority*. Press Release by Uganda National Meteorological Authority. Kampala, Uganda: UNMA.

Mwangi, A. (2007). *Floods continue to ravage Uganda Thousands risk lives as they cross cut off flooded areas*. Press Release by Uganda Meteorological Department. Kampala, Uganda: UMD.

National Environment Management Agency (NEMA). (2002). *Localizing global environmental conventions. A simple guide for Uganda*. Kampala: NEMA.

Ngwomoya, A. (2018). *Why Kampala keeps flooding. The Daily Monitor newspaper report published on 09 March 2018*. Kampala, Uganda: The Nation Group.

Okaka, W. (2010). Developing regional communications campaigns strategy for environment and natural resources management policy awareness for the East African community. *Research Journal of Environmental and Earth Sciences, 2*(2), 106-111.

Okaka, W. (2011). Developing effective climate change policy communication for sustainable development in Africa. *Climate change stakeholders' forum*. Available at www.stakeholderforum.org/…ach/index.php/cif-day2...South Africa

Panos. (2012). *Communicating climate change in Uganda: Challenges and opportunities*. Cardiff University.

Rogers, E. M. (1995). Diffusion of innovations. New York: The Free Press.

Uganda Radio Network - URN. (2018). *Wakiso floods: 2 bodies recovered, 3 rescued. November 21, 2018. A local radio news bulletin clip*. Kampala, Uganda: URN.

Uganda Radio Network - URN. (2018). *Wakiso floods: 2 bodies recovered, 3 rescued. November 21, 2018. A local radio news bulletin clip*. Kampala, Uganda: URN.

United Nations FCCC. (2002). *Report of the conference of the parties on its seventh session*. Retrieved from www.eac.int

Chapter 12
Environmental Hazards Assessment at Pre-Saharan Local Scale:
Case Study From the Draa Valley, Morocco

Ahmed Karmaoui
https://orcid.org/0000-0003-3881-4029
Southern Center for Culture and Sciences, Morocco

Adil Moumane
Ibn Tofail University, Morocco

Jamal Akchbab
Association des Amis de l'Environnement, Morocco

ABSTRACT

Ecosystem management requires biophysical and socio-economic measurement. The intervention of the government and the local community in order to combat the degradation of ecosystems must take into account the effects of the environmental hazards. This can reinforce the inhabitants' ability to adapt at local level. The impact on ecosystem and resources are numerous and complex. Consequently, a multidisciplinary evaluation is needed. In this context, a new approach was proposed, called environmental hazards assessment at local scale. It was used to evaluate the risk of several oasis resources to multiple hazards in the Middle Draa Valley. The findings show that for all resources, desertification is the biggest challenge affecting this area followed by drought, sandstorms, and then floods. This risk assessment approach can provide guidance for future assessments.

DOI: 10.4018/978-1-5225-9771-1.ch012

INTRODUCTION

Today, climate change has become an internationally recognized reality. Climate change has affected several vital sectors, agriculture (Xie et al., 2018; Neset et al., 2018), health (Karmaoui, 2018), economy (Dogru et al., 2019; Zerouali et al., 2019), tourism (Rosselló-Nadal, 2014), and various ecosystems and the associated ecosystem services (Karmaoui et al., 2015a). The population response must focus on reducing the vulnerability of the socio-ecological system through adaptation to change. The arid environment is an example of the most vulnerable systems and supports a socially fragile population. Around the world, the oasis as a wetland in an arid environment that offers an interesting example of what adaptation could be. Indeed, the oasis produces ecosystem services to local populations. While they can reduce the vulnerability of populations to climate change, but unfortunately they are not taken into account in local adaptation projects.

The study area, the Middle Draa Valley (MDV) is among the most affected areas in the country by desertification (Karmaoui et al., 2014a; 2014b) and drought (Karmaoui et al., 2019; 2016a; 2015b), floods (Karmaoui et al., 2016b), and sandstorms. From here comes the idea to carry out an assessment of those environmental hazards on the ecosystem services of water and soil. The impact of these hazards is accelerated by human direct intervention such as overgrazing, urbanization, and overexploitation of resources. All of these factors contribute to the degradation of arid ecosystems and the disappearance of many villages.

An analysis of the impact of these environmental hazards on ecosystem resources and services is lacking. Such an evaluation must be undertaken. In this context, a multidisciplinary approach was proposed to evaluate the impact of the main environmental hazards and disasters on water and soil resources. This new method was developed using information collected from an expert workshop at local scale. It offers a logical process that helps to better understand the links between these hazards and the different types of resources (natural, physical, financial, human, and social). This tool aims to help planners and managers to put in place the most appropriate measures to adapt to climate change and the main environmental hazards at the local level.

MATERIAL AND METHODS

Study Area

MDV covers a total area of 15 000 Km^2 (Figure 1), and is located at an elevation that range from 400 to 1800 m. The climate is pre-Saharan with an average annual

Figure 1. Localization of the study area. MDV, the Middle Draa Valley, Morocco.
Source: (Authors)

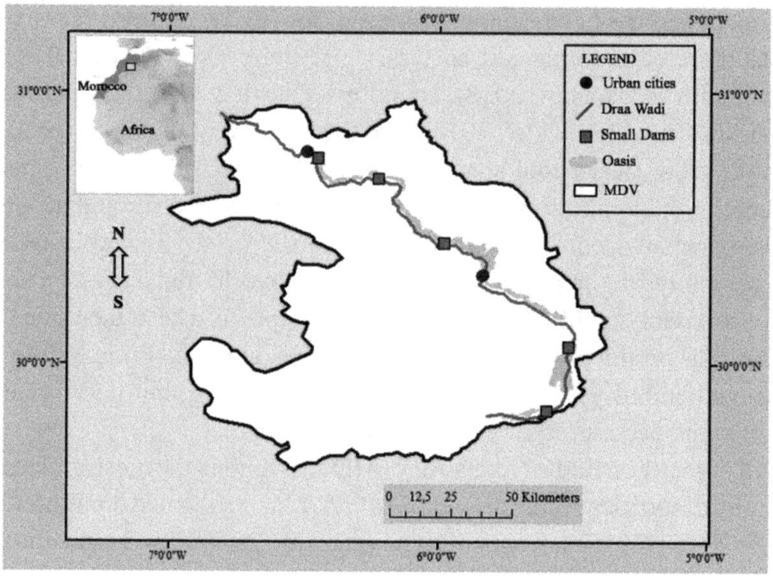

rainfall of about 80 mm; the average annual temperature is 23 °C. In summer, the temperature often exceeds 40 °C and less than two degrees on winter nights.

A large part of the MDV is arid and the most intensive agricultural production is concentrated in oasis ecosystem. The oasis covers 26000 Ha and crossed longitudinally by Draa Wadi (Temporary River). MDV in the south of Morocco provides a perfect opportunity to study the actual effects of environmental risks and adaptation.

Like all oasis ecosystems in Morocco, the oasis of this valley produces ecosystem services such as water, food, and wood that supply and support a large part of population. Other services like regulatory services can reduce the exposure to climate change, allowing the recharge of groundwater, which is very useful in times of drought. The oasis offers several cultural services that attract thousands of tourists (visiting historical sites and exploring oasis biodiversity).

In addition to drought and floods, sand dune encroachment is among the most pressing environmental problems in this region. These hazards are affecting the whole oasis system, which impact consequently, the local socio-economical system. The MDV is a very poor and marginalized area since the Morocco's independence in 1956 (Harakat, 2007). Palm groves are in danger of extinction, in particular under the effect of soil degradation and water scarcity (Chelleri et al., 2014). According to Ait El Hayane (1991) and Benmohammadi, (1998), the causes of the degradation of the natural environment combine extreme climatic conditions and poor management of space. This trend of degradation is often linked to a lack of awareness of vulnerability, adaptation, and conservation. Regarding the groundwater resources, a decreasing trend was recorded in the aquifers of the study area (Figure 2).

Talking about climate change, Figure 3 depicts the evolution of seasonal and annual anomalies temperature and precipitation for three future horizons 2020, 2050 and 2080. It shows an increase in mean temperatures from 1.4 to 3 °C and decrease in precipitation by about 3.9–15%.

Methodology

The approach presented in this study takes into account the impact of environmental hazards on local socio-ecological resources. The proposed tool can help strengthen the adaptive capacity of local communities. This model leads to better understand the links between the environmental risks that affect the socio-ecological systems. Integrating the impact of climate change allows good environmental management for the benefit of the local population. This approach allows also choosing policies that sustain ecosystems for local-long term development.

Figure 4 shows the structure of the produced framework called Environmental Hazards Assessment at Local Scale (EHALS). It was used to evaluate the risk of several oasis resources (palm-grove) to multiple hazards in the MDV. It presents the impact flow of the risks to resources to human well-being components and the human response on resources and reducing or exacerbating the risks effects.

Figure 2. Level of groundwater resources in the MDV
Source: (IMPETUS project)

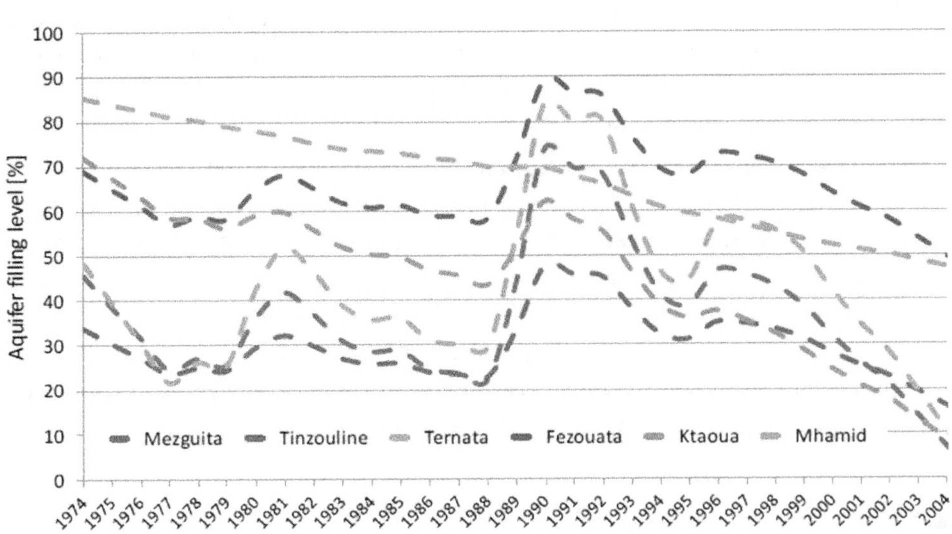

Figure 3. a, Seasonal and annual anomalies (°C) of mean temperature for the three future horizons 2020, 2050 and 2080 and for both A2 and B2 scenarios; b, Percentage change in the level of cumulative seasonal (winter, spring and autumn) and annual precipitation for three future horizons 2020, 2050 and 2080 and for both A2 and B2 scenarios at MDV.
Source: (Karmaoui et al., 2019)

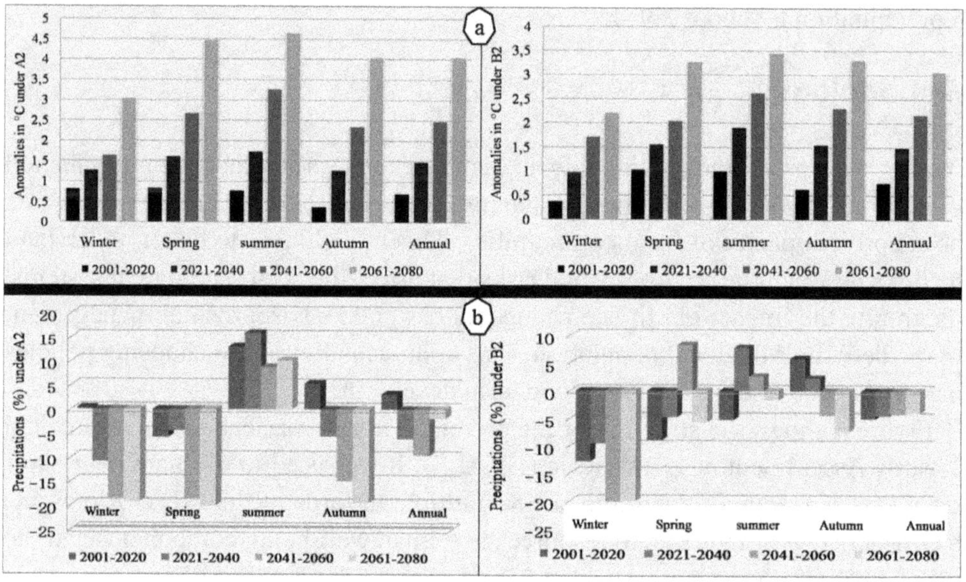

The most productive ecosystem of MDV is a palm-grove (considered wetland). In the last thirty years, its area has been exposed to several socio-ecological challenges. Both the assessment of the impact and the adaptation to these threats are necessary to understand the state of the art. The method allows characterizing scores of several indicators regarding different threats. This approach can be used also to prioritize the impact of climate change on socio-ecological system.

The proposed approach was applied to the MDV, one of the most threatened provinces in Morocco in terms of desertification and drought. Taking this valley as the geographical focus, 40 local experts' participate in a local workshop (Figure 5).

This chapter presents a case study that put us in touch with some local village association leaders confronted with silting, and the scarcity of water resources, with women's association and history-geography teachers, and also heads of centers, and institutions such as the National Institute of Agronomic Research (INRA), the Belgian cooperation, and the association of water irrigation users.

Figure 4. Environmental challenges and their impacts on resources and the well-being of the local population
Source: (Authors)

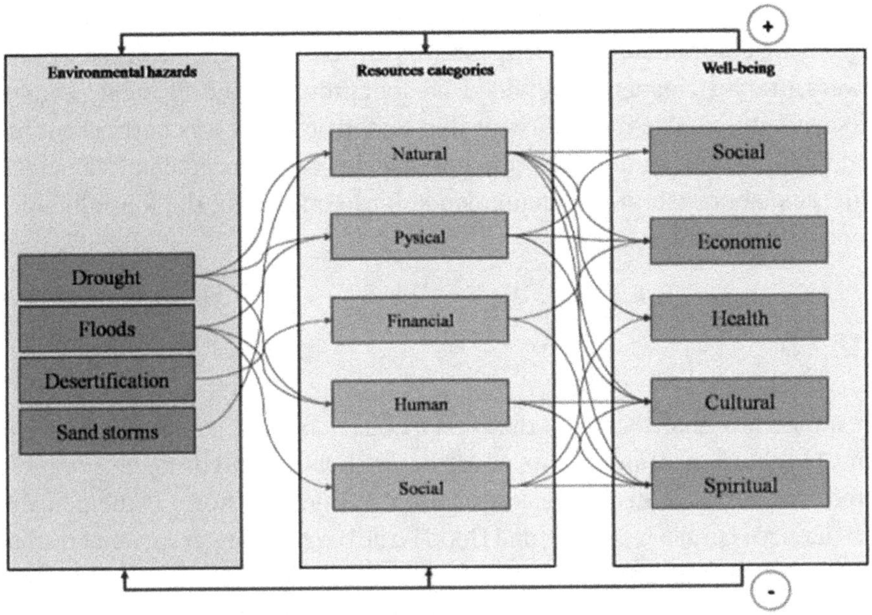

Figure 5. Photos of participant groups in the workshop

The workshop was facilitated by two trainers: Dr. Karmaoui and Mr. Moumane organized by the friends of the environment association (FEA) in Zagora (Draa Valley) on April 02, 2018 from 9:00 pm that lasted all day. After the words of the president of the association who introduced the workshop and thanked the participants and the organizing committee, the two facilitators presented the theme of the workshop. Afterward, the participants were divided into four groups, flood, drought, sand storm, and desertification. The realization of this scientific event was carried out in two phases. First the collection of information in terms of environmental vulnerability, second, the elaboration of the recommendations concerning the four climatic and anthropogenic hazards.

RESULTS

The results show that the risks that most impact natural, financial, and human resources (Table 1 & 2, and Figure 6) are desertification and drought. For physical resources are very sensitive to desertification. While sandstorm is the danger that most influences human resources and flood is the biggest danger to social resources. Generally, for all resources, the results show that desertification is the biggest challenge followed by drought and then sandstorms, and then floods (Figure 7).

For the danger of desertification, participants described traditional and modern irrigation canals, oasis and natural vegetation as the most affected resources. This first group proposed as a response the intervention of the stakeholders and the concerned services of the state for the rehabilitation of these irrigation structures. Concerning the oasis vegetation, the participants proposed to set up systems of mechanical fixation of sand and the conservation of the fertility of the soil. While for natural vegetation, participants proposed to deal with the "overgrazing" and also to educate and raise awareness about the environment. For the danger of drought, the second group of participants selected three negative impacts: The unemployment, the reduction of the cultivated area and the reluctance of the young people of the farms. This group proposed seasonal migration in case of absence of work as an adaptation response and the introduction of new techniques such as the economic irrigation system "drip", the use of greenhouses, and the use of solar energy for pumping irrigation water. However, the reluctance of young people can be remedied through the encouragement of alternative projects. The third group working on sandstorms selected impacts on crops, stopping and disrupting road traffic which caused traffic

Table 1. The main environmental risks and their impacts on socio-ecological variables of the study area

	N	Indicators	Desertification						Drought						Sand Storm						Floods						
		Resources	**0**	**1**	**2**	**3**	**4**	**5**	**0**	**1**	**2**	**3**	**4**	**5**	**0**	**1**	**2**	**3**	**4**	**5**	**0**	**1**	**2**	**3**	**4**	**5**	
Natural resources	1	Water surface					X							X				X			X						
	2	Groundwater		X									X		X						X						
	3	Soil						X						X						X						X	
	4	Palmgrove vegetation						X					X							X			X				
	5	Natural vegetation						X				X			X						X						
Total N.R			0	1	0	0	1	3	0	0	0	1	2	2	2	0	0	1	0	2	3	0	1	0	0	1	20
Physical resources	6	Sand fixation				X					X								X				X				
	7	Canal irrigation						X						X						X						X	
	8	Wells					X							X							X						
	9	Small dams						X					X					X					X				
	10	Roads						X		X										X						X	
Total P.R			0	0	0	1	1	3	0	1	1	0	1	2	0	0	0	1	1	2	1	0	2	0	0	2	19
Financial resources	11	Vegetal Production						X						X						X						X	
	12	Animal production						X						X						X	X						
	13	income					X							X						X			X				
	14	Remittances					X							X												X	
Total F.R			0	0	0	0	2	2	0	0	0	0	0	4	0	0	0	0	0	3	1	0	1	0	0	2	15

continued on the folllowing page

Table 1. Continued

	N	Indicators	Desertification						Drought						Sand Storm						Floods						
		Resources	**0**	**1**	**2**	**3**	**4**	**5**	**0**	**1**	**2**	**3**	**4**	**5**	**0**	**1**	**2**	**3**	**4**	**5**	**0**	**1**	**2**	**3**	**4**	**5**	
Human resources	15	Youth					X							X						X			X				
	16	Farmers						X					X							X			X				
	17	Handicraftsman				X							X					X					X				
	18	Traditional makers				X							X										X				
	19	Tradespeople				X			X										X				X				
	20	Transhumants						X						X						X					X		
	21	Tourism guides	X						X								X									X	
Total H.R			1	0	0	3	1	2	2	0	0	0	3	2	0	0	1	1	1	3	0	0	5	0	1	1	27
Social resources	22	Associations		X								X														X	
	23	Cooperatives and AUEA					X					X							X							X	
	24	Women associations				X					X							X						X			
	25	Qbila and Jmaaa				X							X					X								X	
Total S.R			0	1	0	2	1	0	0	0	0	1	2	0	0	0	0	2	1	0	0	0	0	1	0	3	14
Total General			1	2	0	6	6	10	2	1	1	2	8	10	2	0	1	2	3	10	2	0	9	1	1	9	89

AUEA: Water Users associations

Table 2. Synthesis of the main environmental risks and their impacts on socio-ecological variables

		Resources	Desertification	Drought	Sand Storm	Floods
Natural Resources	1	Water surface	4	5	3	0
	2	Groundwater	1	4	0	0
	3	Soil	5	5	5	5
	4	Palmgrove vegetation	5	4	5	2
	5	Natural vegetation	5	3	0	0
Physical Resources	6	Sand fixation	3	2	4	2
	7	Canal irrigation	5	5	5	5
	8	Wells	4	5		0
	9	Small dams	5	4	3	2
	10	Roads	5	1	5	5
Financial Resources	11	Vegetal Production	5	5	5	5
	12	Animal production	5	5	5	0
	13	Income	4	5	5	2
	14	Remittances	4	5		5
Human Resources	15	Youth	4	5	5	2
	16	Farmers	5	4	5	2
	17	Handicraftsman	3	4	3	2
	18	Traditional makers	3	4		2
	19	Tradespeople	3	0	4	2
	20	Transhumants	5	5	5	4
	21	Tourism guides	0	0	2	5
Social Resources	22	Associations	1	3		5
	23	Cooperatives and AUEA	4	3	4	5
	24	Women associations	3	2	3	3
	25	Qbila and Jmaaa	3	4	3	5
Total			94	92	79	70

Table 3. Negative impacts of the environmental risks and the proposed adaptation strategies

	Negative Impacts	Adaptation Strategies
Desertification	Seguias and traditional canals	• Stakeholders and responsibles intervention
	Palmegrove vegetation	• Mechanic fixation • Soil fertility conservation
	Natural vegetation	• Overgrazing stoping • Spreading the culture of preserving the environment (Education)
Drought	Unemployment	• Seasonal migration
	Crop area decreasing	• Use of technology
	Reluctance of young people from farming	• Encouragement of Alterantive projects
Sand storms	Farms damages	• Use of plants dapted to sand dunes environments
	Stop transport and traffic accidents	• Bulletins and prognostic awareness
	Health problem (respiratory system)	• Providing specializations in respiratory system
Floods	Human and material damages	• Reviewing the method of using the domain • Small dam construction
	Migration	• Preparation of a development strategy of resilience to floods
	Environmental imbalance	• Ecosystem rehabilitaton

accidents. The last impact is on health. Regarding the proposed strategies, this third group has advanced the use of plant species adapted to the conditions of sands, the communication of weather information by radio and television to warn road users. For health, this group proposed to increase the staff and the hospital structures concerning respiratory diseases.

The fourth group discussions on flooding led to three types of impacts: human and material damage, migration and environmental imbalance (erosion and loss of water and soil quality). To prevent human and material damage, the state must prohibit construction in flood-prone areas and the installation of small diversion and storage dams. For the migration impact, participants proposed to prepare development strategies that take into account resilience to floods. The environmental imbalance can be remedied by the rehabilitation of ecosystems as recommended by this fourth group.

Figure 6. Profile of environmental risks on socio-ecological variables

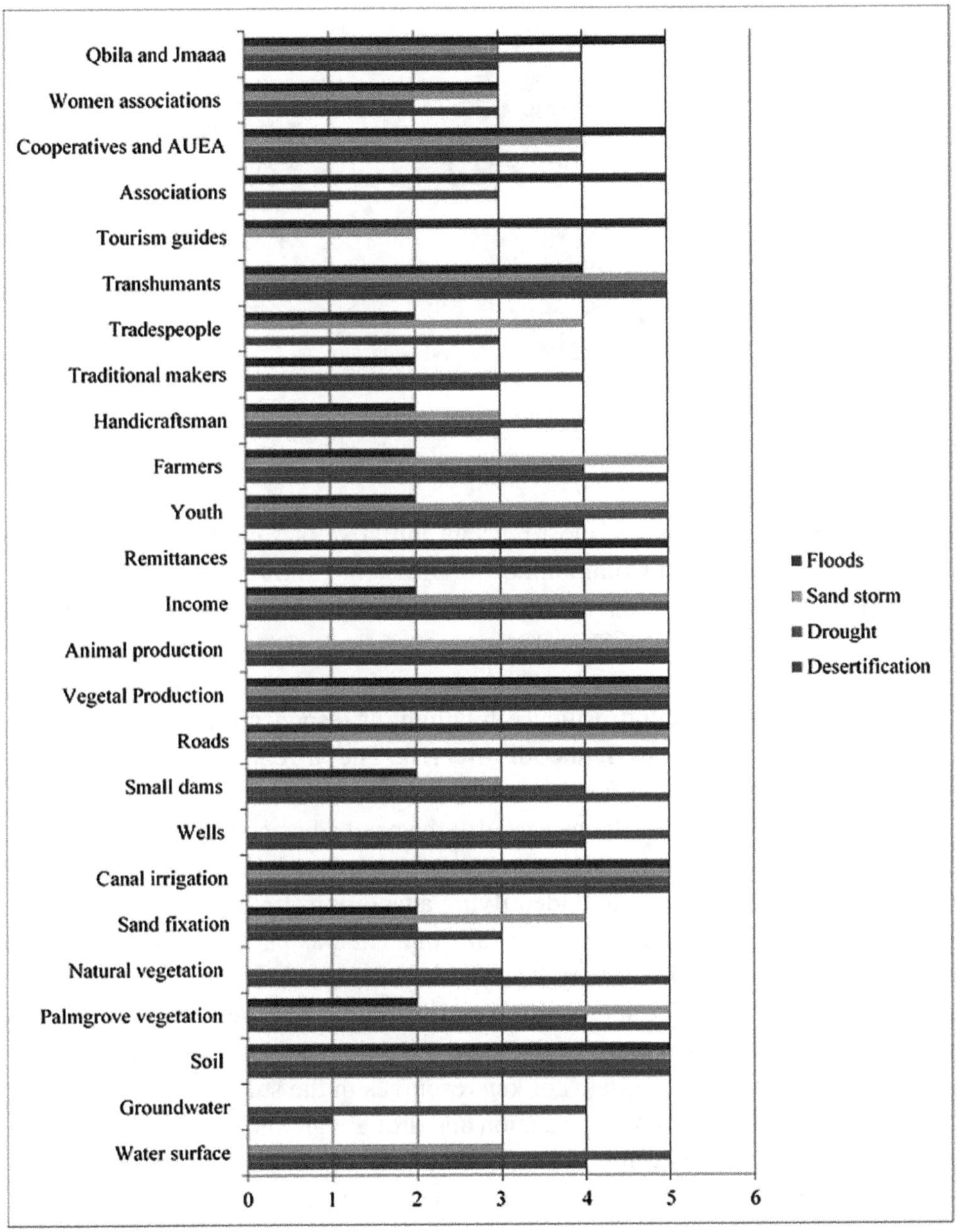

Figure 7. Impact of environmental risks in MDV socio-ecological systems

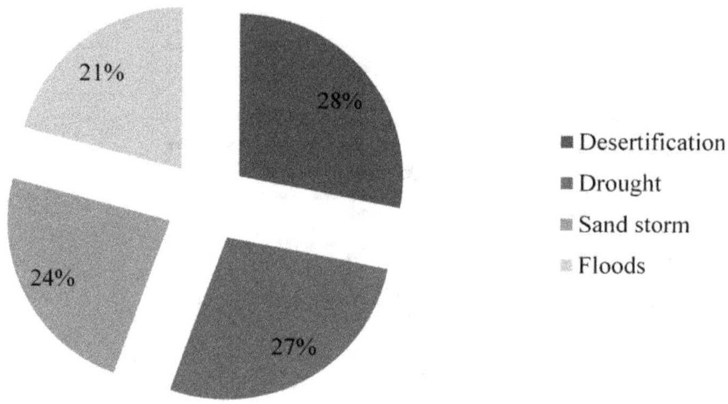

DISCUSSION

In terms of risk reduction, the most relevant initiative was entitled "Building the Resilience of Nations and Communities to Disasters", the Hyogo Framework for Action (HFA) 2005-2015. Next the Sendai Framework for Disaster Risk Reduction 2015-2030 supported by the United Nations Office for Disaster Risk Reduction (UNISDR) is an instrument to reduce of disaster risk, preventing new risk, reducing existing risk and strengthening resilience, highlighting responsibility of states (society and institutions) to prevent and reduce disaster risk. The current article supports these frameworks through the identification and assessment of the potential environmental risk in the south of Morocco. It answers also the need advanced by De Lange et al., (2010) and Hagenlocher et al., (2018) to use indicators of environmental vulnerability when evaluating disaster risk and identifying adaptation strategies.

This case study is based on a workshop conducted as part of a climate change adaptation project in the Draa Valley. It was based on a consultation of all stakeholders in the region (participatory approach). Participating groups were asked to discuss the main environmental issues in the area. Participants also identified, ranked, prioritized, and assessed the status of key resources in the study area.

The results showed that desertification and drought are the main environmental problems. This result was also found by Chelleri et al., (2014) and Karmaoui, (2019). Any development and conservation project must take into account these two main environmental constraints, especially in projects related to water such as agriculture and tourism.

The region can move towards economic projects that saves water consumption. The state has already started to guide these two sectors through the subsidizing of drip systems and renewable energies (solar panels) (Minucci & Karmaoui, 2017). Unfortunately, these initiatives do not affect the small farmers (the majority of the farmers) in the study area.

For the most important impact of each risk participants selected two to three impacts and proposed appropriate strategies. The most important strategies are synthesized as follows:

- The role of the state and its services at the provincial level:
  - Construction and installation of structures to combat floods, desertification and health problems
  - Laws and information on the importance and fragility of ecosystems and oasis heritage
  - Technical and financial support of smallholder initiatives
- The role of the local community, individuals and groups:
  - Environmental awareness and education
  - Conservation and respect of ecosystems through the organization and use of resources (for human and animal consumption).

The approach is based on resources instead of ecosystems.

Given the diversity of resources and risks, a multidisciplinary approach was proposed. It integrates at the same time five components of resources (natural, physical, financial, human, and social) in relation to the four most important risks in the MDV.

The chapter presents the development of a new approach that can be used for multi-hazard risk assessments of oasis social-ecological system of MDV. This risk assessment approach can provide guidance for future assessments in similar areas.

CONCLUSION

MDV is located in the Draa-Tafilalet region, where local ecosystem services support more than 300 000 inhabitants. The intense oasis agricultural impacts increasingly water and soil resources. Consequently, these two ecosystem services are degraded and the oasis system is at disappearance risk. In response to this trend, the government and local population must adapt to this situation. However, the challenges facing the water and soil ecosystem services (ES) are numerous and complex, the most dangerous environmental disasters were investigated. A multidisciplinary approach is needed to evaluate the impact of the main Environmental hazards on these two ES in a vulnerable area. In this context, a new framework was developed using information collected from an expert workshop at provincial scale. The results show that the knowledge from participants allow conceiving the proposed method. The expert information complemented the insufficiency of available data and support decision makers in terms of environmental hazards assessment.

EXPERIMENTATION ON HUMAN SUBJECTS

I confirm that the study complies with all regulations and confirmation that informed consent was obtained.

COMPETING INTERESTS

No competing interests in the manuscript.

ACKNOWLEDGMENT

The research is part of the project “Adaptation to climate change in Zagora”. The workshop was organized by the association FEA, Zagora and funded by the Zagora provincial council. I want to thank all the participants: women’s associations and history-geography teachers, and also heads of centers, and institutions such as the National Institute of Agronomic Research (INRA), the Belgian cooperation, and the association of water irrigation users.

REFERENCES

Ait-Alhayane, K. (1991). Les reṕresentations du désert et de la désertification en Afrique du Nord: contribution à une étude exploratoire au Maroc. Institut agronomique méditerranéen de Montpellier.

Benmohammadi, A. (1998). Ensablement et désertification dans la vallée moyenne de l'oued Drâa: Le mythe de la fluctuation de la limite nord du désert. *Africa Geoscience Review*, *5*, 11–22.

Chelleri, L., Minucci, G., Ruiz, A., & Karmaoui, A. (2014). Responses to Drought and Desertification in the Moroccan Drâa Valley Region: Resilience at the Expense of Sustainability? *International Journal of Climate Change: Impacts & Responses*, *5*(2).

De Lange, H. J., Sala, S., Vighi, M., & Faber, J. H. (2010). Ecological vulnerability in risk assessment—A review and perspectives. *The Science of the Total Environment*, *408*(18), 3871–3879. doi:10.1016/j.scitotenv.2009.11.009 PMID:20004002

Dogru, T., Marchio, E. A., Bulut, U., & Suess, C. (2019). Climate change: Vulnerability and resilience of tourism and the entire economy. *Tourism Management*, *72*, 292–305. doi:10.1016/j.tourman.2018.12.010

Hagenlocher, M., Renaud, F. G., Haas, S., & Sebesvari, Z. (2018). Vulnerability and risk of deltaic social-ecological systems exposed to multiple hazards. *The Science of the Total Environment*, *631*, 71–80. doi:10.1016/j.scitotenv.2018.03.013 PMID:29524904

Harakat, I. (2007). *Les acteurs de la coopération et la dimension socio-économique de la désertification dans le sud du Maroc: cas de Zagora*. Mémoire de maitrise en science politique, Université du Québec. Retrieved from http://archipel.uqam.ca/id/eprint/628

Karmaoui, A. (2018). The cutaneous leishmaniasis vulnerability index (CLVI). *Acta Ecologica Sinica*, *38*(4), 288–295. doi:10.1016/j.chnaes.2018.01.001

Karmaoui, A., Balica, S. F., & Messouli, M. (2016b). Analysis of applicability of flood vulnerability index in Pre-Saharan region, Morocco. *Natural Hazards and Earth System Sciences*, *2016*. doi:10.5194/nhess-2016-96

Karmaoui, A., Ifaadassan, I., Babqiqi, A., Messouli, M., & Khebiza, M. Y. (2016a). Analysis of the Water Supply-demand Relationship in the Middle Draa Valley, Morocco, under Climate Change and Socio-economic Scenarios. *Journal of Scientific Research & Reports*, *9*(4), 1–10. doi:10.9734/JSRR/2016/21536

Karmaoui, A., Ifaadassan, I., Messouli, M., & Khebiza, M. Y. (2015b). Sustainability of the Moroccan oasean system (Case study: Middle Draa Valley). *Global J Technol Optim*, *6*(01), 170. doi:10.4172/2229-8711.1000170

Karmaoui, A., Messouli, M., Ifaadassan, I., & Khebiza, M. Y. (2014b). A multidisciplinary approach to assess the environmental vulnerability at local scale in context of climate change (Pilot study in upper draa valley, South Morocco). *Glob. J. Technol. Optim*, *6*, 1–11.

Karmaoui, A., Messouli, M., Khebiza, Y. M., & Ifaadassan, I. (2014a). Environmental Vulnerability to climate change and anthropogenic impacts in dryland (Pilot study: Middle Draa Valley, South Morocco). *J Earth Sci Clim Change, S11*. Doi:10.4172/2157-7617.S11-002

Karmaoui, A., Messouli, M., & Yacoubi, M. (2015a). Vulnerability of Ecosystem Services to Climate Change and Anthropogenic Impacts in South East of Morocco (Case study: Drying up of Iriki lake). *The International Journal of Climate Change: Impacts and Responses.*, *7*(3), 1835–7156.

Karmaoui, A., Minucci, G., Messouli, M., Khebiza, M. Y., Ifaadassan, I., & Babqiqi, A. (2019). Climate Change Impacts on Water Supply System of the Middle Draa Valley in South Morocco. In *Climate Change, Food Security and Natural Resource Management* (pp. 163–178). Cham: Springer. doi:10.1007/978-3-319-97091-2_8

Maroufi, A. (2008). Désertification et migration: un défi pour le développement durable au Maroc. Le cas de Zagora. *Sustainable Mediterranean, 54.*

Minucci, G., & Karmaoui, A. (2017). Exploring the water-food-energy and climate nexus: insights from the Moroccan Draa Valley. In Peri-Urban Areas and Food-Energy-Water Nexus Sustainablity and Resilience Strategies in the Age of Climate Change. Springer International Publishing AG. doi:10.1007/978-3-319-41022-7_11

Neset, T. S., Wiréhn, L., Opach, T., Glaas, E., & Linnér, B. O. (2018). Evaluation of indicators for agricultural vulnerability to climate change: The case of Swedish agriculture. *Ecological Indicators*. doi:10.1016/j.ecolind.2018.05.042

Rosselló-Nadal, J. (2014). How to evaluate the effects of climate change on tourism. *Tourism Management*, *42*, 334–340. doi:10.1016/j.tourman.2013.11.006

Xie, W., Huang, J., Wang, J., Cui, Q., Robertson, R., & Chen, K. (2018). Climate change impacts on China's agriculture: The responses from market and trade. *China Economic Review*. doi:10.1016/j.chieco.2018.11.007

Zerouali, S., Khebiza, M. Y., & El Qorchi, F. (2019). Monetary Value Change of Some Provisioning Ecosystem Services of Middle Draa Valley, South of Morocco. In *Climate Change and Its Impact on Ecosystem Services and Biodiversity in arid and semi arid regions*. IGI Publisher. Retrieved from https://www.igi-global.com/book/climate-change-its-impact-ecosystem/206548#table-of-contents

Chapter 13
Vulnerability of Forest Vegetation Due to Anthropogenic Disturbances in Western Himalayan Region of India

Akash
Gurukula Kangri University, India

Bhupendra Singh Bhandari
HNB Garhwal University, India

Navneet
Gurukula Kangri University, India

Kamal Bijlwan
SGRR University, India

ABSTRACT

The Western Himalayan zone of India is not only threatened by rapid climatic changes, natural floods, and fires, but also by anthropogenic activities. Himalayan forests are vulnerable due to climatic changes and faced severe ecological deterioration due to anthropogenic pressures. The degradation of forests due to anthropogenic disturbances is increasing because of overgrowth of population, high poverty ratio, as well as the limitations of alternative livelihood options. Further resources from forest makes it inseparable to manage forest stands without considering the importance of socio-economic status and ecological aspects of forest management to the well-being of local communities. Therefore, the Himalayan forests and the communities depending on forests should be seen as a part of an evolving. This chapter will explore the vulnerability of the knowledge towards Western Himalayan forests and community-based management of forests. Additionally, it will sketch potential sites affected through anthropogenic pressures.

DOI: 10.4018/978-1-5225-9771-1.ch013

INTRODUCTION

The forest-dominated mountainous landscapes provide a wide range of ecosystem services for people residing in the mountains as well as for people residing in lower hills and plains (Grêt-Regamey et al., 2013). Many species in the forest have been adapted as the climatic changes occur (Keenan, 2015), but the role of forest is critical in terms of socio-ecological status and provides various ecosystem services to the earth survivals (Pandey & Jha, 2012). As per the report of IPCC (Intergovernmental Panel on Climate Change), continuous changing of the climate has severely affected the ecological and social status (IPCC, 2014). Scientists and researchers argued that due to the climatic changes forest are facing vulnerability through human induced pressures which are mainly responsible for present condition of the plant species (Sharma et al., 2007). Forest of the world are facing severe anthropogenic pressures in various forms of pollutants, grazing, deforestation, trampling, scraping resulting into the loss of biodiversity (Akash and Navneet, 2019). The Western Himalaya, spanning across Jammu and Kashmir, Himachal Pradesh and Uttarakhand is distinctly different from the Eastern Himalaya. Its gentler and wider slope, continental climate conditions with lower humidity and higher snowfall, lower tree line, narrow ranging and alpine scrub zone, and an overall lower primary productivity lends a vast difference in the biological diversity of the two regions (Miller, 1987; Mani, 1994). The Shivalik landscape is also called as sub Himalaya and considered as one of the youngest mountain in India. Its alignment is more or less parallel to lesser Himalaya. The region extends from the basin of Indus to Bhahmaputra with one gap of over 300 km form sapta kosi to Manas river (Kumar et al., 2010). The Shivalik region is categories under Indo-Gangetic plains and has great significance in terms of taxa from Indo- Malyalam and Palaeratic regions. The North-western part of India, representing six major types of zones which are Central, Eastern, Southern, Western as well as the North-eastern of the country covers Uttarakhand, Jammu & Kashmir, Himachal Pradesh, Punjab, Delhi, Haryana and a union territory of Chandigarh. On the basis of the physiographic, it has two zone named as and Indo- Gangetic plains and Himalayan Ecosystem. The Shivalik Landscape is the sandwiched between these two major physiographic region. This region is unstable due to the unstable shape of the landscape, unconsolidated land mass, torrential rains as well as the unscientific management practices (Yadav et al. 2005). Shivalik hills are one the youngest maintains ranges parallel to the other ranges in Himalaya. Himalayan forest has also faced severe vulnerability due to the anthropogenic as well as the natural pressures. The biodiversity and the productivity of forest in Himalayan

Table 1. Average climatic data of the some areas of Western Himalaya

Climatic Zone	Altitude (m)	Mean Annual Tem.(C⁰)	Mean Annual Rainfall (Cm)	Mean Max Tem (june-in C⁰)	Mean Min. Tem (jan- C⁰)
Shivalik Subtropical	400-800	2240	150	31.0	14.0
Doon valley sub tropical	600-1000	21.1	212	29.4	13.3
Outer Himalaya	1200-1500	18.9	298	27.2	11.1
Rajaji tiger reserve	300-1000	23.4	300	32.3	8.0

(Source: Modified from Bhainsora, 1995)

region are under the effects of climatic changes (Sharma et al., 2015; Upgupta et al., 2015). The vulnerability of vegetation due to the biotic pressures are adjudged through the various changes in the phenological characteristics of plants (Pau et al., 2011), distribution of forest type (Chaturvedi et al., 2011; Gopalakrishnan et al., 2011; Upgupta et al., 2015) as well as the productivity of the vegetation. The assessment of vulnerability of different ecosystem of the forest requires descriptive understanding both in terms of abiotic and biotic factors of ecosystem which are mainly responsible for growth of vegetation in forest. The vulnerability of ecosystem of forest is enhanced by stress imposed by the human and natural factors which is beyond the capacity to adapt to stress (Pandey & Bardsley, 2015). It can also be expressed as the differences among adaptability and sensitivity (Zhang et al., 2017). The deviation in steady in ecosystem services are determined through the different variables of the ecosystem in a fixed time (Coulson & Joyce, 2006). On the other hand the forest's vulnerability reveals the alteration in the relative features of carbon and distribution of forest (Kumar et al., 2018). In present, it is noticeable that changes in the habitats and its fragmentation as well as the loosing adaption of plants in forests would make them vulnerable, resulting into the existing alignment between various plants communities.

OBJECTIVES OF THE STUDY

Considering the importance of Western Himalayan region of India, the study aims to enhance vulnerability through various pressures on the floristic vegetation or through development activities by considering the ecological parameters. The chapter will describes the treasure of Western Himalaya and the potentiality of vegetation in the Himalaya. The study also aims to evolve a set of plausible guidelines, and exploration through development guidelines by considering the anthropogenic pressures as the barrier for development.

METHODOLOGY

Vegetation barrier in the development programme and its relevance to Himalaya is to be studied. Various pressures, policies as well as programmes through which the species has faced vulnerability would be evaluated in the terms of present problems in the area and in order to study their effectiveness.

WESTERN HIMALAYAN VEGETATION ANALYSIS AND VULNERABILITY

The forests of Western Himalayan region are known to be multifunctional as they provide a various range of ecosystem services for supporting livelihood options of local communities (Rasul, 2014). Himalaya Mountains are the important ecosystem services and have a significant. In Western Himalayan region of India, most of the protected areas are surrounded by number of Gujjars (tribal community) and other villages which have potentials of promoting wildlife tourism along with cultural tourism, adventurer tourism, natural tourism, pilgrimage tourism playing (Akash et al., 2018 d). At the same time they also imposed significant pressures on the local biodiversity in Western Himalaya.

Forest are the legacies of the interactions of human with the nature which have been going on for a long time and created cultural landscapes and traditional systems of forest resource. Although the balance among the nature and have been warning and further the pressures on the forests arises due to anthropogenic disturbances are increasing day by day due to overpopulation high rate of poverty as well as limitations of other livelihood things (Arya et al. 2012). In Himalaya, there is a very strong relationship between the population and the resources of forest which makes it inseparable to manage forests without considering the importance of social, ecological, economic and aspects of forest management to the well-being of local communities (Baland et al., 2010). On the other way, the forests of Himalaya's communities associated with the region depends on forests can be seen as a part of social-ecological system, whereas an assessing the impacts of climatic alteration and human induced pressures in the era of anthropocene. Various research methodology have been implemented for assessing the rate of vulnerability of natural forests and associated vegetation which usually involves identification of disturbances index, vulnerability of ecosystems, climatic effects, natural calamities. Further it is crucial to know that the vulnerability concept can be summarized through research community based on various disciplines level (Lwasa, 2015). On the other way, classification and measurement of vulnerability is highly debatable, it totally depends on the rationale of identifying patterns of vulnerability (Kok et al., 2016). The ecosystem of natural forest is considered as vulnerable if there are so many changes in changes in the species composition and their structure like migration, phenology / local extinction or reduction in the productivity of the vegetation (Allen et al., 2015). Climatic changes in terms of precipitation, temperature, extreme weather conditions are also responsible for vulnerability.

The Western Himalayan region includes many Wildlife Sanctuary and National parks like Rajaji tiger reserve, Corbett tiger reserve, Nanda Devi wildlife sanctuary etc. which are main repository of biodiversity. Rajaji tiger reserve is one of the most important repository of biodiversity in Western Himalaya. It is the fourthly eighth tiger reserve in the country and second tiger reserve in terms of area in Uttarakhand state after Corbett tiger reserve in Western Himalaya. It was declared tiger reserve because it sustains wide range of tigers in upper Gangetic plains. The tiger reserve is an essentials part of the terai landscape between Sharda and Yamuna river. This area is collectively known as Rajaji – Corbett conservation unit in Shivalik landscape which maintains the viable population of tiger. The River Ganga flows 24 km through the park dividing the tiger reserve into two unequal halves. In summer the temperature raises 40-45^{0}C and in winter 20-25^{0}C. The annual rainfall ranges from 1200-1500 mm. Generally the soil is poor and infertile but in some places accumulation of humus occurs. The Chilla forest range of the tiger reserve lies in the east of the river Ganges and attached to the Garhwal forest Division at an elevation between 302 and 1000 meter above sea level The Chilla range of the reserve is one of the great centre of attractions for tourists (Akash et al., 2018b). Approximately 90% tourist visit in Chillla range every year to enjoy the wildlife and scenic beauty. The dominant plant species area *Mallotus phillipensis, Dalbergia sissoo, Shorea robusta, Acatia catachu, Cassia fistula, Helicteres isora,* and the ground vegetation is mainly comprises of *Ageratun conyziodes, Anagallis arvensis, Cynodon dactylon, Kyllinga monocephala, Abutilon indicum, Sida spinosa* etc. whereas the high altitudinal area of the Rajaji tiger reserve which lie to the Pauri Garhwal is mainly comprises of *Pinus* spp. and mixed forest vegetation (Akash et al., 2018a,b). Some of the faunal species of the reserve are *Panthera pardus, P . tigris, Axis axis, Axis peroconius, Cervus unicolor, Elephus maximus, Naemohaedus goral* etc. The area of the Chilla forest division of Western Himalayan zone comes under the protected area network but undergoing rapid changes in fauna and vegetation pattern due to the large scale anthropogenic forcing at some places in form of lopping, grazing and hydro-power project, scraping, trampling and extraction of non timber (Akash et al., 2019). The Haridwar forest division of Rajaji tiger reserve cover 7304.60 hectare of forest cover in Uttarakhand state (Akash et al., 2018 c). Rajaji tiger reserve appears to be India's one of the most successful protected area and further its development has boost up the viable population of Asiatic elephant and *Panthera tigris* in their natural habitat. The number of tigers in last few years has greatly increases that is why the status of tiger reserve to it is given by government of India.

Factors Responsible for Vulnerability

In Western Himalayan forest ecosystem disturbance occurs in various forms in which locals communities removes a little portion of plant biomass in different form like grazing, lopping, trampling, scraping etc. These disturbances causes severe effects on the plant biodiversity and retards the successional process of the forest. Various forest ecosystem arises on different altitude due to the anthropogenic disturbances as well as the variation in topography, rainfall, soil, rainfall and other climatic conditions are responsible for sustaining the specific types of plant community (Akash, 2018 a). Our study also states that the different plant community like *Shorea robusta - Mallottus phillipensis, Dalbergia sisoo, Holeptelia integrifolia,* and mixed forest community in Western Himalaya has evolved due to the various anthropogenic activities, different temperature at different sites, different soil, rainfall pattern and different climatic condition. It has been also observed in the study area that these different plant communities have different temperature, climatic, different coordinates, soil, rainfall and different environmental conditions.

Forest Fire

Fire is an important factor in the determination of the composition and structure of forest ecosystems. Tropical forests especially the dry forests are highly susceptible to fire during the period of drought (Saha & Hiremath, 2003). The forest fire has both negative as well as the positive impacts on the ecological processes. According to

Table 2. Vegetational attributes studied by different workers in Western Himalayan and adjoining areas of Uttarakhand, Northern India

Location	Density/ha.	TBC (m^2/ha.)	Authors and Year
Chambhi (Nainital)	416	32.71	Ralhen et al., [1982]
Maheshkhan (Nainital)	940	39.4	Saxena and Singh, [1982]
Mussoorie (Dehradun)	640	-	Joshi et al., [1985]
Chakrata (Dehradun)	320	29.19	Singhal et al.,[1986]
Dhanolti (Dehradun)	140	13.55	Singhal and Soni, [1989]
Binsar Mahadev (Pauri)	310-520	31.50-57.33	Baduni and Sharma, [1996]
Askot WLS (Pithoragarh)	550	50.8	Dhar et al., [1997]
Mandal-Tungnath (Chamoli)	337	90.16	Singh and Rawat, [2012]
Kedarnath WLS (Chamoli)	340-810	30.1-62.2	Rai et al., [2012]
Pindari (Bageshwar)	480	73.4	Adhikari et al., [1995]
Kunjakharak (Nainital)	830	30.69	Rikhari et al., [1989]
Shitalakhet (Almora)	880	56.37	Pandey, [2003]
Siahidevi (Almora)	1260	31.7	Rana et al., [1985]
Buvakhal (Pauri Garhwal)	790	35.39	Lata and Bisht, [1991]
Pauri (Pauri Garhwal)	390	17.27	Dhanai et al., [2000]
Mandal-Tungnath (Chamoli)	433	88.06	Singh and Rawat, [2012]
Dhakuri (Pindari) Maheshkhan (Bhawali)	1060- 1300	98.49 83.77	Saxena and Singh, [1982]
Mandal-Tungnath (Chamoli)	433	110.47	Singh and Rawat, [2012]
Dudhatoli region (Dehradun)	250-340	18.45-38.25	Baduni and Sharma, [1996]
Chamoli	10-1300	0.02-12.80	Bhandari et l.,[1997]
Mandal(Chamoli)	5-430	0.08-22.06	Sharma et al., [2009]
Narianbagar (Chamoli)	10-762.5	0.003-35.37	Bhandari et al., [1999]
Chilla forest-Rajaji tiger reserve	2.01- 216.68	0.011-11.6	Akash et al., 2018a
Chilla forest-Rajaji tiger reserve	2.0 - 216.67	0.01- 11.75	Akash et al., 2018b
Rajaji tiger reserve	149.99 - 397.91	3.612 - 46.813	Akash et al., 2018b
Rajaji tiger reserve	2.083- 60.416	0.02- 1.4511	Akash et al., 2019

Du Toit, 1972 and Hardwick, 2000). Forest fire has adverse effects on the seedlings of tree species and growth. Fire verily affects the juvenile and seedling lying on the ground surface (Bond & Keeley, 2005). Forest fire is a major problem in the Western Himalyan region of India as it severely harms the ground flora and animals. It is both natural and anthropogenic in origin. Fire causes profound changes in the habitat of the livings and this has important implications for the survival, population size as well as the structure, composition and diversity of plants and animals (Sukumar, 1989). Forest fire and felling of fodder species are considered to be important influences as far as some frequent movement of elephants is concerned and both factors greatly affect the structure and composition of the plant communities in the Rajaji tiger reserve and allied areas of the Western Himalaya (Joshi, 2002). It has been observed from the previous stuides that the Western Himalaya fires are entirely of anthropogenic origin. If natural disasters are excluded from the forest fire then the forest fire come close to being the worst kind of all disaster. The most important approaches for prevention of forest fire is fire line. Fire-lines are crucial for preventing the forest fire in Rajaji tiger reserve. Fire- line is an important approaches in preventing of Shivalik biodiversity in Northern India. Generally the the fire-lines are of 100 feet, 50 feet, 20 feet, 10 feet and 4 feet (Rasily, 2008). These fire lines are constructed in all the forest range of the tiger reserve. In Fire- lines, burning of grasses and other inflammable materials are collected in controlled manner before the beginning of fire season. Regular maintenance is kept so that L*antana camara* and *Eucalypus* spp cannot overtake the fire-line as these plants are highly inflammable in dried form. Forest fire is one of the most dangerous threats for forests that lead to significant losses of the bio-diversity and different forest ecosystems (Pew & Larsen, 2001; Fearbside, 2005). Fire of Forests destroys billions of hectares of the world's protected and unprotected forest areas every year resulting into the loss of wildlife and human property and apart from release of tons of carbon to the atmosphere, raises costs on fire suppression and prevention heavily, and damage to other environmental (Davidenko & Eritsov, 2003; Fao,2005).Forest fire rate in Western Himalayan region of in India has increased dramatically in last few decades as a result of accumulated *Pinus roxburghii*, needles which are inflammable due to its high resin amount and provide abundant source of fuel for these wildfires. Secondly drought, hot weather condition of the region is another reason responsible for the forest fire. In Western Himalaya, burning of forest, clearing as well as the shifting cultivation is an old traditional practice. On the other way the locals inhabitants have poor socio-economic conditions continue to use or burn forests to support their daily livelihood by implementing various activities like fodder and agriculture or to collect of forest products (Gadgil & Meher, 1985).In Uttarakhand and other region of Western Himalaya have faced human induced forest fire as a regular and historic

Table 3. Affected forests through wildfires in some of the forested regions of Western Himalaya during 2007-2009

S.No.	Area Affected by Wildfire (in hectare)			Burned Area of Forest (%)		
1	Haridwar Forest Division					
	2007	2008	2009	2007	2008	2009
	120	96	105	1.4	1.11	1.23
2	Chilla Forest Division					
	75	95	60	0.5	0.64	0.4

Source: (Rasily, 2008; Joshi and Singh, 2010)

feature. Forest fire has been noticed since 1990 -2016 at a regular basis which has impacted severe biodiversity. In past years have been associated with the against the then the forest policies of Bristish (Bhandari et al, 2012).At the time from 1990-2016, wildfires has destroyed many dense forest and their vegetation in Uttarakhand. Approximately, 3.75 lakh hectares in 1995 and 80,000 hectare forested areas was destroyed in Ganga-Yamuna region of Uttrakhand. Himalachal and Uttarakhand are the top state of the Western Himalaya which have been effected with forest fire. Approximately 4,500 hectares area was affected in Himachal Pradesh, and 40% more than the 3,185 hectares in Uttarakhand state since few years. Foreste fire had destroyed approximately 3500 hectares of land and various losses of flora and fauna which has caused severe effects on the social, economic and ecological impacts (Kavita, 2016). In Western Himalaya, fires are mostly intentional for the purpose of collecting resources of forests like sal seeds, resin,, honey and timber etc. At the same time locals also burnt the forests to improve the growth of various grasses, for protecting them from wild fauna, and accidental or other purposes. Various trees like *Pinus roxburghii*, *Quarcus* spp. are susceptible to fire. Although the mature pine is fire tolerant but there is lack of the regeneration potentials of other trees and fire resistance trees in Western Himalayan region area facing severe threats to the vegetation. The two tourists places of the Western Himalaya are Uttarakhand and Himachal Pradesh Pradesh. These are most tourist attracted centre in summer but fire created them panic. The spreading of Himalayan forest vegetation is more than other with limitation of movement as more depend on local people for control the fire. At present time migration of local peoples for employment and Jobs as well as the government pro conservation plan for wilderness by vacating the people from

Table 4. Details of Lantana eradication in some of the forested areas of Western Himalaya

S.No.	Name of Forest Range	Total Area (hectare)	Lantana Infestation (in 2001)	Lantana Eradicated (in hectare)	Lantana Infestation (in 2011)
1.	Kansarao	7932.70	3695.05	927.00	4745.60
2.	Ramgarh	7703.00	2831.00	900.00	2854.00
3.	Motichur	8042.20	3589.05	580.00	4622.25
4.	Gohri	10177.90	2816.15	370.25	6367.37
5.	Chilla	14829.80	4896.15	580.00	7104.24
6.	Haridwar	8525.50	2625.80	409.00	3408.00
7.	Dholkhand	5995.10	4821.51	629.00	1640.00
8.	Chilawali	11531.39	4239.00	413.00	4894.00
9.	Beribara	7304.60	-	311.00	5605.00

Source: (Rasily, 2008)

forested area without investing in watch and ward of forests resulting into the total isolation of these forested area of Western Himalaya.

Floods

The Western Himalayan regions are also vulnerable to various natural events, which are landslides, earthquakes, floods, drought as well as the cyclones. Amongst all of these natural calamities landslides and floods are most disastrous in Western Himalayan region more especially to the Uttarakhand causing severe loss to plants and property. Floods of other parts of India also has destroyed many natural forest and property like Mumbai floods (Prasad & Singh, 2005), Leh floods and Gujrat floods. The Western Himalayan region is very sensitive to rain-induced hazards which occur in various forms like flash floods, and glacial lake, cloudburst etc. Floods and cloudbursts are so common (Joshi & Kumar 2006).The floods occurs in Garhwal Himalaya (Uttarakhand) which includes the Kedarnath temples has causes severe loss to the biodiversity and other casualties of humans and animals. Geologically, the temples of Kedarnath is located in the southern part of Higher Himalaya. On the other hand, the rocks are found in the temples regions are generally high-grade metamorphic rocks, with granite intrusion at different regions.The disaster of the temple Kedarnath in Uttarakhand is due to the excessive rainfall-induced flash-floods-associated and is now one of the worst flood-related disaster in Western Himalaya, in terms the loss of human lives and biodiversity. As far as the climatic condition in concerned, the topographic of the Himalayan region which is also evident from the river profiles. On the other hand the some regions of Uttarakhand and Himachal area occupies a very important position in terms of seismic belt. The Western and Eastern part of Himalaya faced strong floods in past. Floods has killed approximately 4000 people and has caused unprecedented damage to the roadside vegetation, habitats, business industry and tourism. Due to floods in Uttrakhand in past approximately, 30 governmental establishments including guest houses, paramilitary camps, food outlets, schools and government offices have also been destroyed either completely or partially. An initial assessment made from various survey and reports suggests that more than 5000 hill villages and shelters of locals along with the forest and residing animals have been affected severely.

Soil Erosion

The process in which soil particles detaches from soil surface and their deposition in other areas is known as soil erosion. Soil erosion reduces the productivity of an area in different ways as it reduces the capacity of water in soil, decreases the efficiency of nutrients of plant, destroyed seedlings of trees, shrubs and small plants, reduces plant rooting depth, increase runoff as well as reduces its infiltration rates (Pimentel, 2006). In past soil erosion has impacted severely on the composite of plan species density of seed as well as the seed distribution (Blaikie, 2016; Cheng et al., 2006). Erosion of soil occurs when the soil is left exposed to the rain energy as the raindrops hit bare soil at great amount of energy and their easily displaceable ability from the surface. Raindrops removes thin layers soil from the surface of the land and the soil get eroded (Oldeman, 1997). The energy from the Rainfall is the cause of soil erosion in barren land, and it occurs when the soil lacks protective cover of vegetation. On the other way the main cause of soil erosion is deforestation, improper treatment of catchment and various anthropogenic pressures (Mahabaleshwara & Nagabhushan, 2014). At the time of erosion the process of organic matter and essential plant nutrients and other minerals are detached and soil depth is reduced. The changes changes both to vegetation growth and biota as well as the biodiversity (Pimentel et al., 1995) as the eroded soil carries very important soil nutrients like potassium and calcium nitrogen, phosphorous. Plant cover and vegetation reduces the rate of soil erosion, and its effects depends upon the height and continuity of canopy cover as well as the density of ground cover and density of root. Vegetation are most effective tool in reducing erosion because of their cover and density (Zuazo & Pleguezuelo, 2009).

The vegetation of Western Himalayas is highly destroyed by landslides and erosion of soil and these are mostly linked with diverse activities of human, like construction of roads and buildings as well as the removal of forests. The overall impact of erosion from earthquakes in the area is comparatively minimal mainly as these events are relatively very rare. On the other hand the natural phenomenon which has caused erosion of soil like exceptional rains, glacial lake and earthquakes outburst flooding in the high Himalayan region which are so common (Shrestha, 1997). It was evident from the various study that all of the eroded sites was once dominated with *Justicia adhatoda, Arundo donax, Aristida purpurea, Euphorbia hirta, Pinus roxburgii, Xanthium strumarium, Ficus palmate, Indigofera heterantha,, Conyza bonariensis, Zanthoxylum armatum and Ageratum conyzoides and Cynodon dactylon* These all species has been destroyed from the erosion of from the natural phenomenon as well as the anthropogenic pressures like trampling, grazing, road construction, and browsing. Soil erosion affecting development of vegetation and also has controlled by the response of vegetation (Shihong et al., 2003). In Himalaya, the productivity of soil loss due to soil erosion because the top soil is removed and the depth of roots reduced resulting into the plant nutrients like nitrogen, potassium, calcium are also removed. When these all soil nutrients are removed through soil erosion then growth of plant is stunted and overall productivity of soil decline. So less numbers of plants grown in the soil. Further as a result of eroded soil, vegetation cover reduced, richness of species decreased as well as the succession slow down. Secondly the slope had adverse impact on density of plant and the diversity of forest, with increase in the degree of slopes resulting into the recorded density and diversity was low in the area.

Natural Calamity or a Manmade Disaster

Various studies from Western Himalaya reveals that inadequate consideration of geology and geomorphology of the region whereas the fragile ecosystem in expanding through the urban clusters including the regions of Uttarakhand (Kedarnath) the factors responsible for disturbance of vegetation and biodiversity. On the other hand various constructing projects like hydropower projects, unmanaged road alignment with unfair constructions, without consideration of the stability of slope as well as the faulty engineering techniques are the other major things which are responsible for the more recent fury of the flash flood in Uttarakhand. Secondly the while tampering with the view of Western Himalaya region region, It should be be more careful because the slope which have evolved by and endogenic and exogenic processes are precautionly balanced (Sati et al. 2011). The policy makers are unaware of the sensitivity of the regions and other things is that the awareness is suppressed by the anthropogenic pressure of utilizing the resources of the region for overall growth Further it should be clear that over construction, utilization of excessive natural resources and huge number of power projects could also be the main cause of future calamites in the region (Rana et al. 2007). The various ongoing projects in the Himalayan region has a significant role in enhancing the disasters as they carelessly disposed of muck in the areas which was produced by the ongoing hydro power projects whereas increasing size and destroying capacity of the flooding considerably destroyed the floral and faunal wealth. The 330 MW hydropower project in Garhwal region produced approximately 800,000 m^3 of debris and muck. On the other hand approximately, about 500,000 m^3 of the muck was scraped at the time of recent flood in Garhwal, causing severe damages to the lower settlements in Srinagar Town. High floods level has been recorded at present time in Alaknanda valley. Except all these, in constructing projects, the uses of heavy explosives have caused fractures in the rocks and overlying loose materials resulting into the loss of plants associated with these rocks like *Woodfordia fruticosa*, *Terminalis* spp. etc. further the fractures causes due to the rainwater to go inside, causing the problems of landslides. On the other hand, excess amount of rainfall, some of the surroundings areas have become more vulnerable for future landslides as well as the threat for locals living in the area.

Socio-Economic Aspect of Vulnerability

Western Himalaya is an important habitat of Asiatic elephants and tigers in Northern India. It forms important boundary with Corbett tiger reserve and conserve the viable population of flora and fauna. These two protected areas located at the sites of ecologically valuable faunal corridor where once so many locals completes with each other for their basis needs.. All the forested areas in Western Himlaya protecting approximately more than one thousands number of Asiatic elephants and other globally endangered population of other flora and fauna. Locals residing in the forested areas are mainly depends on forest for various resources such as firewood, fodder, grazing land, medicinal plants and fruits (Badola 1997). So confliction in men and animals communities also occur by interfering habitats which includes

Table 5. Status of households, population of human and cattle as well as agricultural land in some of the forested areas of Western Himalaya

S.No.	Village Name	Agricultural Land (in hactate)	No. of Households	Human Population	Cattle Population
1	Teera	17.00	121	726	180
2	Rusulpur Tongiya	39.50	124	1024	292
3	Hajara Toungiya	68.80	150	1135	258
4	Chilla chaur	3.00	11	45	91
5	Gothiya gujjars	-	989	7418	9283
6	Kunou chaur	7.383	36	130	192
7	Koriya 1	-	-	-	-
8	Koriya 2	-	-	-	-
Total		**135,683**	**1,431**	**10,748**	**10,296**

Source: Information collected from village survey of the reserve, and from various published data. Human population data is based on as per the census of 2011.

- information not available

Table 6. Scenario of Gujjar (Pastoralist community) families in some of the forested areas of Western Himalaya during 1985 and 1998.

Forest Range	Number of Family 1985	1998
Gohri	-	149
Ramgarh	17	99
Haridwar	85	254
Chilla	181	193
Motichur	37	116
Chillawali	65	260
Kansarao	11	85
Dholkhand	116	234
Ramgarh	17	99
Total family rehabilitated	512	1390

Source: (Joshi & Singh, 2009).

predation of livestock by leopards, tigers and other animals harm to crops like elephants, birds, wild boar (Badola 1998; WII 2005; Johnsingh and Negi 2003). Himalayan protected areas have most of the Brahmna, Rajouts and other nomadic community like Gujjars, Tharu, Voksa etc. On the other hand in terms of wealth, approximately 94% of households owned at least three cow or bull. They mainly grows maize, rice, wheat and various seasonal vegetables.

These locals and nomadic communities are responsible for the lopping of Saussurea *atkinsoni, Meconopsis aculeate, Aconitum falconeri, A. balfouri, Betula utilis, Sorbus lanata, Sorbus lanata, Rhododendron campanulatum, Morina longifolia, Corydalis cashmeriana, Polemonium caerulium, Taxus wallichiana* and various medicinal plants like *Polygonatum multiflorum, Picrorhiza kurrooa, Aconitum violaceum.* The major changes primarily comprises of enhancement of wilderness in those places, which were used by Gujjar as their shelters. It was inferred from the results of the study that their abandoned areas are presently replaced by huge variety of vegetation like *Saccharum munja, Saccharum spontaneum, Holarrhena antidysenterica, Trewia nudiflora, Murraya* spp., *Cynodon dactylon, Syzygium cumini* etc. Besides, the water holes are presently completely recharged with natural water and are being used by wildlife. Before the commencement of rehabilitation programme of Gujjars, elephants must scarify the feeding grounds in order to feed on the small plants due to domestic buffaloes being grazed and looping, scraping of trees by Gujjars besides. Secondly at the day time wild animals are unable to drink water as most of the natural water sources are present near to their shelters resulting into the vulnerability of the biodiversity.

CONCLUSION

The present chapter focused on wilderness aspects of vulnerability, Western Himalayan cultures, personal growth and learning new ways to live. In a broad sense irresponsible management which includes the excessive negative effects on traditional vegetation due to pressures as well as on the natural environment, and decreasing the cultural integrity of local people. The analysis above indicates that Western Himalaya have large potentials of eco-tourism policies as well as the biodiversity rich potentialities but there are some difficulties or bottlenecks. If these difficulties are taken care, the place can be an attractive destination for various species and important medicinal and aromatic plants as well. Looking into the present influx in terms of vulnerability in the area It can be concluded that the status of vegetation in the Western Himalaya is enhancing day- by- day. Comparatively dualism in policy like allowing over construction in some regions of the protected areas of Shivalik like Chilla wildlife sanctuary, power projects in Garhwal Himalayas as well as in some sensitive part of Uttarakhand and protected parts of Himachal is another dimension, which has declined the species aspect from conservation efforts. The zone of the construction should be in planned or sustainable manner so that the species can get maximum survivals of the area and the aims and objectives of the development can be achieved. The area of Uttarakhand like Rajaji Tiger reserve, Corbett tiger reserve and valley of flowers has numbers of natural attractions and scenic beauty. On the other hand, Maximum of the tourists are unaware of these protected areas but some area like Kansarso and Motichur of Rajaji tiger reserve and some area of valley of flowers which are also unknown to the tourists, have great potentials of species and biodiversity. The implementation of community- based projects to enhance the forest as well as the wildlife conservation in the Himalaya. On the other hand, Participation of local villagers and tribes in policy making further provide the long-term survival of wildlife and appropriate strategy if we provide alternate to local people. This will be useful in reducing anthropogenic pressure from forest especially from the crucial wildlife corridors of the protected areas in Western Himalaya.

REFERENCES

Adhikari, B. S., Rawat, Y. S., & Singh, S. P. (1995). Structure and function of high altitude forests of Central Himalaya. Dry matter Dynamics. *Annals of Botany*, *72*(3), 237–248. doi:10.1006/anbo.1995.1017

Akash, Navneet, & Bhandari. (2019). A Community Analysis of Woody Species in a Tropical Forest of Rajaji Tiger Reserve. *Environment and Ecology*, *37*(1), 48–55.

Akash & Navneet. (2018). *Eco-Tourism as a Viable Option for Conservation of Wildlife in protected areas under Shivalik Hills of the outer Himalaya, India*. IGI Global.

Akash & Navneet. (2019). Impact of Climatic changes on earth's survivals- an analysis of changing environment and its consequences at global level: Impact of Climatic changes on earth's survivals. In Climate Change and Its Impact on Ecosystem Services and Biodiversity in Arid and Semi-Arid Zones. IGI Global Publishers.

Akash, Navneet, & Bhandari. (2018a). Phytosociological studies, biodiversity conservation in a sub tropical moist deciduous forest of Rajaji tiger reserve. Uttarakhand, India. *Int. J. Res. Anal. Rev.*, *5*(3), 39-51.

Akash., Navneet, & Bhandari. (2018b) Phytosociological investigation, biodiversity conservation and life form pattern in a *Holeptelia integrifolia* community under Rajaji tiger reserve, Uttarakhand, India. *Int. Res. J. Bio. Sci.*, *7*(7), 1-8.

Akash, Navneet, & Bhandari. (2018c). Tree diversity, stand structure and community composition in tropical forest of Rajaji tiger reserve, Northern India. *J. of App. and Nat. Sci.*, *10*(3), 945 – 953.

Allen, C. D., Breshears, D. D., & McDowell, N. G. (2015). On underestimation of global vulnerability to tree mortality and forest die-off from hotter drought in the Anthropocene. *Ecosphere*, *6*(8), 1–55. doi:10.1890/ES15-00203.1

Arya, N., Tewari, B., & Ram, J. (2012). The effect of natural and anthropogenic disturbance in forest canopy and its effect on species richness in forests of Uttarakhand Himalaya, India. *Russian Journal of Ecology*, *43*(2), 117–121. doi:10.1134/S1067413612020099

Baduni, N. P., & Sharma, C. M. (1997). Flexibility fitness compromise in some moist temperate forests of Garhwal Himalaya. *Annal of Forest*, *5*, 126–135.

Baland, J. M., Bardhan, P., Das, S., & Mookherjee, D. (2010). Forests to the people: Decentralization and forest degradation in the Indian Himalayas. *World Development*, *38*(11), 1642–1656. doi:10.1016/j.worlddev.2010.03.007

Bhandari, B. S., Mehta, J. P., Nautiyal, B. P., & Tiwari, S. C. (1997). Structure of a chir- pine community along an altitudinal gradient in Garhwal Himalaya. *International Journal of Ecology and Environmental Sciences*, *23*, 67–74.

Bhandari, B. S., Mehta, J. P., & Semwal, R. L. (2012). Forest Fire in Uttarakhand Himalaya: An Overview. In *Glimpses of Forestry Research in the Indian Himalayan Region. Almora* (pp. 167–175). G.B. Pant Institute of Himalayan Environment and Development.

Bhandari, B. S., Nautiyal, D. C., & Gaur, R. D. (1999). Structural attributes and productivity potential of an alpine pasture of Garhwal Himalaya. *Journal of the Indian Botanical Society*, *78*, 321–329.

Blaikie, P. (2016). *The political economy of soil erosion in developing countries*. Routledge.

Chaturvedi, R. K., Gopalakrishnan, R., Jayaraman, M., Bala, G., Joshi, N. V., Sukumar, R., & Ravindranath, N. H. (2011). Impact of climate change on Indian forests: A dynamic vegetation modeling approach. *Mitigation and Adaptation Strategies for Global Change*, *16*(2), 119–142. doi:10.100711027-010-9257-7

Cheng, J., Wan, H., & Hu, X. (2006). Soil seed bank and meadow renewal in the grassland on Loess Plateau. *Turang Xuebao*, *43*, 679–683.

Coulson, D., & Joyce, L. (2006). Indexing variability: A case study with climate change impacts on ecosystems. *Ecological Indicators*, *6*(4), 749–769. doi:10.1016/j.ecolind.2005.09.008

Davidenko, E. P., & Eritsov, A. (2003). The fire season in 2002 in Russia. Report of the Aerial Forest Fire Service, Avialesookhrana. *International Forest Fire News*, *28*, 15–17.

Dhanai, C. S., Panwar, V. P., & Sharma, C. M. (2000). Effect of aspect and soil on the structure of *Quercus leucotrichophora* natural stands in Western Himalaya. *Indian Journal of Forestry*, *23*(3), 349–356.

FAO. (2005). Global forest resources assessment. *Forest Health and Vitality*, 60–65.

Fearnside, P. M. (2005). Deforestation in Brazilian Amazonia: History, rates, and consequences. *Conservation Biology, 19*(3), 680–688. doi:10.1111/j.1523-1739.2005.00697.x

Gadgil, M., & Meher-Homji, V. M. (1985). Ecology and management of World's Savannas. In Land use and productive potential of Indian savanna. Canberra: Australian Academy of Science.

Gopalakrishnan, R., Jayaraman, M., Bala, G., & Ravindranath, N. H. (2011). Climate change and Indian forests. *Current Science, 101*, 348–355.

Grêt-Regamey, A., Brunner, S. H., & Kienast, F. (2013). Mountain ecosystem services: Who cares? *Mountain Research and Development, 32*(S1), S23–S34. doi:10.1659/MRD-JOURNAL-D-10-00115.S1

IPCC. (2014). Summary for Policymakers. In Climate Change 2014: Impacts, Adaptation, and Vulnerability. Part A: Global and Sectoral Aspects. Contribution of Working Group II to the Fifth Assessment Report of the Intergovernmental Panel on Climate Change. Cambridge University Press.

Joshi, S. P., Raizada, A., & And Srivastava, M. M. (1985). Vegetational analysis of oak forest at Mussoorie (Barlowganj), India. *Indian Journal of Forestry, 8*(1), 78–84.

Joshi, V., & Kumar, K. (2006). Extreme rainfall events and associated natural hazards in Alaknanda valley, Indian Himalayan region. *Journal of Mountain Science, 3*(3), 228–236. doi:10.100711629-006-0228-0

Keenan, R. J. (2015). Climate change impacts and adaptation in forest management: A review. *Annals of Science, 72*, 1–23. PMID:26104086

Kok, M., Lüdeke, M., Lucas, P., Sterzel, T., Walther, C., Jansses, P., ... de Soysa, I. (2016). A new method for analysing socio-ecological patterns of vulnerability. *Regional Environmental Change, 16*(1), 229–243. doi:10.100710113-014-0746-1

Kumar, M., Rawat, S., & Singh, H. (2018). Dynamic forest vegetation models for predicting impacts of climate change on forests: An Indian perspective. *Indian Journal of Forestry, 41*, 1–12.

Lata, K., & Bisht, N. S. (1991). Quantitative analysis and regeneration potential of moist temperature forest in Garhwal Himalayas. *Indian Journal of Forestry, 14*(2), 98–106.

Lwasa, S. (2015). A systematic review of research on climate change adaptation policy and practice in Africa and South Asia deltas. *Regional Environmental Change*, *15*(5), 815–824. doi:10.100710113-014-0715-8

Mahabaleshwara, H., & Nagabhushan, H. (2014). A study on soil erosion and its impacts on floods and sedimentation. *International Journal of Research in Engineering and Technology*, *3*(3), 443–451.

Oldeman, L. (1997). *Soil Degradation: A Threat to Food Security?* Paper presented at the International Conference on Time Ecology: "Time for Soil Culture-Temporal Perspectives on Sustainable Use of Soil", Tutzing.

Pandey, J. C. (2003). Vegetation analysis in a mixed Oak-Conifer forest of Central Himalaya. *Indian Journal of Forestry*, *26*(1), 66–74.

Pandey, R., & Bardsley, D. K. (2015). Social-ecological vulnerability to climate change in the Nepali Himalaya. *Applied Geography (Sevenoaks, England)*, *64*, 74–86. doi:10.1016/j.apgeog.2015.09.008

Pandey, R., & Jha, S. (2012). Climate vulnerability index-measure of climate change vulnerability to communities: A case of rural lower Himalaya, India. *Mitigation and Adaptation Strategies for Global Change*, *17*(5), 487–506. doi:10.100711027-011-9338-2

Pew, K. L., & Larsen, C. P. S. (2001). GIS analysis of spatial and temporal patterns of human-caused wildfires in the temperate rain forest of Vancouver Island, Canada. *Forest Ecology and Management*, *140*(1), 1–18. doi:10.1016/S0378-1127(00)00271-1

Pimentel, D. (2006). Soil erosion: A food and environmental threat. *Environment, Development and Sustainability*, *8*(1), 119–137. doi:10.100710668-005-1262-8

Prasad, A. K., & Singh, R. P. (2005). Extreme rainfall event of July 25–27, 2005 over Mumbai, West Coast, *India. Photonirvachak (Dehra Dun)*, *33*(3), 365–370. doi:10.1007/BF02990007

Rai, L. K., Prasad, P., & Sharma, E. (2000). Conservation threats to some important medicinal plants of the Sikkim Himalaya. *Biological Conservation*, *93*(1), 27–33. doi:10.1016/S0006-3207(99)00116-0

Ralhan, P. K., Saxena, A. K., & Singh, J. S. (1982). Analysis of forest vegetation at and around Nainital in Kumaun Himalaya. *Proceding of Indian National Academy of Sciecnes. B*, *48*(1), 121–137.

Rana, B. S., Verma, K. R., & Pandey, A. N. (1985). Analysis of forest vegetation at Siahi Devi Hill of West Almora Division in Kumaun Himalayas. *Indian Forester, 111*(9), 745–759.

Rasul, G. (2014). Food, water, and energy security in South Asia: A nexus perspective from the Hindu Kush Himalayan region. *Environmental Science & Policy, 39*, 35–48. doi:10.1016/j.envsci.2014.01.010

Rikhari, H. C., Ralhan, P. K., & Singh, S. P. (1989). Phytosociology and population structure of chir-pine forests of Kumoan Himalayas. *Annals of Biology, 5*(2), 129–140.

Saha, S., & Howe, H. F. (2001). The bamboo fire cycle hypothesis: A comment. *American Naturalist, 158*(6), 659–663. doi:10.1086/323593 PMID:18707360

Saxena, A. K., & Singh, J. S. (1982). A phytosociological analysis of woody species in forest communities of a part of Kumaun Himalaya. *Vegetation, 50*(1), 3–22. doi:10.1007/BF00120674

Sharma, E., Bhuchar, S., Xing, M., & Kcithyari, B. P. (2007). Land use change and its impact on hydro-ecological linkages in Himalayan watersheds. *Tropical Ecology, 48*, 151–161.

Sharma, J., Jayaraman, M., Kumar, V., & Ravindranath, N. H. (2015). Climate change impact and vulnerability assessment of forests in the Indian western Himalayan region: A case study of Himachal Pradesh, India. *Climate Risk Management, 10*, 63–76. doi:10.1016/j.crm.2015.08.002

Sharma, S. K., Deml, K., Dangal, S., Rana, E., & Madigan, S. (2015). REDD+ framework with integrated measurement, reporting and verification system for community based Forest management systems (CBFMS) in Nepal. *Current Opinion in Environmental Sustainability, 14*, 17–27. doi:10.1016/j.cosust.2015.01.003

Shihong, L., Wusheng, X., Xiankun, L., & Runqin, T. (2003). A review of vegetation restoration in eroded area of red soil. *Guangxi Zhi Wu, 23*(1), 83–89.

Shrestha, D. P. (1997). Assessment of soil erosion in the Nepalese Himalaya: A case study in Likhu Khola Valley, Middle Mountain Region. *Land Husbandry, 2*(1), 59–80.

Singh, G., & Rawat, G. S. (2012). Quantitative analysis of tree species diversity in different oak (Quercus spp.) dominated forests in Garhwal Himalaya, India. *Notulae Scientia Biologicae, 4*(4), 132. doi:10.15835/nsb448200

Singhal, R. M., & Soni, S. (1989). Quantitative ecological analysis of some woody species of Mussoorie Himalayas (UP). *Indian Forester, 115*(5), 327–336.

Upadhyay, K. (2016, April 30). Uttarakhand battles fire crisis. *The Hindu*.

Upgupta, S., Sharma, J., Jayaraman, M., Kumar, V., & Ravindranath, N. H. (2015). Climate change impact and vulnerability assessment of forests in the Indian western Himalayan region: A case study of Himachal Pradesh, India. *Climate Risk Management*, *10*, 63–76. doi:10.1016/j.crm.2015.08.002

Zhang, Y., Qi, W., Zhou, C., Ding, M., Liu, L., Gao, J., ... Zheng, D. (2015). Spatial and temporal variability in the net primary production of alpine grassland on the Tibetan Plateau since 1982. *Journal of Geographical Sciences*, *24*(2), 269–287. doi:10.100711442-014-1087-1

Zuazo, V. C. H. D., & Pleguezuelo, C. R. O. R. (2009). Soil-erosion and runoff prevention by plant covers. A review. In Sustainable agriculture. Springer.

Compilation of References

Abbas, A., Amjath-Babu, T. S., Kächele, H., & Müller, K. (2015). Non-structural flood risk mitigation under developing country conditions: An analysis on the determinants of willingness to pay for flood insurance in rural Pakistan. *Natural Hazards*, *75*(3), 2119–2135. doi:10.100711069-014-1415-x

Abbaspour, K. C., Rouholahnejad, E., Vaghefi, S., Srinivasan, R., Yang, H., & Kløve, B. (2015). A continental-scale hydrology and water quality model for Europe: Calibration and uncertainty of a high-resolution large-scale SWAT model. *Journal of Hydrology (Amsterdam)*, *524*, 733–752. doi:10.1016/j.jhydrol.2015.03.027

Abdechiri, M., Meybodi, M. R., & Bahrami, H. (2013). Gases brownian motion optimization: An algorithm for optimization (GBMO). *Appl. Soft Comput. J.*, *13*(5), 2932–2946. doi:10.1016/j.asoc.2012.03.068

Abdullah, A., Rahman, A., & Vojinovic, Z. (2009). LiDAR filtering algorithms for urban flood application: Review on current algorithms and filters test. *Laser Scanning, 9*(38), 30-36.

Abdullah, A. F., Vojinovic, Z., Price, R. K., & Aziz, N. A. A. (2012). Improved methodology for processing raw LiDAR data to support urban flood modelling–accounting for elevated roads and bridges. *Journal of Hydroinformatics*, *14*(2), 253–269. doi:10.2166/hydro.2011.009

Abdulrazzak, M., Elfeki, A., Kamis, A., Kassab, M., Chaabani, A., & Noor, K. (2019). Flash flood risk assessment in urban arid environment : Case study of Taibah and Islamic universities' campuses, Medina, Kingdom of Saudi Arabia. *Geomatics, Natural Hazards & Risk*, *10*(1), 780–796. doi:10.1080/19475705.2018.1545705

ABHS. (2010). *Etude de schéma directeur de protection de la ville de Fès contre les inondations*. ABHS.

ABHS. (2015). *Data*. Agence du bassin hydraulique de Fès.

ACAPS Briefing Note,. (2018). *Uganda - Flooding and Landslides in Bududa District*. Kampala, Uganda: ACAPS.

Adebayo, B., & Ntale, S. (2018). *Uganda mudslides, floods spur deaths, destruction*. Mbale, Uganda: CNN.

Adger, W. N. (2006). Vulnerability. *Global Environmental Change*, *16*(3), 268–281. doi:10.1016/j.gloenvcha.2006.02.006

Adger, W. N., Hughes, T. P., Folke, C., Carpenter, S. R., & Rockstrom, J. (2005). Social- ecological resilience to coastal disasters. *Science*, *309*(5737), 1036–1039. doi:10.1126cience.1112122 PMID:16099974

Adhikari, B. S., Rawat, Y. S., & Singh, S. P. (1995). Structure and function of high altitude forests of Central Himalaya. Dry matter Dynamics. *Annals of Botany*, *72*(3), 237–248. doi:10.1006/anbo.1995.1017

Ahmad, I., Abbas, A., Saghir, A., & Fawad, M. (2016). Regional Frequency Analysis of Annual Peak Flows in Pakistan Using Linear Combination of Order Statistics. *Polish Journal of Environmental Studies*, *25*(6), 2255–2264. doi:10.15244/pjoes/63782

Ahmad, I., Fawad, M., & Mahmood, I. (2015). At-Site Flood Frequency Analysis of Annual Maximum Stream Flows in Pakistan Using Robust Estimation Methods. *Polish Journal of Environmental Studies*, *24*(6), 2345–2353. doi:10.15244/pjoes/59585

Ahmad, I., Shah, S. F., Mahmood, I., & Ahmad, Z. (2013). Modeling of monsoon rainfall in Pakistan based on Kappa distribution. *Science International Lahore.*, *25*(2), 333–336. doi:10.24949/njes.v10il.1570.g94

Ahmed, I. (2015). *People of many rivers: Tales from the riverbanks*. Dhaka, Bangladesh: The University Press Limited.

Ait-Alhayane, K. (1991). Les représentations du désert et de la désertification en Afrique du Nord: contribution à une étude exploratoire au Maroc. Institut agronomique méditerranéen de Montpellier.

Akash & Navneet. (2018). *Eco-Tourism as a Viable Option for Conservation of Wildlife in protected areas under Shivalik Hills of the outer Himalaya, India*. IGI Global.

Akash & Navneet. (2019). Impact of Climatic changes on earth's survivals- an analysis of changing environment and its consequences at global level: Impact of Climatic changes on earth's survivals. In Climate Change and Its Impact on Ecosystem Services and Biodiversity in Arid and Semi-Arid Zones. IGI Global Publishers.

Akash, Navneet, & Bhandari. (2018a). Phytosociological studies, biodiversity conservation in a sub tropical moist deciduous forest of Rajaji tiger reserve. Uttarakhand, India. *Int. J. Res. Anal. Rev.*, *5*(3), 39-51.

Akash, Navneet, & Bhandari. (2018c). Tree diversity, stand structure and community composition in tropical forest of Rajaji tiger reserve, Northern India. *J. of App. and Nat. Sci.*, *10*(3), 945 – 953.

Akash., Navneet, & Bhandari. (2018b) Phytosociological investigation, biodiversity conservation and life form pattern in a *Holeptelia integrifolia* community under Rajaji tiger reserve, Uttarakhand, India. *Int. Res. J. Bio. Sci.*, *7*(7), 1-8.

Akash, Navneet, & Bhandari. (2019). A Community Analysis of Woody Species in a Tropical Forest of Rajaji Tiger Reserve. *Environment and Ecology*, *37*(1), 48–55.

Alam, G. M. M. (2016). *An assessment of the livelihood vulnerability of the riverbank erosion hazard and its impact on food security for rural households in Bangladesh* (PhD thesis). School of Commerce, University of Southern Queensland, Australia.

Alam, G. M. M., Alam, K., & Shahbaz, M. (2016). Influence of institutionalaccess and social capital on adaptation choices: Empirical evidence from vulnerable rural households in Bangladesh. *Ecological Economics*, *130*(C), 243–251. doi:10.1016/j.ecolecon.2016.07.012

Alam, G. M. M., Alam, K., Shahbaz, M., & Clarke, M. L. (2017). Drivers of vulnerability to climatic change in riparian char and river-bank households in Bangladesh: Implications for policy, livelihoods and social development. *Ecological Indicators*, *72*, 23–32. doi:10.1016/j.ecolind.2016.06.045

Alaoui, M. (2013). Water Sector in Morocco: Situation and Perspectives. *Journal of Water Resources and Ocean Science.*, *2*(5), 108–114. doi:10.11648/j.wros.20130205.18

Al-Betar, M. A., Awadallah, M. A., Farris, H., Yang, X. S., Khader, A. T., & Alomari, O. A. (2018). Bat-inspired algorithms natural selection mehanisms for global optimization. *Neurocomputing*, *273*, 448–465. doi:10.1016/j.neucom.2017.07.039

Alho, P., Hyyppä, H. & Hyyppä, J. (2009). Consequence of DTM precision for flood hazard mapping: A case study in SW Finland. *Nordic Journal of Surveying and Real Estate Research*, *6*, 21-39.

Ali, A. (2013). *Indus basin floods: Mechanisms, impacts, and management.* Mandaluyong City, Philippines: Asian Development Bank.

Ali, I., Hatta, Z., & Azman, A. (2014). Transforming the local capacity on natural disaster risk reduction in Bangladeshi communities: A social work perspective. *Asian Social Work and Policy Review*, *8*(1), 34–42. doi:10.1111/aswp.12023

Ali, I., Hatta, Z., Azman, A., Islam, S., Mazumder, A. H., & Hossain, K. (2016). Community participation in disaster management: A case study of Bangladesh. *Indian Journal of Ecology*, *43*(2), 463–472.

Allen, C. D., Breshears, D. D., & McDowell, N. G. (2015). On underestimation of global vulnerability to tree mortality and forest die-off from hotter drought in the Anthropocene. *Ecosphere*, *6*(8), 1–55. doi:10.1890/ES15-00203.1

Allison, E. H., & Seeley, J. A. (2004). HIV and AIDS among fisherfolk: A threat to responsible fisheries? *Fish and Fisheries*, *5*(3), 215–234. doi:10.1111/j.1467-2679.2004.00153.x

Ambroise, B. (1991). *Hydrologie des petits bassins versants ruraux en milieu tempéré, processus et modèles.* Séminaire du Conseil Scientifique du département « Science du sol » de l'Institut National de Recherches Agronomiques de Dijon.

Ambroise, B. (1998). La dynamique du cycle de l'eau dans un bassin versant. Processus, Facteur, Modèles.

Anees, M. T., Abdullah, K., Nawawi, M. N. M., Ab Rahman, N. N. N., Piah, A. R. M., Zakaria, N. A., ... Omar, A. M. (2016). Numerical modeling techniques for flood analysis. *Journal of African Earth Sciences*, *124*, 478–486. doi:10.1016/j.jafrearsci.2016.10.001

Armenakis, C., Du, E. X., Natesan, S., Persad, R. A., & Zhang, Y. (2017). Flood Risk Assessment in Urban Areas Based on Spatial Analytics and Social Factors. *Geosciences, 7*(4), 123.

Arnold, J. G., Kiniry, J. R., Srinivasan, R., Williams, J. R., Haney, E. B., & Neitsch, S. L. (2012). *Soil & Water Assessment Tool.* Academic Press.

Arora, K., & Singh, V. P. (1989). A Comparative Evaluation of the Estimators of the Log Pearcon Type (LP) 3 Distribution. *Journal of Hydrology (Amsterdam)*, *105*(1-2), 19–37. doi:10.1016/0022-1694(89)90094-2

Arya, N., Tewari, B., & Ram, J. (2012). The effect of natural and anthropogenic disturbance in forest canopy and its effect on species richness in forests of Uttarakhand Himalaya, India. *Russian Journal of Ecology*, *43*(2), 117–121. doi:10.1134/S1067413612020099

ASCE. (2000). *Artificial Neural Networks in Hydrology. J.* Australian Government.

Attorney General's Department Flood. (2009). *Preparedness Australian emergency manual series.* Commonwealth of Australia.

Atyang, A. (2014). *Study on Early Warning Systems in Uganda. A consultancy report submitted on 3October 2014 tor the Office of the Prime Minister of Uganda, Department for Disaster Preparedness and Management with UNDP funding*. Kampala, Uganda: Government of Uganda.

Australian Disaster Resilience Guideline (ADRG). (2014). *Guideline 7-3: Technical flood risk management guideline: Flood hazard, Australian Institute for Disaster Resilience CC BY-NC.* Australian Institute for Disaster Resilience.

Avanzia, A., Franka, E., Righetto, M., & Fattorellia, S. (2013). LIDAR Data Resolution Versus Hydro-Morphological Models for Flood Risk Assessment. ISPRS-International Archives of the Photogrammetry. *Remote Sensing and Spatial Information Sciences*, *1*, 119–124.

Babu, R., Tejwani, K. K., Agrawal, M. C., & Bhusan, L. S. (1979). Rainfall Intensity-Duration-Return Period Equations and Nomographs of India. CSWCRTI, ICAR, Dehradun, India.

Baduni, N. P., & Sharma, C. M. (1997). Flexibility fitness compromise in some moist temperate forests of Garhwal Himalaya. *Annal of Forest*, *5*, 126–135.

Baeck, S. H., Choi, S. J., Choi, G. W., & Lee, D. R. (2014). A study of evaluating and forecasting watersheds using the flood vulnerability assessment index in Korea. *Geomatics, Natural Hazards & Risk*, *5*(3), 208–231. doi:10.1080/19475705.2013.803268

Baland, J. M., Bardhan, P., Das, S., & Mookherjee, D. (2010). Forests to the people: Decentralization and forest degradation in the Indian Himalayas. *World Development*, *38*(11), 1642–1656. doi:10.1016/j.worlddev.2010.03.007

Balica, S. F., Douben, N., & Wright, N. G. (2009). Flood vulnerability indices at varying spatial scales. *Water Science and Technology*, *60*(10), 2571–2580. doi:10.2166/wst.2009.183 PMID:19923763

Balica, S. F., Popescu, I., Beevers, L., & Wright, N. G. (2013). Parametric and physically based modelling techniques for flood risk and vulnerability assessment: A comparison. *Environmental Modelling & Software*, *41*, 84–92. doi:10.1016/j.envsoft.2012.11.002

Balica, S. F., Wright, N. G., & van der Meulen, F. (2012). A flood vulnerability index for coastal cities and its use in assessing climate change impacts. *Natural Hazards*, *64*(1), 73–105. doi:10.100711069-012-0234-1

Balica, S., & Wright, N. G. (2010). Reducing the complexity of the flood vulnerability index. *Environmental Hazards*, *9*(4), 321–339. doi:10.3763/ehaz.2010.0043

Barati, R. (2011). Parameter Estimation of Nonlinear Muskingum Models Using Nelder-Mead Simplex Algorithm. *J. Hydrol. Eng.*, 946–954. .1943-5584.0000379 doi:10.1061/(ASCE)HE

Barati, R. (2013). Application of Excel Solver for Parameter Estimation of the Nonlinear Muskingum Models. *KSCE Journal of Civil Engineering*, *17*(5), 1139–1148. doi:10.100712205-013-0037-2

Barcellos, P. D. C. L., da Costa, M. S., Cataldi, M., & Soares, C. A. P. (2017). Management of non-structural measures in the prevention of flash floods: A case study in the city of Duque de Caxias, state of Rio de Janeiro, Brazil. *Natural Hazards*, *89*(1), 313–330. doi:10.100711069-017-2966-4

Barroca, B., Mouchel, J. M., Bonierbale, T., & Hubert, G. (2008). *Flood Vulnerability Assessment Tool (FVAT). In Day Water: An Adaptive Decision Support System for Urban Stormwater Management* (pp. 121–127). London, UK: Published by IWA Publishing.

Base, C. D. E., Modelisation, D. E. L. A., & Et, H. (2013). *Compétitivité économique du maroc*. Academic Press.

Bater, C. W., & Coops, N. C. (2009). Evaluating error associated with lidar-derived DEM interpolation. *Computers & Geosciences*, *35*(2), 289–300. doi:10.1016/j.cageo.2008.09.001

Beniston, M. (2003). Climatic change in mountain regions: A review of possible impacts. *Climatic Change*, *59*(1/2), 5–31. doi:10.1023/A:1024458411589

Benjiamin, J. R., & Cornell, C. A. (1970). *Probability, Statistics and Decision for Civil Engineers*. New York: McGraw-Hill.

Benmohammadi, A. (1998). Ensablement et désertification dans la vallée moyenne de l'oued Drâa: Le mythe de la fluctuation de la limite nord du désert. *Africa Geoscience Review*, *5*, 11–22.

Berterreche, L., Champredonde, R., & Dartus, D. (2008). *Etude de la génération de crues rapides par la confrontation de deux logiciels*. Academic Press.

Beven, K. (1989). Changing ideas in hydrology, the case of physically-based models. *Journal of Hydrology (Amsterdam)*, (105): 157–172.

Bhandari, B. S., Mehta, J. P., Nautiyal, B. P., & Tiwari, S. C. (1997). Structure of a chir- pine community along an altitudinal gradient in Garhwal Himalaya. *International Journal of Ecology and Environmental Sciences*, *23*, 67–74.

Bhandari, B. S., Mehta, J. P., & Semwal, R. L. (2012). Forest Fire in Uttarakhand Himalaya: An Overview. In *Glimpses of Forestry Research in the Indian Himalayan Region. Almora* (pp. 167–175). G.B. Pant Institute of Himalayan Environment and Development.

Bhandari, B. S., Nautiyal, D. C., & Gaur, R. D. (1999). Structural attributes and productivity potential of an alpine pasture of Garhwal Himalaya. *Journal of the Indian Botanical Society*, *78*, 321–329.

Bicheron, P., Defourny, P., Brockmann, C., Schouten, L., Vancutsem, C., Huc, M., … Arino, O., (2008). *GlobCover: Products Description and Validation Report*. ESA GlobCover project. Available online at: http://www.esa.int/due/ionia/globcover

Biringer, J. L., Hansen, L., & Hoffman, J. (2003). *Buying Time: A User's Manual to Building Resistance and Resilience to Climate Change in Natural Systems*. World Wildlife Fund.

Birkmann, J. (2006). Measuring vulnerability to promote disaster-resilient societies: Conceptual frameworks and definitions. *Measuring vulnerability to natural hazards: Towards disaster resilient societies, 1*, 9-54.

Blaikie, P. (2016). *The political economy of soil erosion in developing countries*. Routledge.

Bobee, B. (1975). The Log Pearson Type 3 Distribution and Its Application in Hydrology. *Water Resources Research, 11*(5), 681–689. doi:10.1029/WR011i005p00681

Bodoque, J. M., Guardiola-Albert, C., Aroca-Jimenez, E., Eguibar, M. A., & Martinez-Chenoll, M. L. (2016). Flood Damage Analysis: First Floor Elevation Uncertainty Resulting from LiDAR-Derived Digital Surface Models. *Remote Sensing, 8*(7), 604. doi:10.3390/rs8070604

Bollin, C., Hidajat, R., & Birkmann, J. (2006). Community-based risk index: Pilot implementation in Indonesia. In Measuring vulnerability to natural hazards: Towards disaster resilient societies (pp. 271-289). United Nation University Press.

Borden, K. A., Schmidtlein, M. C., Emrich, C. T., Piegorsch, W. W., & Cutter, S. L. (2007). Vulnerability of US cities to environmental hazards. *Journal of Homeland Security and Emergency Management, 4*(2), 1547–7355. doi:10.2202/1547-7355.1279

Boussaïd, I., Lepagnot, J., & Siarry, P. (2013). A survey on optimization metaheuristics. Information Sciences, 82–117. doi:10.1016/j.ins.2013.02.041

Bouvard, Gaross, & Berthet. (1994). Les crues de projet des barrages méthode du gradex, editors. *Bulletin du comité français des grands barrages*, 17-23. Available on: https://hydrologie.org/BIB/Gradex/Gradex.pdf

Bouwer, L. M., Crompton, R. P., Faust, E., Hoppe, P., & Pielke, R. A. Jr. (2007). Confronting disaster losses. *Science, 318*(5851), 753. doi:10.1126cience.1149628 PMID:17975052

Brown, K. M., Hambidge, C. H., & Brownett, J. M. (2016). Progress in operational flood mapping using satellite synthetic aperture radar (SAR) and airborne light detection and ranging (LiDAR) data. *Progress in Physical Geography, 40*(2), 196–214. doi:10.1177/0309133316633570

Candela, L., Tamoh, K., Olivares, G., & Gomez, M. (2012). Modelling Impacts of Climate Change on Water Resources in Ungauged and Data-Scarce Watersheds. Application to the Siurana Catchment (NE Spain). *The Science of the Total Environment, 440*, 253–260. doi:10.1016/j.scitotenv.2012.06.062 PMID:22795259

Cazenave, A., & Llovel, W. (2010). Contemporary sea level rise. *Annual Marine Science Review, 2*(1), 145–173. doi:10.1146/annurev-marine-120308-081105 PMID:21141661

CCA. (2010). *Coastal regeneration handbook*. CCA.

Cerny, V. (1985). Thermodynamical approach to the traveling salesman problem: An efficient simulation algorithm. *Journal of Optimization Theory and Applications*, *45*(1), 41–51. doi:10.1007/BF00940812

Chatterjee, R., & Huq, S. (2002). A report on the inter-regional conference on adaptation to climate change. *Mitigation and Adaptation Strategies for Global Change*, *7*(4), 403–406.

Chaturvedi, R. K., Gopalakrishnan, R., Jayaraman, M., Bala, G., Joshi, N. V., Sukumar, R., & Ravindranath, N. H. (2011). Impact of climate change on Indian forests: A dynamic vegetation modeling approach. *Mitigation and Adaptation Strategies for Global Change*, *16*(2), 119–142. doi:10.100711027-010-9257-7

Chelleri, L., Minucci, G., Ruiz, A., & Karmaoui, A. (2014). Responses to Drought and Desertification in the Moroccan Drâa Valley Region: Resilience at the Expense of Sustainability? *International Journal of Climate Change: Impacts & Responses*, *5*(2).

Cheng, J., Wan, H., & Hu, X. (2006). Soil seed bank and meadow renewal in the grassland on Loess Plateau. *Turang Xuebao*, *43*, 679–683.

Chu, H. J., & Chang, L. C. (2009). Applying Particle Swarm Optimization to Parameter Estimation of the Nonlinear Muskingum Model. *Journal of Hydrologic Engineering*, *14*(9), 1024–1027. doi:10.1061/(ASCE)HE.1943-5584.0000070

Church, J. A., & White, N. J. (2006). A 20th century acceleration in global sea-level rise. *Geophysical Research Letters*, *33*(1), 1–4. doi:10.1029/2005GL024826

Clarke, M., & Rendell, H. (2007). Climate, extreme events and land degradation. In Climate and Land Degradation. Springer. doi:10.1007/978-3-540-72438-4_7

Coles, S. G., & Powell, E. A. (1996). Bayesian methods in extreme value modelling: a review and new developments. *International Statistical Review/Revue Internationale de Statistique*, 119-136, Doi:10.2307/1403426

Comer, Crist, Reid, Hak, Hamilton, Unnasch, … Kutner. (2012). A *Rapid Ecoregional Assessment of the Mojave Basin and Range Ecoregion*. Bureau of Land Management.

Connor, R. F., & Hiroki, K. (2005). Development of a method for assessing flood vulnerability. *Water Science and Technology*, *51*(5), 61–67. doi:10.2166/wst.2005.0109 PMID:15918359

Cook, A., & Merwade, V. (2009). Effect of topographic data, geometric configuration and modeling approach on flood inundation mapping. *Journal of Hydrology (Amsterdam)*, *377*(1-2), 131–142. doi:10.1016/j.jhydrol.2009.08.015

Costabile, P., Macchione, F., Natale, L., & Petaccia, G. (2015). Flood mapping using LIDAR DEM. Limitations of the 1-D modeling highlighted by the 2-D approach. *Natural Hazards*, *77*(1), 181–204. doi:10.100711069-015-1606-0

Coulson, D., & Joyce, L. (2006). Indexing variability: A case study with climate change impacts on ecosystems. *Ecological Indicators*, *6*(4), 749–769. doi:10.1016/j.ecolind.2005.09.008

CRED. (2015). *Centre for Research on the Epidemiology of Disasters Emergency Database*. CRED.

Curtis, S., Salahuddin, A., Adler, R. F., Huffman, G. J., Gu, G., & Hong, Y. (2007). Precipitationextremes estimated by GPCP and TRMM: ENSO relationships. *Journal of Hydrometeorology*, *8*(4), 678–689. doi:10.1175/JHM601.1

Cutter, S. L., Mitchell, J. T., & Scott, M. S. (2000). Revealing the vulnerability of people and places: A case study of Georgetown County, South Carolina. *Annals of the Association of American Geographers*, *90*(4), 713–737. doi:10.1111/0004-5608.00219

DARA. (2012). DARA and the Climate Vulnerable Forum, Climate vulnerability monitor 2nd edition. In *A guide to the cold calculus of a hot planet*. DARA.

Das, A. (2004). Parameter Estimation for Muskingum Models. *Journal of Irrigation and Drainage Engineering*, *130*(2), 140–147. doi:10.1061/(ASCE)0733-9437(2004)130:2(140)

Das, A. (2007). Chance-Constrained optimization-based parameter estimation for Muskingum models. *Journal of Irrigation and Drainage Engineering*, *133*(5), 487–494. doi:10.1061/(ASCE)0733-9437(2007)133:5(487)

Dasgupta, S. (2016). *Climate change poses urgent threat to poor of coastal Bangladesh*. Academic Press.

Davidenko, E. P., & Eritsov, A. (2003). The fire season in 2002 in Russia. Report of the Aerial Forest Fire Service, Avialesookhrana. *International Forest Fire News*, *28*, 15–17.

de Brito, M. M., Evers, M., & Höllermann, B. (2017). Prioritization of flood vulnerability, coping capacity and exposure indicators through the Delphi technique: A case study in Taquari-Antas basin, Brazil. *International Journal of Disaster Risk Reduction*, *24*, 119–128. doi:10.1016/j.ijdrr.2017.05.027

De Lange, H. J., Sala, S., Vighi, M., & Faber, J. H. (2010). Ecological vulnerability in risk assessment—A review and perspectives. *The Science of the Total Environment*, *408*(18), 3871–3879. doi:10.1016/j.scitotenv.2009.11.009 PMID:20004002

Dell, V. K., Arthur, S., & Pender, G. (2008). *The use of a LiDAR DEM in modelling sewerage derived urban flood wave routing. 11th International Conference on Urban Drainage*, Edinburgh, UK.

Demir & Kisi. (2016). Flood Hazard Mapping by Using Geographic Information System and Hydraulic Model: Mert River, Samsun, Turkey Hindawi Publishing Corporation. *Advances in Meteorology*, *2016*, 9. doi:10.1155/2016/4891015

Dhanai, C. S., Panwar, V. P., & Sharma, C. M. (2000). Effect of aspect and soil on the structure of *Quercus leucotrichophora* natural stands in Western Himalaya. *Indian Journal of Forestry, 23*(3), 349–356.

Di Baldassarre, G., Kooy, M., Kemerink, J. S., & Brandimarte, L. (2013). Towards understanding the dynamic behavior of floodplains as human-water systems. *Hydrology and Earth System Sciences Discussions, 10*(3), 3869–3895. doi:10.5194/hessd-10-3869-2013

Di Baldassarre, G., & Uhlenbrook, S. (2012). Is the current flood of data enough? A treatise on research needs for the improvement of flood modelling. *Hydrological Processes, 26*(1), 153–158. doi:10.1002/hyp.8226

Dincer, H., & Hacioglu, U. (2017). Investigating the Economic Vulnerability Factors of Emerging Markets after the Global Financial Crisis of 2008 with a Hybrid Multi-Criteria Decision-Making Approach. In Handbook of Research on Economic, Financial, and Industrial Impacts on Infrastructure Development (pp. 313-333). IGI Global. doi:10.4018/978-1-5225-2361-1.ch016

Dogan, B., & Olmez, T. (2015). A new metaheuristic for numerical function optimization: Vortex Search Algorithm. *Information Sciences, 293*, 125–145. doi:10.1016/j.ins.2014.08.053

Dogru, T., Marchio, E. A., Bulut, U., & Suess, C. (2019). Climate change: Vulnerability and resilience of tourism and the entire economy. *Tourism Management, 72*, 292–305. doi:10.1016/j.tourman.2018.12.010

Dorigo, M., Maniezzo, V., & Colorni, A. (1996). Ant System: Optimization by a colony of cooperating agents. *IEEE Trans. Syst. Man. Cybern. –Part B, 26*(1), 29–41. doi:10.1109/3477.484436 PMID:18263004

Dottori, F., Martina, M. L. V., & Figueiredo, R. (2016). A methodology for flood susceptibility and vulnerability analysis in complex flood scenarios. *Journal of Flood Risk Management, 11*, S632–S645. doi:10.1111/jfr3.12234

Douben, J. K. (2006). Characteristics of River floods and Flooding: A Global Overview, 1985-2003. *Journal, 59*, 59–521.

Douben, N. (2006). Characteristics of river floods and flooding: A global overview, 1985– 2003. *Irrigation and Drainage, 55*(1), S9–S21. doi:10.1002/ird.239

Dupont, B. S., & Allen, D. L. (2000). *Revision of the Rainfall Intensity Duration Curves for the Commonwealth of Kentucky*. Kentucky Transportation Research Center Research Report, 292.

East African Community Secretariat. (1999). *The Treaty for the Establishment of the East African Community*. Arusha, Tanzania: East African Governments.

Easterling, D. R., Evans, J. L., Groisman, P. Y., Karl, T. R., Kunkel, K. E., & Ambenje, P. (2000). Observed variability and trends in extreme climate events: A brief review. *Bulletin of the American Meteorological Society, 81*(3), 417–425. doi:10.1175/1520-0477(2000)081<0417:OVATIE>2.3.CO;2

Elsebaie, I. H. (2012). Developing Rainfall Intensity-Duration-Frequency Relationship for Two Regions in Saudi Arabia. *Journal of King Saud University. Engineering and Science, 24*, 131–140.

EMDAT. (2017). *OFDA/CRED International Disaster Database, Universitécatholique de Louvain – Brussels – Belgium*. Retrieved from https://ourworldindata.org/natural-disasters

Emergency Management Australia and Australian Bureau of Meteorology. (1998). *Flood-warning, preparedness and safety*. Emergency Management Australia.

Enarson, E. (2007). Identifying and addressing social vulnerabilities. *Emergency Management: Principles and Practice for Local Government, 2*, 257-278.

Eskandar, H., Sadollah, A., Bahreininejad, A., & Hamdi, M. (2012). Water cycle algorithm - A novel metaheuristic optimization method for solving constrained engineering optimization problems. *Computers & Structures, 110–111*, 151–166. doi:10.1016/j.compstruc.2012.07.010

Eyotaru, O. (2013, July 23). East African integration hindered by information gaps – minister. Daily Monitor Newspaper, p. 1.

Fadil, A., Rhinane, H., Kaoukaya, A., Kharchaf, Y., & Bachir, O. A. (2011). Hydrologic Modeling of the Bouregreg Watershed (Morocco) Using GIS And SWAT Model. *Journal of Geographic Information System, 3*(04), 279–289. doi:10.4236/jgis.2011.34024

FAO. (2005). Global forest resources assessment. *Forest Health and Vitality*, 60–65.

Fearnside, P. M. (2005). Deforestation in Brazilian Amazonia: History, rates, and consequences. *Conservation Biology, 19*(3), 680–688. doi:10.1111/j.1523-1739.2005.00697.x

Fernandez, P., Mourato, S., Moreira, M., & Pereira, L. (2016). A new approach for computing a flood vulnerability index using cluster analysis. *Physics and Chemistry of the Earth Parts A/B/C, 94*, 47–55. doi:10.1016/j.pce.2016.04.003

Filahi, S., Tanarhte, M., Mouhir, L., El Morhit, M., & Tramblay, Y. (2016). Trends in indices of daily temperature and precipitations extremes in Morocco. *Theoretical and Applied Climatology, 124*(3), 959–972. doi:10.100700704-015-1472-4

Flood, M. (2001). Laser Altimetry: From Science to Commercial Lidar Mapping. *Photogrammetric Engineering and Remote Sensing*, 67.

Frigerio, I., Ventura, S., Strigaro, D., Mattavelli, M., De Amicis, M., Mugnano, S., & Boffi, M. (2016). A GIS-based approach to identify the spatial variability of social vulnerability to seismic hazard in Italy. *Applied Geography (Sevenoaks, England)*, *74*, 12–22. doi:10.1016/j.apgeog.2016.06.014

Furevik, T., Drange, H., & Sorteberg, A. (2002). *Anticipated changes in the Nordic Seas marine climate: Scenarios for 2020, 2050 and 2080*. Academic Press.

Gadgil, M., & Meher-Homji, V. M. (1985). Ecology and management of World's Savannas. In Land use and productive potential of Indian savanna. Canberra: Australian Academy of Science.

Gao, J., Nickum, J. E., & Pan, Y. (2007). An assessment of flood hazard vulnerability in the Dongting Lake Region of China. *Lakes and Reservoirs: Research and Management*, *12*(1), 27–34. doi:10.1111/j.1440-1770.2007.00318.x

Gassman, P. W., Arnold, J. G., Srinivasan, R., & Reyes, M. (2010). The worldwide use of the SWAT Model: Technological drivers, networking impacts, and simulation trends. In *Proceedings of the Watershed Technology Conference*. American Society of Agricultural and Biological Engineers, Earth University.

Gaume, E. (2009). Hydrologie de versants et de bassins versants et modélisation pluie-débit. Master2 Sciences et Génie de l'environnement, Ecole Nationale des Ponts et Chaussées.

Geem, Z. W. (2006). Parameter Estimation for the Nonlinear Muskingum Model Using the BFGS Technique. *Journal of Irrigation and Drainage Engineering*, *132*(5), 474–478. doi:10.1061/(ASCE)0733-9437(2006)132:5(474)

Geem, Z. W. (2011). Parameter Estimation of the Nonlinear Muskingum Model Using Parameter-Setting-Free Harmony Search. *Journal of Hydrologic Engineering*, *16*(8), 684–688. doi:10.1061/(ASCE)HE.1943-5584.0000352

Gelman, A., Carlin, J. B., Stern, H. S., & Rubin, D. B. (1997). *Bayesian Data Analysis*. New York: Chapman and Hall.

Gerland, P., Raftery, A. E., Sevčíková, H., & Wilmoth, J. (2014). World population stabilization unlikely this century. *Science*, *346*(6206), 234–237. doi:10.1126cience.1257469 PMID:25301627

Ghimire, S. (2013). Application of a 2D hydrodynamic model for assessing flood risk from extreme storm events. *Climate (Basel)*, *1*(3), 148–162. doi:10.3390/cli1030148

Gleick, P. H. (1993). *Igor Shiklomanov's chapter "World fresh water resources", Water in Crisis: A Guide to the World's Fresh Water Resources*. New York: Oxford University Press.

Glover, F. (1986). Future paths for integer programming and links to artificial intellegience. *Computers & Operations Research*, *13*(5), 533–549. doi:10.1016/0305-0548(86)90048-1

Gopalakrishnan, R., Jayaraman, M., Bala, G., & Ravindranath, N. H. (2011). Climate change and Indian forests. *Current Science*, *101*, 348–355.

Government of Rwanda. (2012). *Rwanda and UNDP climate change project*. Government of the Republic of Rwanda and UNDP Rwanda.

Government of Tanzania. (2006). *Tanzania NAPA guiding principles*. Government of the Republic of Tanzania.

Government of Uganda. (2005). *Uganda National Report And Information On Disaster Risk Reduction Efforts For The World Conference On Disaster Reduction*. Government of Uganda.

Government of Uganda. (2010a). *Climate change policy draft 2012*. Kampala, Uganda: Ministry of Water and Environment.

Government of Uganda. (2010b). *The National Policy for Disaster Preparedness and Management*. Kampala, Uganda: Government of Uganda.

Government of Uganda. (2015a). *Economic assessment of the impacts of climate change in Uganda: Key results*. Government of Uganda.

Government of Uganda. (2015b). *The National Development Plan II (2016-2020)*. Kampala, Uganda: Government of Uganda.

Grayson, R. A. (2000). *Spatial Patterns in Catchment Hydrology*. Cambridge University.

Grêt-Regamey, A., Brunner, S. H., & Kienast, F. (2013). Mountain ecosystem services: Who cares? *Mountain Research and Development*, *32*(S1), S23–S34. doi:10.1659/MRD-JOURNAL-D-10-00115.S1

Grinsted, A., Moore, J. C., & Jevrejeva, S. (2009). Reconstructing sea level from paleo and projected temperatures 200 to 2100 AD. *Climate Dynamics*, *34*(2), 461–472.

Groisman, Knight, Easterling, Karl, Hegerl, & Razuvaev. (2005). Trends in intense precipitation in the climate record. *J. Climate, 18*, 1326–1350.

Guhathakurta, P., Sreejith, O. P., & Menon, P. A. (2011). Impact of Climate Change on Extreme Rainfall Events and Flood Risk in India. *Journal of Earth System Science*, *120*(3), 359–373. doi:10.100712040-011-0082-5

Guillot, P., & Duband, D. (1967). La méthode du gradex pour le calcul de la probabilité des crues à partir des pluies. *Flood and Their Computation. Proceedings of the Leningrad Symposium*, *84*, 560–569. Available on: https://iahs.info/uploads/dms/084063.pdf

Gumbel, E. J. (1941). The return period of flood flows. *Annals of Mathematical Statistics*, *12*(2), 163–190. doi:10.1214/aoms/1177731747

Gumbel, E. J. (1954). *Statistical Theory Of Extreme Values and Some Practical Applications*. U.S. Department of Commerce, National Bureau of Standards.

Gupta, H. V., Sorooshian, S., & Yapo, P. O. (1999). Status of automatic calibration for hydrologic models: Comparison with multilevel expert calibration. *Journal of Hydrologic Engineering*, *4*(2), 135–143. doi:10.1061/(ASCE)1084-0699(1999)4:2(135)

Hagenlocher, M., Renaud, F. G., Haas, S., & Sebesvari, Z. (2018). Vulnerability and risk of deltaic social-ecological systems exposed to multiple hazards. *The Science of the Total Environment*, *631*, 71–80. doi:10.1016/j.scitotenv.2018.03.013 PMID:29524904

Haile, A. T., & Rientjes, T. H. M. (2005). Effects of LiDAR DEM resolution in flood modelling: A model sensitivity study for the city of Tegucigalpa, Honduras. *Isprs*, *3*, 12–14.

Hallegatte, S., Green, C. H., Nicholls, R. J., & Corfee-Morlot, J. (2013). Future flood losses in major coastal cities. *Nature Climate Change*, *3*(9), 802–806. doi:10.1038/nclimate1979

Hansen, P. & Mladenovic, J.A.M. (2008). Variable neighbourhood search: methods and applications. *4OR, 6*, 319–360.

Han, W., Meehl, G. A., Rajagopalan, B., Fasullo, J. T., Hu, A., Lin, J., ... Yeager, S. (2010). Patterns of Indian Ocean sea-level change in a warming climate. *Nature Geoscience*, *3*(8), 546–550. doi:10.1038/ngeo901

Harakat, I. (2007). *Les acteurs de la coopération et la dimension socio-économique de la désertification dans le sud du Maroc: cas de Zagora*. Mémoire de maitrise en science politique, Université du Québec. Retrieved from http://archipel.uqam.ca/id/eprint/628

Hashim, M. S., Hassan, S., & Bakar, A. A. (2018) Developing a Household Flood Vulnerability Index: A Case Study of Kelantan. *Sch. J. Econ. Bus. Manag.* Doi:10.21276jebm.2018.5.7.4

Helga-Jane, S., & Richard, L. (2004). *Risque d'inondation et aménagement durable des territoires*. Available at http://books.openedition.org/septentrion/15682?lang=fr

Hengl, T., Mendes de Jesus, J., Heuvelink, G. B. M., Ruiperez Gonzalez, M., Kilibarda, M., Blagotić, A., ... Kempen, B. (2017). SoilGrids250m: Global gridded soil information based on machine learning. *PLoS One*, *12*(2). doi:10.1371/journal.pone.0169748 PMID:28207752

Hirabayashi, Y., Mahendran, R., Koirala, S., & Kanae, S. (2013). Global flood risk under climate change. *Nature Climate Change*, *3*(9), 1–6. doi:10.1038/nclimate1911

Hohenthal, J., Alho, P., Hyyppä, J., & Hyyppä, H. (2011). Laser scanning applications in fluvial studies. *Progress in Physical Geography*, *35*(6), 782–809. doi:10.1177/0309133311414605

Ho, L. T. K., Umitsu, M., & Yamaguchi, Y. (2010). Flood Hazard Mapping by Satellite Images and SRTM DEM in the Vu Gia- Thu Bon Alluvial Plain, Central Vietnam. *The International Archives of the Photogrammetry, Remote Sensing and Spatial Information Sciences*, *38*, 8.

Holland, J. H. (1975). *Adaptation in Natural and Artificial Systems*. Ann Arbor, MI: University of Michigan Press.

Hong, L., Ouyang, M., Peeta, S., He, X., & Yan, Y. (2015). Vulnerability assessment and mitigation for the Chinese railway system under floods. *Reliability Engineering & System Safety*, *137*, 58–68. doi:10.1016/j.ress.2014.12.013

Horton, R. E. (1933). The role of infiltration in the hydrologic cycle. *Transactions of the American Geophysics Union, 14th Annual Meeting*, 446–460. 10.1029/TR014i001p00446

Horton, R. E. (1945). Erosional development of streams and their drainage basins; Hydrophysical approach to quantitative morphology. *Geological Society of America Bulletin*, *56*(3), 275–370. doi:10.1130/0016-7606(1945)56[275:EDOSAT]2.0.CO;2

Hossain, K., Quaik, S., Ismail, N., Rafatullah, M., Ali, I., Hatta, Z., ... Rameeja, S. (2016). Climate change-perceived impacts on agriculture, vulnerability and response strategies for improving adaptation practice in developing countries (South Asian Region). *International Journal of Agricultural Research*, *11*(1), 1–12. doi:10.3923/ijar.2016.1.12

Ibrahim, N. F., Zardari, N. H., Shirazi, S. M., Mohd Haniffah, M. R., Mat Talib, S., Yusop, Z., & Mohd Yusoff, S. M. A. (2017). Identification of Vulnerable Areas to Floods in Kelantan River Sub-basins by using Flood Vulnerability Index. *International Journal of GEOMATE*, *12*(29), 107–114.

IFRC/RCS. (2009). *East Africa: Preparedness for El Floods Emergence Appeal of 13 October 2009*. IFRC/RCS.

Intergovernmental Panel on Climate Change (IPCC). (2012). *Managing the risks of extreme events and disasters to advance climate change adaptation. A special report of Working Groups I and II of the Intergovernmental Panel on Climate Change*. New York: Cambridge University Press.

International Council for Science Regional Office for Africa (ICSU-ROA). (2007). *Science plan: Sustainable energy in sub-Saharan Africa*. Pretoria, South Africa: ICSU-ROA.

International Council for Science Regional Office for Africa (ICSU-ROA). (2008). *Science plan: Global environmental change (including climate change and adaptation) in sub-Saharan Africa*. Pretoria, South Africa: ICSU-ROA.

IPCC. (2007). *Impacts, adaptation, and vulnerability. Contribution of working group*. IPCC.

IPCC. (2007). *Synthesis Report*. IPCC.

IPCC. (2012). *Managing the Risks of Extreme Events and Disasters to Advance Climate Change Adaptation. Cambridge University Press.*

IPCC. (2013). Climate Change 2013: The Physical Science Basis. Cambridge University Press.

IPCC. (2014). Summary for Policymakers. In Climate Change 2014: Impacts, Adaptation, and Vulnerability. Part A: Global and Sectoral Aspects. Contribution of Working Group II to the Fifth Assessment Report of the Intergovernmental Panel on Climate Change. Cambridge University Press.

ISDR. (2009). *Global assessment report on disaster risk reduction*. Geneva, Switzerland: United Nations.

Islam, M. S., Hasan, T. M., Chowdhury, S. I. R., Rahaman, M. H., & Tusher, T. R. (2012). Coping techniques of local people to flood and river erosion in char areas of Bangladesh. *Journal of Environmental Science & Natural Resources*, *5*(2), 251–261. doi:10.3329/jesnr.v5i2.14827

ISRIC – World Soil Information. (2013). *Soil property maps of Africa at 1 km*. Available for download at www.isric.org

Jamal, P., Stitou, E., Messari, E., Dridri, P. A., Abdel, P., Chaouni, A., … Fès, F. S. T. (2012). *UFR: Chimie de l ' Environnement Discipline*. Géologie Présentée par Jamal Naoura N aoura.

Jongman, B., Ward, P. J., & Aerts, J. C. J. H. (2012). Global exposure to river and coastal flooding: Long-term trends and changes. *Global Environmental Change*, *22*(4), 823–835. doi:10.1016/j.gloenvcha.2012.07.004

Jonkman, S. N. (2005). Global perspectives on loss of human life caused by floods. *Natural Hazards*, *34*(2), 151–175. doi:10.100711069-004-8891-3

Jorgensen, S. E. (1992). *Integration of ecosystem theories: a pattern*. Kluwer Academic Publishers Dordrecht. doi:10.1007/978-94-011-2682-3

Joshi, S. P., Raizada, A., & And Srivastava, M. M. (1985). Vegetational analysis of oak forest at Mussoorie (Barlowganj), India. *Indian Journal of Forestry*, *8*(1), 78–84.

Joshi, V., & Kumar, K. (2006). Extreme rainfall events and associated natural hazards in Alaknanda valley, Indian Himalayan region. *Journal of Mountain Science*, *3*(3), 228–236. doi:10.100711629-006-0228-0

Joy, S. (2005). *Application of GIS and remote sensing in flood management: A case study of West Bengal, India* (Doctoral dissertation). National University of Singapore. Retrieved from ScholarBank@NUS Repository.

Kannan, N., Santhi, C., White, M. J., Mehan, S., Jeffrey, G., & Gassman, P. W. (2019). *Some Challenges in Hydrologic Model Calibration for Large-Scale Studies : A Case Study of SWAT Model Application to Mississippi-Atchafalaya River Basin*. Academic Press. doi:10.3390/hydrology6010017

Karaboga, D., & Basturk, B. (2007). A powerful and efficient algorithm for numerical function optimization: Artificial bee colony (ABC) algorithm. *Journal of Global Optimization*, *39*(3), 459–471. doi:10.100710898-007-9149-x

Karahan, H., Gurarslan, G., & Geem, Z. W. (2013). Parameter Estimation of the Nonlinear Muskingum Flood-Routing Model Using a Hybrid Harmony Search Algorithm. *Journal of Hydrologic Engineering*, *18*(3), 352–360. doi:10.1061/(ASCE)HE.1943-5584.0000608

Karmaoui, A. & Balica, S. (2019). A New Flood Vulnerability Index Adapted for the Pre-Saharan Region. *International Journal of River Basin Management*. Doi:10.1080/15715124.2019.1583668

Karmaoui, A., Messouli, M., Khebiza, Y. M., & Ifaadassan, I. (2014a). Environmental Vulnerability to climate change and anthropogenic impacts in dryland (Pilot study: Middle Draa Valley, South Morocco). *J Earth Sci Clim Change, S11*. Doi:10.4172/2157-7617.S11-002

Karmaoui, A. (2018). The cutaneous leishmaniasis vulnerability index (CLVI). *Acta Ecologica Sinica*, *38*(4), 288–295. doi:10.1016/j.chnaes.2018.01.001

Karmaoui, A., Balica, S. B., & Messouli, M. (2016). Analysis of applicability of flood vulnerability index in Pre-Saharan region, Morocco. *Natural Hazards and Earth System Sciences*, *2016*. doi:10.5194/nhess-2016-96

Karmaoui, A., Ifaadassan, I., Babqiqi, A., Messouli, M., & Khebiza, M. Y. (2016a). Analysis of the Water Supply-demand Relationship in the Middle Draa Valley, Morocco, under Climate Change and Socio-economic Scenarios. *Journal of Scientific Research & Reports*, *9*(4), 1–10. doi:10.9734/JSRR/2016/21536

Karmaoui, A., Ifaadassan, I., Messouli, M., & Khebiza, M. Y. (2015b). Sustainability of the Moroccan oasean system (Case study: Middle Draa Valley). *Global J Technol Optim*, *6*(01), 170. doi:10.4172/2229-8711.1000170

Karmaoui, A., Messouli, M., Ifaadassan, I., & Khebiza, M. Y. (2014b). A multidisciplinary approach to assess the environmental vulnerability at local scale in context of climate change (Pilot study in upper draa valley, South Morocco). *Glob. J. Technol. Optim*, *6*, 1–11.

Karmaoui, A., Messouli, M., & Yacoubi, M. (2015a). Vulnerability of Ecosystem Services to Climate Change and Anthropogenic Impacts in South East of Morocco (Case study: Drying up of Iriki lake). *The International Journal of Climate Change: Impacts and Responses.*, *7*(3), 1835–7156.

Karmaoui, A., Minucci, G., Messouli, M., Khebiza, M. Y., Ifaadassan, I., & Babqiqi, A. (2019). Climate Change Impacts on Water Supply System of the Middle Draa Valley in South Morocco. In *Climate Change, Food Security and Natural Resource Management* (pp. 163–178). Cham: Springer. doi:10.1007/978-3-319-97091-2_8

Kaspersen, P.S., HøeghRavn, N., Arnbjerg-Nielsen, K., Madsen, H., & Drews, M. (2015). Influence of urban land cover changes and climate change for the exposure ofEuropean cities to flooding during high-intensity precipitation. *Proc. Int. Assoc. Hydrol. Sci. (IAHS), 370*, 21-27.

Kaveh, A., & Mahdavi, V. R. (2014). Colliding bodies optimization : A novel meta-heuristic method. *Computers & Structures*, *139*, 18–27. doi:10.1016/j.compstruc.2014.04.005

Keenan, R. J. (2015). Climate change impacts and adaptation in forest management: A review. *Annals of Science*, *72*, 1–23. PMID:26104086

Kellermann, P., Schobel, A., Kundela, G., & Thieken, A.H. (2015). Estimating flood damage to railway infrastructure - the case study of the March River flood in 2006 at the Austrian Northern Railway. *Nat. Hazards Earth Syst. Sci.*, *15*(11), 2485-2496. . doi:10.5194/nhess-15-2485-2015

Kelman, I. (2003). *Physical flood vulnerability of residential properties in coastal, eastern England* (Doctoral dissertation). University of Cambridge.

Kenbl, M. R., Yang, Z. L., Hutchison, K., & Maidment, D. R. (2005). Regional Scale Flood Modelling Using NEXRAD Rainfall, GIS and HEC-HMS/RAS: A Case Study for the San Antonio River Basin Summer 2002 Storm Event. *Journal of Environmental Management*, *75*(4), 325–336. doi:10.1016/j.jenvman.2004.11.024 PMID:15854726

Kerala Government. (2018). *Stand With Kerala Kerala Flood Relief*. Retrieved from https://donation.cmdrf.kerala.gov.in

Kew, S. F., Selten, F. M., Lenderink, G., & Hazeleger, W. (2013). The simultaneous occurrence of surge and discharge extremes for the Rhine delta. *Natural Hazards and Earth System Sciences*, *13*(8), 2017–2029. doi:10.5194/nhess-13-2017-2013

Keynes, J. M. (1921). *A Treatise on Probability*. Macmillan and Co.

Khaddor, I. (2016). *Simulation numérique des écoulements à surface libre du bassin versant de Mghogha en moyen de SIG, Télédétection, HEC-HMS et HEC-RAS: Études des protections contre les inondations dans la ville de Tanger, nord de Maroc*. Universite Abdelmalek Essaadi Faculte Des Sciences Et Techniques Tanger.

Khan, S. (2012). Vulnerability assessments and their planning implications: A case study of the Hutt Valley, New Zealand. *Natural Hazards*, *64*(2), 1587–1607. doi:10.100711069-012-0327-x

Kheder, K. (2014). Application of Remote Sensing and GIS for Flood Risk Assessment: A case study of Al Kharj Valley Saudi Arabia. *International Journal of Scientific & Engineering Research*, *5*(5), 452–459. doi:10.14299/ijser.2014.05.001

Kim, J. H., Geem, Z. W., & Kim, E. S. (2001). Parameter estimation of the nonlinear Muskingum model using Harmony Search. *Journal of the American Water Resources Association*, *37*(5), 1131–1138. doi:10.1111/j.1752-1688.2001.tb03627.x

Kirkpatrick, S., Gellat, C.D., Jr., & Vecchi, M.P. (1983). Optimization by simulated annealing. *Science*, (80), 220.

Kiva, K. (2018). *September-December Rains to Trigger Floods, Landslides - Metrological Authority*. Press Release by Uganda National Meteorological Authority. Kampala, Uganda: UNMA.

Klein, R. J. T., & Nicholls, R. J. (1999). Assessment of Coastal Vulnerability to Climate Change. *Ambio*, *28*, 182–187.

Klemas, V. (2014). Remote sensing of floods and flood-prone areas: An overview. *Journal of Coastal Research*, *31*, 1005–1013.

Knapp, A. K., Beier, C., Briske, D. D., Classen, A. T., Luo, Y., Reichstein, M., ... Weng, E. (2008). Consequences of more extreme precipitation regimes forterrestrial ecosystems. *Bioscience*, *58*(9), 811–821. doi:10.1641/B580908

Kok, M., Lüdeke, M., Lucas, P., Sterzel, T., Walther, C., Jansses, P., ... de Soysa, I. (2016). A new method for analysing socio-ecological patterns of vulnerability. *Regional Environmental Change*, *16*(1), 229–243. doi:10.100710113-014-0746-1

Kotzee, I., & Reyers, B. (2016). Piloting a social-ecological index for measuring flood resilience: A composite index approach. *Ecological Indicators*, *60*, 45–53. doi:10.1016/j.ecolind.2015.06.018

Kuczera, G. (1982). Combining site-specific and regional information: An empirical Bayes approach. *Water Resources Research*, *18*(2), 306–314. doi:10.1029/WR018i002p00306

Kuczera, G. (1999). Comprehensive at-site flood frequency analysis using Monte Carlo Bayesian inference. *Water Resources Research*, *35*(5), 1551–1557. doi:10.1029/1999WR900012

Kumar, M., Rawat, S., & Singh, H. (2018). Dynamic forest vegetation models for predicting impacts of climate change on forests: An Indian perspective. *Indian Journal of Forestry*, *41*, 1–12.

Kumar, P. K. D. (2006). Potential vulnerability implications of sea level rise for the coastal zones of Cochin, Southwest Coast of India. *Environmental Monitoring and Assessment*, *123*(103), 333–344. doi:10.100710661-006-9200-2 PMID:16738753

Kumar, P., Tiwart, K. N., & Pal, D. K. (1991). Establishing SCS Runoff Curve Number from IRS Digital Data Base. *Photonirvachak (Dehra Dun)*, *19*(4), 245–252. doi:10.1007/BF03023971

Kundzewicz, Z.W., Pinskwar, I., & Brakenridge, G.R. (2003). Large floods in Europe, 1985-2009. *Hydrol. Sci. J.*, *58*, 736.

Lahlaoi, H., Rhinane, H., & Hilali, A. (2015). Potentialerosion risk calculation using Remote Sensing and GISin Oued El Maleh Watershed, Morocco. *Journal of Geographic Information Systems*, *7*(2), 128–139.

Lata, K., & Bisht, N. S. (1991). Quantitative analysis and regeneration potential of moist temperature forest in Garhwal Himalayas. *Indian Journal of Forestry*, *14*(2), 98–106.

Leary, N., Adequwon, J., Barros, V., Burton, I., Kukarni, J., & Lasco, R. (Eds.). (2008). *Climate change and adaptation*. London, UK: Earthscan.

Lee, J. S., & Choi, H. I. (2018). Comparison of Flood Vulnerability Assessments to Climate Change by Construction Frameworks for a Composite Indicator. *Sustainability, 10*(3), 768.

Leiter, A.M., Oberhofer, H., & Raschky, P.A. (2009). Creative disasters? Flooding effects on capital, labour and productivity within European firms. Environ. *Resour. Econ., 43*(3), 333-350. . doi:10.100710640-009-9273-9

Lenderink, G., Mok, H. Y., Lee, T. C., & van Oldenborgh, G. J. (2011). Scaling and trends of hourly precipitation extremes in two different climate zones—Hong Kong and the Netherlands. *Hydrology and Earth System Sciences*, *15*(9), 3033–3041. doi:10.5194/hess-15-3033-2011

Lenderink, G., & van Meijgaard, E. (2008). Increase in hourly precipitation extremes beyond expectations from temperature changes. *Nature Geoscience*, *1*(8), 511–514. doi:10.1038/ngeo262

Li, J., & Wong, D. W. (2010). Effects of DEM sources on hydrologic applications. *Computers, Environment and Urban Systems*, *34*(3), 251–261. doi:10.1016/j.compenvurbsys.2009.11.002

Likens, G. E. (2013). *Biogeochemistry of a forested ecosystem*. Springer Science & Business Media. doi:10.1007/978-1-4614-7810-2

Liu, X. (2008). Airborne LiDAR for DEM generation: Some critical issues. *Progress in Physical Geography*, *32*(1), 31–49. doi:10.1177/0309133308089496

Luo, J. G., & Xie, J. C. (2010). Parameter Estimation for Nonlinear Muskingum Model Based on Immune Clonal Selection Algorithm. *Journal of Hydrologic Engineering*, *15*(10), 844–851. doi:10.1061/(ASCE)HE.1943-5584.0000244

Luzio, M., Srinivasan, R., & Arnold, J. G. (2003). *Integration of watershed tools and SWAT model into basins*. Academic Press.

Lwasa, S. (2015). A systematic review of research on climate change adaptation policy and practice in Africa and South Asia deltas. *Regional Environmental Change*, *15*(5), 815–824. doi:10.100710113-014-0715-8

Magnuss, D. R., Morton, J. M., Huettmann, F., Chapin, F. S. III, & McGuire, D. (2011). A climate-change adaptation framework to reduce continental-scale vulnerability across conservation reserves. *Ecosphere*, *2*(10), 112. doi:10.1890/ES11-00200.1

Mahabaleshwara, H., & Nagabhushan, H. (2014). A study on soil erosion and its impacts on floods and sedimentation. *International Journal of Research in Engineering and Technology*, *3*(3), 443–451.

Maison, P. (2000). *Un modèle hydrologique de suivi de la pollution diffuse en bassin versant. Approche mécaniste simplifiée de la zone non saturée*. Thèse de Doctorat, L'institut National Polytechnique de Toulouse en Science de la terre et environnement, 303

Mantegna, R. N. (1994). Fast, Accurate Algorithm for Numerical Simulation of Lévy Stable Stochastic Process. *Physical Review*, *49*, 4677–4683. PMID:9961762

Ma, R. (2005). DEM generation and building detection from lidar data. *Photogrammetric Engineering and Remote Sensing*, *71*(7), 847–854. doi:10.14358/PERS.71.7.847

Marchandise, A. (2007). *Modélisation hydrologique distribuée sur le gardon d'Anduze: étude comparative de différents modèles pluie-débit, extrapolation de la normale à l'extrême et tests d'hypothèses sur les processus hydrologiques*. Thèse de Doctorat, Univ.

Maria, A., Evangelos, A., & Vassilios, A. (2015). *Hydrology and Water Resource Systems Analysis*. Academic Press.

Maroufi, A. (2008). Désertification et migration: un défi pour le développement durable au Maroc. Le cas de Zagora. *Sustainable Mediterranean*, *54*.

Mason, D. C., Cobby, D. M., Horritt, M. S., & Bates, P. D. (2003). Floodplain friction parameterization in two-dimensional river flood models using vegetation heights derived from airborne scanning laser altimetry. *Hydrological Processes*, *17*(9), 1711–1732. doi:10.1002/hyp.1270

Masselink, G., & Russell, P. (2013). Impacts of climate change on coastal erosion. *MCCIP Science Review*, *2013*, 71–86.

McCarthy, J. J., Canziani, O., Leary, N. A., Dokken, D. J., & White, K. S. (Eds.). (2001). *Climate change 2001: Impacts, adaptation and vulnerability. IPCC working group II*. Cambridge, UK: Cambridge University Press.

McGranahan, G., Balk, D., & Anderson, B. (2007). The rising tide: Assessing the risks of climate change and human settlements in low elevation coastal zones. *Environment and Urbanization*, *19*(1), 17–37. doi:10.1177/0956247807076960

McLachlan, J. S., Hellmann, J. J., & Schwartz, M. W. (2007). A Framework for Debate of Assisted Migration in an Era of Climate Change. *Conservation Biology*, *21*(2), 297–302. doi:10.1111/j.1523-1739.2007.00676.x PMID:17391179

Meehl, G. A. (2007). Global climate projections. In Climate change 2007: The Physical Science Basis. Contribution of Working Group I to the Fourth Assessment Report of the Intergovernmental Panel on Climate Change (pp. 433-497). Cambridge University Press.

Menoni, S., & Pergalani, F. (1996). An attempt to link risk assessment with land use planning: A recent experience in Italy. *Disaster Prevention and Management: An International Journal*, *5*(1), 6–21. doi:10.1108/09653569610109523

Meraj, G., Romshoo, S. A., Yousuf, A. R., Altaf, S., & Altaf, F. (2015). Assessing the influence of watershed characteristics on the flood vulnerability of Jhelum basin in Kashmir Himalaya. *Natural Hazards*, *77*(1), 153–175. doi:10.100711069-015-1605-1

Messner, F., & Meyer, V. (2006). Flood damage, vulnerability and risk perception – Challenges for flood damage research. Part 4: Vulnerability and flood damages. In *Flood risk management: hazards, vulnerability and mitigation measures* (pp. 149–168). Amsterdam: Springer with NATO Public Diplomacy Division. doi:10.1007/978-1-4020-4598-1_13

Metropolis, N., Rosenbluth, A., Rosenbluth, M., Teller, A., & Teller, E. (1953). Equation of state calcula-tions by fast computing machines. *The Journal of Chemical Physics*, *21*(6), 1087–1092. doi:10.1063/1.1699114

Meyer, V., & Schwarze, R. (2019). The Economics and Management of Flood Risk in Germany. *Volker Meyer and Reimund Schwarze*, 473–495.

Meyer, V., Priest, S., & Kuhlicke, C. (2012). Economic evaluation of structural and non-structural flood risk management measures: Examples from the Mulde River. *Natural Hazards*, *62*(2), 301–324. doi:10.100711069-011-9997-z

Micevski, T., & Kuczera, G. (2009). Combining site and regional flood informationusing a Bayesian Monte Carlo approach. *Water Resources Research*, *45*(4), W04405. doi:10.1029/2008WR007173

Mileti, D. (1999). *Disasters by design: A reassessment of natural hazards in the United States.* Joseph Henry Press.

Millar, C. I., Stephenson, N. L., & Stephens, S. L. (2007). Climate change and forests of the future: Managing in the face of uncertainty. *Ecological Applications*, *17*(8), 2145–2151. doi:10.1890/06-1715.1 PMID:18213958

Miller & Hutchins. (2017). The impacts of urbanisation and climate change on urban flooding and urban water quality: A review of the evidence concerning the United Kingdom. *Journal of Hydrology: Regional Studies*, *12*, 345–362.

Miller, J. D., & Hess, T. (2017). Urbanization Impact on Storm Runoff Along a Rural-Urban Gradient. *Journal of Hydrology (Amsterdam)*, *552*, 474–489. doi:10.1016/j.jhydrol.2017.06.025

Miller, N., & Cronshey, R. (1989). Runoff Curve Numbers, the next Step. In B. C. Yen (Ed.), *channel Flow and Catchment Runoff, Departement of civil Engineering, Universtity of Virginia* (pp. 910–916). Charlottesvilles, VA: Academic Press.

Milly, P. C. D., Betencourt, J., Falkenmark, M., Hirsch, R. M., Kundzewicz, Z. W., Lettenmair, D. P., & Stouffer, R. J. (2008). Stationarity is Dead: Whither Water Management? *Science*, *319*(5863), 573–574. doi:10.1126cience.1151915 PMID:18239110

Minamikawa, K., Takahashi, M., Makino, T., Tago, K., & Hayatsu, M. (2015). Irrigation with oxygen-nanobubble water can reduce methane emission and arsenic dissolution in a flooded rice paddy. *Environmental Research Letters, 10*(8), 1-11.

Ministère de l'Intérieur Direction Générale des Collectivités Locale (MIDGCL). (2015). *La région de Drâa Tafilalet, Monographie générale*. Author.

Minucci, G., & Karmaoui, A. (2017). Exploring the water-food-energy and climate nexus: insights from the Moroccan Draa Valley. In Peri-Urban Areas and Food-Energy-Water Nexus Sustainablity and Resilience Strategies in the Age of Climate Change. Springer International Publishing AG. doi:10.1007/978-3-319-41022-7_11

Mohan, S. (1997). Parameter estimation of nonlinear Muskingum models using genetic algorithm. *Journal of Hydraulic Engineering, 123*(2), 137–142. doi:10.1061/(ASCE)0733-9429(1997)123:2(137)

Moriasi, Arnold, Van Liew, Bingner, Harmel, & Veith. (2007). *Model evaluation guidelines for systematic quantification of accuracy in watershed simulation*. Academic Press.

Mujere, N. (2011). Flood Frequency Analysis Using the Gumbel Distribution. *International Journal on Computer Science and Engineering, 3*, 2774–2778.

Musy, A., & Higy, C. (2004). Hydrologie, une science de la nature. *Coll. Gérer l'environnement, Presses polytechniques et universitaires romandes*, 314.

Musy, A. (1998). *Hydrologie appliquée*. Bucharest: Edition HGA.

Mwangi, A. (2007). *Floods continue to ravage Uganda Thousands risk lives as they cross cut off flooded areas*. Press Release by Uganda Meteorological Department. Kampala, Uganda: UMD.

Myers, R. A., & Worm, B. (2003). Rapid worldwide depletion of predatory fish communities. *Nature, 423*(6937), 280–283. doi:10.1038/nature01610 PMID:12748640

Nasiri, H., Yusof, M. J. M., & Ali, T. A. M. (2016). An overview to flood vulnerability assessment methods. *Sustainable Water Resources Management, 2*(3), 331–336. doi:10.100740899-016-0051-x

National Aeronautics and Space Administration (NASA). (2018). *Dramatic Flooding in Eastern Africa.* Retrieved from https://earthobservatory.nasa.gov/images/92130/dramatic-flooding-in-eastern-africa

National Environment Management Agency (NEMA). (2002). *Localizing global environmental conventions. A simple guide for Uganda*. Kampala: NEMA.

National Geographic News (NGN). (2015). *Impact of flooding*. Retrieved from https://www.nationalgeographic.com/environment/natural-disasters/floods

Nelson. (2018). *Natural Disasters & Assessing Hazards and Risk*. Retrieved from https://www.tulane.edu/~sanelson/Natural_Disasters/introduction.htm

Neset, T. S., Wiréhn, L., Opach, T., Glaas, E., & Linnér, B. O. (2018). Evaluation of indicators for agricultural vulnerability to climate change: The case of Swedish agriculture. *Ecological Indicators*. doi:10.1016/j.ecolind.2018.05.042

Ngwomoya, A. (2018). *Why Kampala keeps flooding. The Daily Monitor newspaper report published on 09 March 2018*. Kampala, Uganda: The Nation Group.

Nicholls, R.J. (2011). Sea-level rise and its possible impacts given a 'beyond 4o C world' in the twenty-first century. *Philosophical Transaction of Royal Society A, 369*(1934), 161-181.

Nicholls, R.J., & Cazanave, A. (2010). *Sea level rise and its impact on coastal zones*. Academic Press.

Nicholls, R. J. (2004). Coastal flooding and wetland loss in the 21st century: Changes under the SRES climate and socio-economic scenarios. *Global Environmental Change, 14*(1), 69–86. doi:10.1016/j.gloenvcha.2003.10.007

Nichols, J. D., Koneff, M. D., Heglund, P. J., Knutson, M. G., Seamans, M. E., Lyons, J. E., ... Williams, B. K. (2011). Climate change, uncertainty, and natural resource management. *The Journal of Wildlife Management, 75*(1), 6–18. doi:10.1002/jwmg.33

Nikoo, M., Ramezani, F., Hadzima-Nyarko, M., Nyarko, E. K., & Nikoo, M. (2016). Flood-routing modeling with neural network optimized by social-based algorithm. *Natural Hazards, 82*(1), 1–24. doi:10.100711069-016-2176-5

NOAA. (2013). *State of the coast*. National Oceanic and Atmospheric Administration. Retrieved 25 October 2018, from http://stateofthecoast.noaa.gov/

O'Connell, D. R. H., Ostenaa, D. A., Levish, D. R., & Klinger, R. E. (2002). Bayesian flood frequency analysis with paleo hydrologic bound data. *Water Resources Research, 38*(5), 1058. doi:10.1029/2000WR000028

OECD. (2017). *OECD Review of risk Management Policies Morocco*. Paris: OECD Publishing.

Ohl, C. A., & Tapsell, S. (2000). Flooding and human health: The dangers posed are not always obvious. *BMJ: British Medical Journal, 321*(7270), 1167–1168. doi:10.1136/bmj.321.7270.1167 PMID:11073492

Okaka, W. (2010). Developing regional communications campaigns strategy for environment and natural resources management policy awareness for the East African community. *Research Journal of Environmental and Earth Sciences, 2*(2), 106-111.

Okaka, W. (2011). Developing effective climate change policy communication for sustainable development in Africa. *Climate change stakeholders' forum*. Available at www.stakeholderforum.org/…ach/index.php/cif-day2...South Africa

Okkan & Kirdemir. (2018). Investigation of the Behavior of an Agricultural-Operated Dam Reservoir Under RCP Scenarios of AR5-IPCC. *Water Resour Manage, 32*, 2847–2866.

Oldeman, L. (1997). *Soil Degradation: A Threat to Food Security?* Paper presented at the International Conference on Time Ecology: "Time for Soil Culture-Temporal Perspectives on Sustainable Use of Soil", Tutzing.

Pandey, A., Himanshu, S. K., Mishra, S., & Singh, V. P. (2016). Physically based soil erosion and sediment yield models revisited. *Catena, 147*, 595–620. doi:10.1016/j.catena.2016.08.002

Pandey, J. C. (2003). Vegetation analysis in a mixed Oak-Conifer forest of Central Himalaya. *Indian Journal of Forestry, 26*(1), 66–74.

Pandey, R., & Bardsley, D. K. (2015). Social-ecological vulnerability to climate change in the Nepali Himalaya. *Applied Geography (Sevenoaks, England), 64*, 74–86. doi:10.1016/j.apgeog.2015.09.008

Pandey, R., & Jha, S. (2012). Climate vulnerability index-measure of climate change vulnerability to communities: A case of rural lower Himalaya, India. *Mitigation and Adaptation Strategies for Global Change, 17*(5), 487–506. doi:10.100711027-011-9338-2

Panos. (2012). *Communicating climate change in Uganda: Challenges and opportunities*. Cardiff University.

Papaioannou, P., Vasiliades, L., Loukas, A., & Aronica, G. T. (2017). Probabilistic flood inundation mapping at ungauged streams due to roughness coefficient uncertainty in hydraulic modelling. *Advances in Geosciences, 44*, 23–34. doi:10.5194/adgeo-44-23-2017

Pappenberger, F., Beven, K., Frodsham, K., Romanowicz, R., & Matgen, P. (2007). Grasping the unavoidable subjectivity in calibration of flood inundation models: A vulnerability weighted approach. *Journal of Hydrology (Amsterdam), 333*(2-4), 275–287. doi:10.1016/j.jhydrol.2006.08.017

Park, Y. S. (2017). Development of Korean soil loss estimation model. *Geoenviron. Eng., 18*, 14–17.

Patel, D. P., & Srivastava, P. K. (2013). Flood hazards mitigation analysis using remote sensing and GIS: Correspondence with town planning scheme. *Water Resources Management, 27*(7), 2353–2368. doi:10.100711269-013-0291-6

Penning-Rowsell, E. C., Sultana, P., & Thompson, P. M. (2013). The 'last resort'? Population movement in response to climate-related hazards in Bangladesh. *Environmental Science & Policy, 27*(S1), 44–59. doi:10.1016/j.envsci.2012.03.009

Perrin, C. (2000). *Vers une amélioration d'un modèle global pluie-débit au travers d'une approche comparative*. Thèse de Doctorat, INPG (Grenoble) / Cemagref (Antony).

Perrin, C., Samie, R., & Hendrickx, F. (2009). Les outils du projet: modélisation hydrologique et représentation des usages. Journée de restitution du projet.

Pew, K. L., & Larsen, C. P. S. (2001). GIS analysis of spatial and temporal patterns of human-caused wildfires in the temperate rain forest of Vancouver Island, Canada. *Forest Ecology and Management, 140*(1), 1–18. doi:10.1016/S0378-1127(00)00271-1

Philippart, C. J. M., Anadón, R., Danovaro, R., Dippner, J. W., Drinkwater, K. F., Hawkins, S. J., ... Reid, P. C. (2011). Impacts of climate change on European marine ecosystems: Observations, expectations and indicators. *Journal of Experimental Marine Biology and Ecology, 400*(1-2), 52–69. doi:10.1016/j.jembe.2011.02.023

Phine, H. N., & Hira, M. A. (1983). Log Pearson Type-3 Distribution: Parameter Estimation. *Journal of Hydrology (Amsterdam), 64*(1-4), 25–37. doi:10.1016/0022-1694(83)90058-6

Pierre, U., Doctorale, E., Naturelles, R., & Moussa, M. R. (2009). *Quels apports hydrologiques pour les modèles hydrauliques ? Vers un modèle intégré de simulation des crues*. Academic Press.

Pimentel, D. (2006). Soil erosion: A food and environmental threat. *Environment, Development and Sustainability, 8*(1), 119–137. doi:10.100710668-005-1262-8

Plan Bleu. (2003). *Les menaces sur les sols dans les pays méditerranéens Etude bibliographique*. Author.

Plate, E. J. (2002). Flood risk and flood management. *Journal of Hydrology (Amsterdam), 267*(1-2), 2–11. doi:10.1016/S0022-1694(02)00135-X

Pourghasemi, H. R., Moradi, H. R., Aghda, S. F., Gokceoglu, C., & Pradhan, B. (2014). GIS-based landslide susceptibility mapping with probabilistic likelihood ratio and spatial multi-criteria evaluation models (North of Tehran, Iran). *Arabian Journal of Geosciences, 7*(5), 1857–1878. doi:10.100712517-012-0825-x

Prakasa, R., & Murty, B.S. (2005) Estimation of flood vulnerability index for delta. *Geoscience and Remote Sensing Symposium, 5*(25–29), 3611–3614.

Prasad, A. K., & Singh, R. P. (2005). Extreme rainfall event of July 25–27, 2005 over Mumbai, West Coast, *India. Photonirvachak (Dehra Dun), 33*(3), 365–370. doi:10.1007/BF02990007

Pulvirenti, L., Chini, M., Pierdicca, N., Guerriero, L., & Ferrazzoli, P. (2011). Flood monitoring using multi-temporal COSMO-SkyMed data: image segmentation and signature interpretation. *Rem. Sens. Environ., 115*, 990-1002.

Rahmati, O., & Pourghasemi, H. R. (2017). Identification of critical flood prone areas in data-scarce and ungauged regions: A comparison of three data mining models. *Water Resources Management*, *31*(5), 1473–1487. doi:10.100711269-017-1589-6

Rahmstorf, S., Perrette, M., & Vermeer, M. (2012). Testing the robustness of semi-empirical sea level projections. *Climate Dynamics*, *39*(3-4), 861–875. doi:10.100700382-011-1226-7

Rai, L. K., Prasad, P., & Sharma, E. (2000). Conservation threats to some important medicinal plants of the Sikkim Himalaya. *Biological Conservation*, *93*(1), 27–33. doi:10.1016/S0006-3207(99)00116-0

Ralhan, P. K., Saxena, A. K., & Singh, J. S. (1982). Analysis of forest vegetation at and around Nainital in Kumaun Himalaya. *Proceding of Indian National Academy of Sciecnes. B*, *48*(1), 121–137.

Rallison, R. E. (1980). Origin and Evolution of the SCS Runoff Equation. *Proc. symp. on watershed management 1980, 2*, 914-924.

Rana, B. S., Verma, K. R., & Pandey, A. N. (1985). Analysis of forest vegetation at Siahi Devi Hill of West Almora Division in Kumaun Himalayas. *Indian Forester*, *111*(9), 745–759.

Rashetnia, S. (2016). *Flood vulnerability assessment by applying a fuzzy logic method: a case study from Melbourne* (Doctoral dissertation). Victoria University.

Rasul, G. (2014). Food, water, and energy security in South Asia: A nexus perspective from the Hindu Kush Himalayan region. *Environmental Science & Policy*, *39*, 35–48. doi:10.1016/j.envsci.2014.01.010

Refsgaard, J. C. (1997). Parametrisation, calibration and validation of distributed hydrological models. *Journal of Hydrology (Amsterdam)*, *198*(1-4), 69–97. doi:10.1016/S0022-1694(96)03329-X

Reis, D. S. Jr, & Stedinger, J. R. (2005). Bayesian MCMC flood frequency analysis with historical information. *Journal of Hydrology (Amsterdam)*, *313*(1-2), 97–116. doi:10.1016/j.jhydrol.2005.02.028

Reis, D. S. Jr, Stedinger, J. R., & Martins, E. S. (2005). Bayesian generalized least squares regression with application to log Pearson type 3 regional skew estimation. *Water Resources Research*, *41*(10), W10419. doi:10.1029/2004WR003445

Renard, K. G., Foster, G. A., Weesies, G. A., McCool, D. K., & Yoder, D. C. (1997). *Predicting Soil Erosion by Water: A Guide to Conservation Planning with the Revised Universal Soil Loss Equation (RUSLE)*. U.S. Department of Agriculture, Agricultural Research Service.

Resio, D. T., & Westerink, J. J. (2008). Modeling the physics of storm surges. *Physics Today*, *61*(9), 33–38. doi:10.1063/1.2982120

Rhazi, Essahlaou El Hmaidi, & El Ouali. (2017). Etude De La Vulnérabilité Aux Risques D'inondations Dans La Ville De Meknès. Apport Des SIG, Du MNT et des Modèles Empiriques. *European Scientific Journal, 13*. doi:10.19044/esj.2017.v13n36p102

Ribatet, M., Sauquet, E., Gresillon, J. M., & Ouarda, T. B. M. J. (2007). A regional Bayesian POT model for flood frequency analysis. *Stochastic Environmental Research and Risk Assessment, 21*(4), 327–339. doi:10.100700477-006-0068-z

Ribatet, M., Sauquet, E., Gresillon, J. M., & Ouarda, T. B. M. J. (2007). Usefulness of the reversible jump Markov Chain Monte Carlo model in regional flood frequency analysis. *Water Resources Research, 43*(8), W08403. doi:10.1029/2006WR005525

Rikhari, H. C., Ralhan, P. K., & Singh, S. P. (1989). Phytosociology and population structure of chir-pine forests of Kumoan Himalayas. *Annals of Biology, 5*(2), 129–140.

Rogers, E. M. (1995). Diffusion of innovations. New York: The Free Press.

Rooke, D. (2009). What is flood risk management? Why is it needed and how is it alleviated and managed? *Environmental Law & Management, 21*(3), 140–141.

Roose, E-O. (1991). Conservation des sols en zones méditerranéennes. *Synthèse et proposition d'une nouvelle stratégie de lutte antiérosive: la GCES-Cahiers ORSTOM, Série Pédologie, 26*(2), 145-181.

Rosselló-Nadal, J. (2014). How to evaluate the effects of climate change on tourism. *Tourism Management, 42*, 334–340. doi:10.1016/j.tourman.2013.11.006

Saha, S., & Howe, H. F. (2001). The bamboo fire cycle hypothesis: A comment. *American Naturalist, 158*(6), 659–663. doi:10.1086/323593 PMID:18707360

Saha, S., Moorthi, S., Pan, H.-L., Wu, X., Wang, J., Nadiga, S., ... Goldberg, M. (2010). The NCEP Climate Forecast System Reanalysis. *Bulletin of the American Meteorological Society, 91*(8), 1015–1057. doi:10.1175/2010BAMS3001.1

Salami, R. O., von Meding, J. K., & Giggins, H. (2017). Urban settlements' vulnerability to flood risks in African cities: A conceptual framework. *Jàmbá: Journal of Disaster Risk Studies, 9*(1), 1–9. doi:10.4102/jamba.v9i1.370 PMID:29955335

Samanta, S., Pal, D. K., & Palsamanta, B. (2018). Flood susceptibility analysis through remote sensing, GIS and frequency ratio model. *Applied Water Science, 8*(2), 66.

Santangelo, N., Santo, A., Di Crescenzo, G., Foscari, G., Liuzza, V., Sciarrotta, S., & Scorpio, V. (2011). Flood susceptibility assessment in a highly urbanized alluvial fan: The case study of Sala Consilina (southern Italy). *Natural Hazards and Earth System Sciences, 11*(10), 2765–2780. doi:10.5194/nhess-11-2765-2011

Sanyal, J., & Lu, X. X. (2004). Application of remote sensing in flood management with special reference to monsoon Asia: A review. *Natural Hazards*, *33*(2), 283–301. doi:10.1023/B:NHAZ.0000037035.65105.95

Sarwar, M. G. M. (2005). Impacts of sea level rise on the coastal zone of Bangladesh. *Science, 328*(5985), 1517-1520.

Saxena, A. K., & Singh, J. S. (1982). A phytosociological analysis of woody species in forest communities of a part of Kumaun Himalaya. *Vegetation*, *50*(1), 3–22. doi:10.1007/BF00120674

Schanze, J. (2006). Flood Risk Management-A Basic Framework. *Flood Risk Management: Hazards, Vulnerability and Mitigation Measures,* 1-20.

Schanze, J. (2006). Flood risk management–a basic framework. In *Flood risk management: hazards, vulnerability and mitigation measures* (pp. 1–20). Amsterdam: Springer with NATO Public Diplomacy Division.

Schumann, G., Matgen, P., Cutler, M. E. J., Black, A., Hoffmann, L., & Pfister, L. (2008). Comparison of remotely sensed water stages from LiDAR, topographic contours and SRTM. *ISPRS Journal of Photogrammetry and Remote Sensing*, *63*(3), 283–296. doi:10.1016/j.isprsjprs.2007.09.004

Seidou, O., Ouarda, T. B. M. J., Barbet, M., Bruneau, P., & Bobée, B. (2006). A parametric Bayesian combination of local and regional information in flood frequency analysis. *Water Resources Research*, *42*(11), W11408. doi:10.1029/2005WR004397

Sharma, E., Bhuchar, S., Xing, M., & Kcithyari, B. P. (2007). Land use change and its impact on hydro-ecological linkages in Himalayan watersheds. *Tropical Ecology*, *48*, 151–161.

Sharma, J., Jayaraman, M., Kumar, V., & Ravindranath, N. H. (2015). Climate change impact and vulnerability assessment of forests in the Indian western Himalayan region: A case study of Himachal Pradesh, India. *Climate Risk Management*, *10*, 63–76. doi:10.1016/j.crm.2015.08.002

Sharma, S. K., Deml, K., Dangal, S., Rana, E., & Madigan, S. (2015). REDD+ framework with integrated measurement, reporting and verification system for community based Forest management systems (CBFMS) in Nepal. *Current Opinion in Environmental Sustainability*, *14*, 17–27. doi:10.1016/j.cosust.2015.01.003

Sherman, C. W. (1931). Frequency and intensity of excessive rainfall at Boston, Massachusetts. *Transactions of the American Society of Civil Engineers*, *95*, 951–960.

Shihong, L., Wusheng, X., Xiankun, L., & Runqin, T. (2003). A review of vegetation restoration in eroded area of red soil. *Guangxi Zhi Wu*, *23*(1), 83–89.

Shrestha, D. P. (1997). Assessment of soil erosion in the Nepalese Himalaya: A case study in Likhu Khola Valley, Middle Mountain Region. *Land Husbandry*, *2*(1), 59–80.

Singhal, R. M., & Soni, S. (1989). Quantitative ecological analysis of some woody species of Mussoorie Himalayas (UP). *Indian Forester*, *115*(5), 327–336.

Singh, G., & Rawat, G. S. (2012). Quantitative analysis of tree species diversity in different oak (Quercus spp.) dominated forests in Garhwal Himalaya, India. *Notulae Scientia Biologicae*, *4*(4), 132. doi:10.15835/nsb448200

Singh, V. P. (1995). Accuracy of kinematic and diffusion wave approximations for space independent flows on infiltrating surfaces. *Hydrological Processes*, *9*(1), 1–18. doi:10.1002/hyp.3360090102

Small, C., & Nicholls, R. J. (2003). A global analysis of human settlement in coastal zones. *Journal of Coastal Research*, *19*(3), 584–599.

Sodhi, M. S. (2016). Natural disasters, the economy and population vulnerability as a vicious cycle with exogenous hazards. *Journal of Operations Management*, *45*(1), 101–113. doi:10.1016/j.jom.2016.05.010

Spalding, M. D., Ruffo, S., Lacambra, C., Meliane, I., Hale, L. Z., Shepard, C. C., & Beck, M. W. (2014). The role of ecosystems in coastal protection: Adapting to climate change and coastal hazards. *Ocean and Coastal Management*, *90*(March), 50–57. doi:10.1016/j.ocecoaman.2013.09.007

Storn, R., & Price, K. (1997). Differential evolution–a simple and efficient heuristic for global optimization over continuous spaces. *Journal of Global Optimization*, *11*(4), 341–359. doi:10.1023/A:1008202821328

Surendran, S., Gibbs, G., Wade, S., & Udale-Clarke, H. (2008). *Supplementary note on flood hazard ratings and thresholds for development and planning control purpose–Clarification of Table 13.1 of FD2320/TR2 and Figure 3.2 of FD2321*. Environment Agency and HR Wallingford.

Suriya, S., & Mudgal, B. V. (2011). Impact of Urbanization on Flooding: The Thirusoolam Sub Watershed- A Case Study. *Journal of Hydrology (Amsterdam)*, *412*, 210–219.

Tahiri, M., Tabyaoui, H., & El Hammichi, F. (2017). *Quantification of water erosion and sedimentation using empirical models in the Tahaddart watershed*. Northwestern Rif.

Tang, W. H., Hydraul, J., & Div-ASCE. (1980). Bayesian frequency analysis. *J Hydraul Div-ASCE*, *106*, 1203–1218.

Tanoue, M., Hirabayashi, Y., & Ikeuchi, H. (2016). Global-scale river flood vulnerability in the last 50 years. *Scientific Reports*, *6*(36021), 1–9. PMID:27782160

Tate, E. (2012). Social vulnerability indices: A comparative assessment using uncertainty and sensitivity analysis. *Natural Hazards*, *63*(2), 325–347. doi:10.100711069-012-0152-2

Tingsanchali, T. (2012). Urban flood disaster management. *Procedia Engineering*, *32*, 25–37. doi:10.1016/j.proeng.2012.01.1233

Tompkins, E. L. (2005). Planning for climate change in small islands: Insights from national hurricane preparedness in the Cayman Islands. *Global Environmental Change, 15*(2), 139–149. doi:10.1016/j.gloenvcha.2004.11.002

Trapp, R. J., Diffenbaugh, N. S., Brooks, H. E., Baldwin, M. E., Robinson, E. D., & Pal, J. S. (2007). Changes in severe thunderstorm environment frequency during the 21st century caused by anthropogenically enhanced global radiative forcing. *Proceedings of the National Academy of Sciences of the United States of America, 104*(50), 19719–19723. doi:10.1073/pnas.0705494104

Tschakert, P., & Dietrich, K. A. (2010). Anticipatory learning for climate change adaptation and resilience. *Ecology and Society, 15*(2), 1–23. doi:10.5751/ES-03335-150211

Tung, Y. (1985). River flood routing by nonlinear Muskingum method. *Journal of Hydraulic Engineering, 111*(12), 1447–1460. doi:10.1061/(ASCE)0733-9429(1985)111:12(1447)

Turner, A. B., Colby, J. D., Csontos, R. M., & Batten, M. (2013). Flood modeling using a synthesis of multi-platform LiDAR data. *Water (Basel), 5*(4), 1533–1560. doi:10.3390/w5041533

U.S Soil Conservation Service. (1985). *Natioanl Engineering Handbook, Sec. 4, Hydrollogy.* Washington, DC: U.S. Department of Aggricylture.

U.S. Army Corps of Engineers. (2008). *HEC-RAS (Version 4.1) Hydraulic Reference Manuel.* Davis, CA: Hydrologic Engineering Center, U.S. Army Corps of Engineers. Available on: http://www.hec.usace.army.mil/software/hec-ras/documentation/hec-ras_4.1_reference_manual.pdf)

Uganda Radio Network - URN. (2018). *Wakiso floods: 2 bodies recovered, 3 rescued. November 21, 2018. A local radio news bulletin clip.* Kampala, Uganda: URN.

UNFCCC. (2011). *Report of the Conference of the Parties on Its Sixteenth Session, Held in Cancun from 29 November to 10 December 2010. Part Two: Action taken by the Conference of the Parties at its sixteenth session* (FCCC/CP/2010/7/Add.1). UNFCCC.

UNISDR. (2009). *Terminology on Disaster Risk Reduction.* Available at http://www.unisdr.org/we/inform/publications/7817

United Nations (UN). (1982). Assessment of Floodplain Vulnerability during Extreme Mississippi River Flood 2011. *Proceedings of the seminars on flood vulnerability analysis and on the principles of floodplain management for flood loss prevention*, 48(5), 2619-2625.

United Nations (UN). (2018). *Deeply Saddened by Flash Flooding Deaths in Jordan, Secretary-General Pledges United Nations Support for Rescue, Relief Efforts.* Retrieved from https://www.un.org/press/en/2018/sgsm19318.doc.htm

United Nations FCCC. (2002). *Report of the conference of the parties on its seventh session.* Retrieved from www.eac.int

United Nations Office for Disaster Risk Reduction (UNISDR). (2009). *2009 UNISDR terminology on disaster risk reduction*. Retrieved from https://www.unisdr.org/we/inform/publications/7817

United Nations Office for Disaster Risk Reduction (UNISDR). (2017). *Vulnerability*. Retrieved from https://www.unisdr.org/we/inform/terminology

United Nations Office for Disaster Risk Reduction (UNISDR). (2018). *Magnitude of Japan rains and floods sign of growing challenge to disaster risk management*. Retrieved from https://www.unisdr.org/archive/59297

United States Geological Survey. (2016). *Water science school*. Author.

Upadhyay, K. (2016, April 30). Uttarakhand battles fire crisis. *The Hindu*.

Van den Honert, R. C., & McAneney, J. (2011). The 2011 Brisbane floods: Causes, impacts, and implications. *Water (Basel)*, *3*(4), 1149–1173. doi:10.3390/w3041149

Van der Sande, C. J., De Jong, S. M., & De Roo, A. P. J. (2003). A segmentation and classification approach of IKONOS-2 imagery for land cover mapping to assist flood risk and flood damage assessment. *International Journal of Applied Earth Observation and Geoinformation*, *4*(3), 217–229. doi:10.1016/S0303-2434(03)00003-5

Van Oost, K., Govers, G., & Desmet, P. (2000). Evaluating the effects of changes in landscape structure on soil erosion by water and tillage. *Landscape Ecology*, *15*(6), 577–589. doi:10.1023/A:1008198215674

Van Rompaey, A. J., Verstraeten, G., Van Oost, K., Govers, G., & Poesen, J. (2001). Modelling mean annual sediment yield using a distributed approach. *Earth Surface Processes and Landforms*, *26*(11), 1221–1236. doi:10.1002/esp.275

Van, G. P. H. A. J. M., Van, N. J. M., & Duits, M. T. (1999). Selection of probability distribution with a case study on extreme Oder River discharges. *The Tenth European Conference on Safety and Reliability*, 1475–1480. DOI:10.1.1.74.9802&rep:rep1

Vermeer, M., & Rahmstorf, S. (2009). Global sea level linked to global temperature. *Proceedings of the National Academy of Sciences of the United States of America*, *106*(51), 21527–21532. doi:10.1073/pnas.0907765106 PMID:19995972

Vicens, G. J., Rodriguez, I., & Schaake, J. C. Jr. (1975). A Bayesian framework for the use of regional information in hydrology. *Water Resources Research*, *11*(3), 405–414. doi:10.1029/WR011i003p00405

Villordon, M. B. B. L., & Gourbesville, P. (2016). Community-based flood vulnerability index for urban flooding: Understanding social vulnerabilities and risks. In *Advances in Hydroinformatics* (pp. 75–96). Singapore: Springer. doi:10.1007/978-981-287-615-7_6

Vojtek, M., & Vojteková, J. (2016). Flood hazard and flood risk assessment at the local spatial scale: A case study. *Geomatics, Natural Hazards & Risk*, *7*(6), 1973–1992. doi:10.1080/19475705.2016.1166874

Wahl, T., Jain, S., Bender, J., Meyers, S. D., & Luther, M. E. (2015). Increasing risk of compound flooding from storm surge and rainfall for major US cities. *Nature Climate Change*, *5*(12), 1093–1097. doi:10.1038/nclimate2736

Wang, G., Yang, H., Wang, L., Xu, Z., & Xue, B. (2014). *Using the SWAT model to assess impacts of land use changes on runoff generation in headwaters*. doi:10.1002/hyp.9645

Wang, Y. (2002). Mapping extent of floods: What we have learned and how we can do better. *Natural Hazards Review*, *3*(2), 68–73. doi:10.1061/(ASCE)1527-6988(2002)3:2(68)

Wedajo, G. K. (2017). LiDAR DEM Data for Flood Mapping and Assessment; Opportunities and Challenges: A Review. *J Remote Sensing & GIS*, *6*(04), 211. doi:10.4172/2469-4134.1000211

White, G. F., Kates, R. W., & Burton, I. (2001). Knowing better and losing even more: The use of knowledge in hazards management. *Global Environmental Change Part B: Environmental Hazards*, *3*(3), 81–92. doi:10.1016/S1464-2867(01)00021-3

Willmott, C. J., & Legates, D. R. (1991). Rising estimates of terrestrial and global precipitation. *Climate Research*, *1*, 179–186. doi:10.3354/cr001179

Wilson, E. M. (1974). *Engineering Hydrology*. Hampshire, UK: MacMillan. doi:10.1007/978-1-349-02417-9

Winsemius, H. C., Aerts, J. C. J. H., van Beek, L. P. H., & Ward, P. J. (2015). Global drivers of future river flood risk. *Nature Climate Change*, *6*(4), 381–385. doi:10.1038/nclimate2893

Wischmeier, W. H., & Smith, D. D. (1978). *Predicting Rainfall Erosion Losses - A Guide to Conservation Planning*. Hyattsville, MD: USDA, Science and Education Administration.

Wischmeier, W., & Smith, D. (1978). *Predicting Rainfall Erosion Losses: A Guide to Conservation Planning*. Washington, DC: U.S. Department of Agriculture.

Wisetjindawat, W., Ito, H., Fujita, M., & Eizo, H. (2014). Planning disaster relief operations. *Procedia: Social and Behavioral Sciences*, *125*, 412–421. doi:10.1016/j.sbspro.2014.01.1484

Wisner, B., Blaikie, P., Cannon, T., & Davis, I. (2004). The challenge of disasters and our approach. In At Risk: Natural Hazards, people s vulnerability and disasters (pp. 3-48). Routledge.

Wood, E. F., & Rodriguez, I. (1975). A Bayesian approach to analyzing uncertainty among flood frequency models. *Water Resources Research*, *11*(6), 839–843. doi:10.1029/WR011i006p00839

Wood, E. F., & Rodriguez, I. (1975). Bayesian inference and decision making for extreme hydrologic events. *Water Resources Research*, *11*(4), 533–542. doi:10.1029/WR011i004p00533

World Bank. (2009). *Convenient solutions to an inconvenient truth: ecosystem based approaches to climate change*. Washington, DC: The World Bank.

World Health Organization (WHO). (2002). *Vulnerable groups*. Retrieved from https://www.who.int/environmental_health_emergencies/vulnerable_groups/en

Xie, W., Huang, J., Wang, J., Cui, Q., Robertson, R., & Chen, K. (2018). Climate change impacts on China's agriculture: The responses from market and trade. *China Economic Review*. doi:10.1016/j.chieco.2018.11.007

Xu, D. M., Qiu, L., & Chen, S. Y. (2012). Estimation of Nonlinear Muskingum Model Parameter Using Differential Evolution. *Journal of Hydrologic Engineering*, *17*(2), 348–353. doi:10.1061/(ASCE)HE.1943-5584.0000432

Yang, S., He, S., Du, J., & Sun, X. (2015). Screening of social vulnerability to natural hazards in China. *Natural Hazards*, *76*(1), 1–18. doi:10.100711069-014-1225-1

Yang, X. S., Karamanoglu, M., & He, X. (2014). Flower pollination algorithm: A novel approach for multiobjective optimization. *Engineering Optimization*, *46*(9), 1222–1237. doi:10.1080/0305215X.2013.832237

Yonaba, H. (2009). *Modélisation hydrologique hybride: réseau de neurones - Modèle conceptuel*. Thèse de Doctorat, Univ. Laval, Québec.

Yoon, B. J., & Padmanabhan, G. (1993). Parameter estimation of linear and nonlinear Muskingum models. *Journal of Water Resources Planning and Management*, *119*(5), 600–610. doi:10.1061/(ASCE)0733-9496(1993)119:5(600)

Zerouali, S., Khebiza, M. Y., & El Qorchi, F. (2019). Monetary Value Change of Some Provisioning Ecosystem Services of Middle Draa Valley, South of Morocco. In *Climate Change and Its Impact on Ecosystem Services and Biodiversity in arid and semi arid regions*. IGI Publisher. Retrieved from https://www.igi-global.com/book/climate-change-its-impact-ecosystem/206548#table-of-contents

Zhang, K., Chen, S. C., Whitman, D., Shyu, M. L., & Yan, J. (2003). A progressive morphological filter for removing nonground measurements from airborne LIDAR data. *IEEE Transactions on Geoscience and Remote Sensing*, *41*(4), 872–882. doi:10.1109/TGRS.2003.810682

Zhang, Y., Qi, W., Zhou, C., Ding, M., Liu, L., Gao, J., ... Zheng, D. (2015). Spatial and temporal variability in the net primary production of alpine grassland on the Tibetan Plateau since 1982. *Journal of Geographical Sciences*, *24*(2), 269–287. doi:10.100711442-014-1087-1

ZhongMin, L., BinQuan, L., ZhongBo, Y., & WenJuan, C. (2011). Application of Bayesian approach to hydrological frequency analysis. *Science China Technological Sciences, 54*(5), 1183-92. Doi:10.100711431-010-4229-4

Zhou, F., Xu, Y., Chen, Y., Xu, C., Gao, Y., & Du, J. (2013). *Hydrological response to urbanization at different spatio-temporal scales simulated by coupling of CLUE-S and the SWAT model in the Yangtze River Delta region*. doi:10.1016/j.jhydrol.2012.12.040

Zolina, O., Simmer, C., Belyaev, K., Kapala, A., & Gulev, S. (2009). Improving estimates of heavyand extreme precipitation using daily records from European rain gauges. *Journal of Hydrometeorology*, *10*(3), 701–716. doi:10.1175/2008JHM1055.1

Zope, P. E., Eldho, T. I., & Jothiprakash, V. (2016). Impact of Land Use-Land Cover Change and Urbanization on Flooding: A Case Study of Oshiwara River Basin in Mumbai, India. *Catena*, *145*, 142–154. doi:10.1016/j.catena.2016.06.009

Zsamboky, M., Fernández-Bilbao, A., Smith, D., Knight, J., & Allan, J. (2011). *Impacts of climate change on disadvantaged UK coastal communities*. Academic Press.

Zuazo, V. C. H. D., & Pleguezuelo, C. R. O. R. (2009). Soil-erosion and runoff prevention by plant covers. A review. In Sustainable agriculture. Springer.

About the Contributors

Ahmed Karmaoui started his academic career as Research Fellow at Cadi Ayyad University and awarded Ph.D. from Faculty of Semlalia; and as an associate editor at the International Journal of Climate Change: Impacts and Responses, editor at several International Journals such as Journal of Research in Chemistry & Environment (SRI: JRCE), the International Journal of Earth & Environmental Sciences, Nanomedicine and Nanoscience Research Journal; as International Advisory Board member of the International Journal of ICT Research and Development in Africa (IJICTRDA), and the International Journal of Social Ecology and Sustainable Development (IJSESD). He is a Member of the Asian Council of Science Editors. KARMAOUI has reviewed for 61 international journals, graduated from Publons Academy, and awarded Top 1% of international Reviewers (2018) by Publons (part of CLARIVATE, supported by WEB OF SCIENCE). He has published various peer-reviewed international journal articles and book chapters on a range of environmental topics. He is currently the president of the Southern Center for Culture and Science, Morocco. Broad areas of his research interests include ecosystem services assessment.

* * *

Nik Norulaini Nik Ab Rahman is a Professor from School of Distance Education, Universiti Sains Malaysia. She has completed her PhD in Forest Science from Michigan Technological University, USA. She has around 120 research articles and currently supervising 3 projects. Her area of expertise are environment, environmental impact assessment, environmental analysis and management.

Mohd Omar Abdul Kadir joined Universiti Sains Malaysia in 1989 and is teaching there since then. He is also a professional chemical engineer and is active in engineering as well as research on Supercritical Fluid Technology. 18 Masters and 22 Doctoral students have graduated under his supervision. He is currently supervising 9 PhD and 8 MSc students. He has over 150 papers in refereed journals and in international and national conferences. He is a member of National Air Quality Committee, American Institute of Chemical Engineers (AIChe) USA, Specialist Group on Small Wastewater Treatment Plants (IAWQ) United Kingdom and several others.

Khiruddin Abdullah is a Professor from School of Physics, Universiti Sains Malaysia. He has around 262 research articles and has attended many National and International conferences. Currently, he is supervising 10 research projects. Several postgraduate and graduate students working under his supervision. His research areas of interest are environment, geographic information system, satellite image analysis and satellite image processing.

Ahmad Farid Abu Bakar is Senior Lecturer in Department of Geology of Universiti of Malaya. He had completed his B.Sc (Hons.) in Geology, Masters and PhD from Universiti of Malaya. He is member of Geological Society Of Malaysia, Institut Geologi Malaysia and Board of Geologist Malaysia. He has been appointed in several administrative positions such as Coordinator, Chairperson and Member of Research Centre in the Universiti of Malaya. He has published several national and international journals. His areas of expertise are Environmental Protection, Soil & Sediment Environmental Geochemistry, Water Pollution and Environmental Geology. He also supervised several Master and PhD students. He was awarded with Eager Awards, Coordinating Committee for Geoscience Programmes in East and Southeast Asia (Ccop), 2012, (International).

Mohd Talha Anees has completed his PhD in the area of Environmental remote sensing (land and water) from the School of Physics, Universiti Sains Malaysia, Pulau Penang, Malaysia in 2018. Currently he is working as a Post Doctoral Fellow in Department of Geology, Faculty of Science, University of Malaya, Kuala Lumpur. He has completed his Masters and Bachelors in Applied Geology from Aligarh Muslim University, Aligarh in year 2011 and 2013 respectively. He has published several international journal and conference papers and one book chapter. His research areas of interests are hydrology, remote sensing and GIS, hydrological modeling and soil erosion.

Mirza Barjees Baig is working as a Professor of Agricultural Extension and Rural Society at the King Saud University, Riyadh, Saudi Arabia. He received his education in both social and natural sciences from the USA. He completed his Ph.D. in Extension Education for Natural Resource Management from the University of Idaho, Moscow, Idaho, USA. During his doctoral program, he was honored with "1995 Outstanding Graduate Student Award". He earned his MS degree in International Agricultural Extension in 1992 from the Utah State University, Logan, Utah, USA and was placed on the "Roll of Honor". Dr. Baig has published extensively in the national and international journals. He has also presented extension education and natural resource management extensively at various international conferences and fora. Particularly issues like degradation of natural resources, deteriorating environment and their relationship with society/community are his areas of interest. He has attempted to develop strategies for conserving natural resources, promoting environment and developing sustainable communities through rural development programs. Dr. Baig started his scientific career in 1983 as a researcher at the Pakistan Agricultural Research Council, Islamabad, Pakistan and in that capacity, he worked on the various issues related to soils and crops. He has been associated with the University of Guelph, Ontario, Canada as the Special Graduate Faculty in the School of Environmental Design and Rural Planning from 2000-2005. He also served as a Foreign Professor at the Allama Iqbal Open University (AIOU) through Higher Education Commission of Pakistan, from 2005-2009. Dr. Baig is the member of the Assessment Committee of the Intergovernmental Education Organization, United Nations, EDU Administrative Office, Brussels - Belgium. He is also the member of livelihood commission of IUCN. He served as a Visiting Fellow at the Sustainable Development Policy Institute, Islamabad, Pakistan. He serves on the editorial Boards of many International Journals including: International Journal of Agriculture and Biology; TC- Biodiversity Journal, Canada; Journal of Agricultural Extension and Rural Development, International Journal of Social Forestry etc. He is the member of many national and international professional organizations including Nova Scotia Institute of Agrologists. Dr. Baig has travelled in Asia, North America, Europe and the Middle East and has got a vast experience of working with the international community. He has travelled to Sri Lanka through World Agroforestry Centre, Kenya as the Representative of Pakistan. The Asian Productivity Organization – Japan invited him to the Philippines as a Resource Person to educate the Asian scientists. Dr. Baig established professional and academic ties with the Plymouth University, UK and was instrumental in bringing Pakistani students on scholarships to the King Saud University, Saudi Arabia. In Pakistan, he served as the Coordinator for M.Sc. Rural Development degree program at the Allama Iqbal Open University (AIOU), Islamabad. Along with many publications, he has produced books on "Natural Resource Management in Pakistan; Social Forestry in Pakistan. He served as the member of committee, revising curricula for M.Sc. (Hons.) Agricultural Extension and M.Sc. (Hons.) Forestry Extension degree programs at the AIOU.

Abdelkrim Ben Salem, Ph.D., is a Secondary Education Professor in the Tafilalet High School at Daraa Tafilalet Academy and associate researcher in the Hydrobiology, Ecotoxicology, Sanitation & Global Change Lab, Faculty of Science Semlalia,

Cadi Ayyad University Marrakech. Broadly, her methodological research focuses on vulnerability and adaptation to climate change. Within Environment policy, Dr. Ben Salem currently works on socioeconomic vulnerability of agriculture and water sector to climate change in Morocco. He has published interdisciplinary projects across varied outlets, including, Parasites & Vectors Journal, International Journal Of Water Resources & Arid Environments and International Publisher of Information Science and Technology Research (IGI-Global).

Souad Ben Salem has a master's degree in Sciences and Technics at the Faculty of Sciences and Technics in Errachidia Morocco. They are a student at the Faculty of Sciences Semlalia Marrakech, Doctoral Training: Science of Life and Environment, laboratory of hydrobiology, ecotoxicology, sanitation and global change.

Muhammad Izzuddin Syakir Ishak is an Environmental Geoscientist and a Council Member of Centre for Global Sustainability Studies, Universiti Sains Malaysia (CGSS). He has completed his PhD in Earth Science from University of Ottawa, Canada. His research interests are isotope geochemistry, watersheds management and sustainability studies.

Alam Khan contributed to analyzing the data. He is a Research Scholar in the Department of Mathematics and Statistics at the International Islamic University in Islamabad, Pakistan.

Ishfaq Khan is an Associate Professor in the Department of Mathematics and Statistics at the International Islamic University in Pakistan.

Umut Kirdemir is an academician who carries out his academic studies at Dokuz Eylul University, Izmir, Turkey. He is a PhD student at present and his expertise fields are are hydrology, water resources, climate change, statistics, and machine learning techniques.

Adil Moumane is a Researcher in the Department of Geography; Lab. Environment, Societies, Territories at Ibn Tofail University. They are working on climate change and Environmental Hazards Assessment at Pre-Saharan regions of Morocco.

Zineb Moumen is a PhD student at the University of Sidi Mohamed Ben Abdellah, Fez- Morocco Thematic of research: Assessment of climate and land use changes Impact on Water quality and quantity, study case (Innaouene watershed). 2014-2016: Geo-resources & Environment Master in the University of Sidi Mohamed

Ben Abdellah, Fez-Morocco 2013-2014: Water and Environment license in the University of Sidi Mohamed Ben Abdellah, Fez-Morocco 2009-2010: Bachelor in Experimental Sciences, Option: sciences of Earth and Life in High School Moulay Idriss, Fez- Morocco.

Soummaya Nabih is a PhD student at Sidi Mohamed Ben Abdellah University. Her research interests include: Climate change, Hydrological modeling, Hydrology, Floods, etc.

Mohd Nawawi Mohd Nordin is working as a Professor from School of Physics, Universiti Sains Malaysia. He has completed his PhD in Geophysics from Birmingham University, England. He has around 120 research articles and has attended many national and international conferences. Currently, he has 9 research projects under his supervision and has 3 on going grants. His research area of interest is Geology.

Wilson Truman Okaka currently works as Lecturer of Research and Communication Education at, Kyambogo University. Wilson does research in Quantitative Social Research, Qualitative Social Research and Communication and Media. Wilson Okaka is a Ugandan dedicated to public awareness communication and international volunteering with a focus on gender equality He was educated and trained in Uganda, UK, Italy, Sweden, and the USA. He has presented conference papers, published books, book chapters, and peer-reviewed articles for international journals. He has volunteered with VSO International, in Asia. In addition, he belongs to professional interdisciplinary research, media, scientific, and education networks. He has taught at Kyambogo University (Kampala, Uganda) since its inception.

Umut Okkan received his BSc, MSc, and PhD degrees in Civil Engineering Department from Dokuz Eylul University, Izmir, Turkey, in 2007, 2009, and 2013, respectively. He works in Civil Engineering Department at Balikesir University. His research areas include hydrology, climatology, statistics, and soft computing algorithms.

Mohd Rafatullah is a Senior Lecturer in the Environmental Technology Division at the School of Industrial Technology within Universiti Sains Malaysia.

Lim Hwee San has been a lecturer in Unversiti Sains Malaysia (USM) after obtaining a PhD degree from USM in 2006. He obtained their B.Sc. from USM in Geophysics in 2001. After receiving his B.Sc. (Hons), he continued his M.Sc. and PhD under the supervisions of Prof. Dr. Mohd. Zubir Mat Jafri and Professor Dr. Khiruddin Abdullah. Finally, he obtained his M.Sc. degree and PhD degree from USM in environmental remote sensing in 2003 and 2006 respectively. His research interests lie generally in the field of optical remote sensing - both passive and active - and digital image processing, particularly as it applies to spectral image data. In both cases, the primary applications are water quality monitoring, air quality monitoring, land cover/change detection, land surface properties and digital images classification. He is also interested in modeling of the optical properties of atmospheric aerosols. His current effort focus on the applications of ground based LIDAR and satellite based LIDAR (e.g., CALIPSO, AIRS) data for air pollution and green house effects study.

Vartika Singh is an Assistant Professor. She holds a Ph.D. degree in Hydrogeology and Remote Sensing, M.Phil. degree in Hydrogeology from Vikram University and Master degree in App. Geology from Kurukshtera University. She has GATE Qualified all India rank 47. She has 7 years of research experience with Defence Terrain Research Laboratory (DTRL), DRDO, Delhi. Where she is associated with various important projects of Defence Research & Development Organization, DTRL lab like GIM, Thar and PISTA. Her specialization of research is in Hydrogeology, Remote Sensing, Environmental Geology & Artificial Intelligence. She has more than 30 publications in various national, international journals. She is one of the founder member of Computer Intelligence Research Group and Magazine. She is in Editorial Board of several national and international journals.

Index

M

N

O

Q

R

S

V

W

Z

www.ingramcontent.com/pod-product-compliance
Lightning Source LLC
LaVergne TN
LVHW080558300926
849162LV00005B/985